Sustainable Development Goals Series

The **Sustainable Development Goals Series** is Springer Nature's inaugural cross-imprint book series that addresses and supports the United Nations' seventeen Sustainable Development Goals. The series fosters comprehensive research focused on these global targets and endeavours to address some of society's greatest grand challenges. The SDGs are inherently multidisciplinary, and they bring people working across different fields together and working towards a common goal. In this spirit, the Sustainable Development Goals series is the first at Springer Nature to publish books under both the Springer and Palgrave Macmillan imprints, bringing the strengths of our imprints together.

The Sustainable Development Goals Series is organized into eighteen subseries: one subseries based around each of the seventeen respective Sustainable Development Goals, and an eighteenth subseries, "Connecting the Goals," which serves as a home for volumes addressing multiple goals or studying the SDGs as a whole. Each subseries is guided by an expert Subseries Advisor with years or decades of experience studying and addressing core components of their respective Goal.

The SDG Series has a remit as broad as the SDGs themselves, and contributions are welcome from scientists, academics, policymakers, and researchers working in fields related to any of the seventeen goals. If you are interested in contributing a monograph or curated volume to the series, please contact the Publishers: Zachary Romano [Springer; zachary.romano@springer.com] and Rachael Ballard [Palgrave Macmillan; rachael.ballard@palgrave.com].

More information about this series at http://link.springer.com/series/15486

Azizur Rahman Siddiqui • Avijit Sahay
Editors

Climate Change, Disaster and Adaptations

Contextualising Human Responses to Ecological Change

Springer

Editors
Azizur Rahman Siddiqui
Department of Geography
University of Allahabad
Prayagraj, Uttar Pradesh, India

Avijit Sahay
Department of Humanities
and Social Sciences
Indian Institute of Technology Bombay
Mumbai, Maharashtra, India

ISSN 2523-3084 ISSN 2523-3092 (electronic)
Sustainable Development Goals Series
ISBN 978-3-030-91009-9 ISBN 978-3-030-91010-5 (eBook)
https://doi.org/10.1007/978-3-030-91010-5

This Springer imprint is published by the registered company Springer Nature Switzerland AG
The registered company address is: Gewerbestrasse 11, 6330 Cham, Switzerland

Foreword

I am delighted to write Foreword for the book *Climate Change, Disaster and Adaptations: Contextualising Human Responses to Ecological Change*. In a year that has seen us make unprecedented adaptations to our lifestyle, I think the book is a very timely publication. The world today is facing threats due to climate change and associated extreme natural events. Many of these events are causing large-scale displacement of population termed as environment-induced migration. Such migration is going to increase exponentially in the coming decades of this century as sea level rise will lead to the submergence of some of the most densely populated urban areas in the world. Sea level rise is, of course, only one aspect of climate change. Other aspects of climate change like desertification will also lead to a large-scale displacement of the population. It is therefore, essential to understand the phenomenon of environment-induced migration.

The impacts of climate change like sea level rise, land and forest degradation and desertification can all be grouped under the category of slow-onset event. Previously, studies on hazards and disasters had excluded slow-onset events, but now there is a growing recognition that slow-onset events will be a major driver of global environmental change in the future. This fact was recognised by the Sendai Framework for Disaster Risk Reduction (2015–2030) paragraph 15 which defined disasters as both sudden and slow-onset events. By focusing on climate change adaptations and migration caused by slow-onset events, this book is a major contribution in environmental studies and is highly recommended to not just researchers and academicians but also to policymakers for enhancing their understanding of and linkages between slow-onset events and environment-induced migration.

Given the challenges the world is currently facing, and in light of the urgent need to ensure a more sustainable future, we are of the view that Geography is a key subject and that it is vital. The volume will be useful for undergraduate and postgraduate students in the field of Geography, Geology and Climate Sciences. As 'the science of sustainability', Geography has played and continues to play a central role in developing generations of students who have the qualities and skills necessary to address the diverse array of environmental, economic, social, and health challenges that the world is currently experiencing. On behalf of the International Geographical Union (IGU), I welcome such initiative and do hope that students, researchers, and teachers from Geography and Environmental Sciences together with policy makers will be greatly benefited by the volume. I would like to congratulate Prof. Professor A.R. Siddiqui and Dr Avijit Sahay for bringing out this volume.

Secretary-General and Treasurer R. B. Singh
5th February 2021

Preface

The inspiration for writing the book *Climate Change, Disaster and Adaptations: Contextualising Human Responses to Ecological Change* was presented to us while working on riverbank erosion in Majuli Island of Assam, India. We were surprised to find out that even though riverbank erosion has led to the displacement of thousands of people in Majuli, it is not considered a hazard under Indian laws. One of the reasons for this is that erosion is a slow-onset event, while we generally consider hazards to be sudden destructive events. We were intrigued by this hierarchy in hazard classification, how much such a hierarchy is justified, and how this hierarchy affects communities struggling with different hazards and disasters.

The selection of chapters in the book reflects our above-mentioned interest in different categories of hazards. Thus, instead of focusing the book on predefined hazards and disasters, we have broadened our area of interest and have included chapters that deal with the change in *living environment* of a region. Thus, climate change, which in itself is a slow-onset event, is also dealt from different perspectives in this book. Moreover, many different effects of climate change like sea-level rise, increased variability in climate, extreme weather events, desertification, and droughts alter the *living environment*, and hence, separate chapters have been dedicated to them.

Another issue that is dealt with in detail in this book is *environment motivated migration*, also called *environmentally induced migration*. There is already a significant number of people who have been displaced from their land because of changes in their *living environment*, and this is expected to grow exponentially in the coming decades of this century. While an exact number of *environmentally induced migrants* is difficult to calculate, several estimates by the Intergovernmental Panel on Climate Change (IPCC) suggest there will be more than 150–200 million migrants displaced because of environmental change. Thus, it is important to understand the problems of migrants and the challenges they face to better prepare for future surges in migrant numbers.

Based on the perspectives provided by the book on issues discussed above, we feel that this book is an important contribution to the field of climate change and hazard studies.

We would like to acknowledge with thanks the help, support, and guidance provided to us by Prof. M.H. Quereshi, Prof. Savindra Singh, Prof. Salahuddin Quereshi, Prof. Abha Lakshmi Singh, and Dr. Suraj Mal.

We are also thankful to Prof. Vijoy S Sahay, Dr. Vatsala Srivastava, and Mrs. Nuzhat Fatima Siddiqui for the constant motivation they have provided to us.

We would also like to thank Nikhil Roy and Suraj Prasad of the Department of Geography, University of Delhi, for providing us with clear and concise conceptual understanding of various aspects of slow-onset disasters.

Most importantly, we would like to express our sincere thanks to the contributors who have provided us with such valuable chapters.

Last but not least, we would like to thank our publisher, Springer Nature, and their staff, who have gone through various stages of the manuscript patiently, persistently, and diligently, and invariably pointed us in the right direction.

Prayagraj, Uttar Pradesh, India Azizur Rahman Siddiqui
Mumbai, Maharashtra, India Avijit Sahay

Contents

About the Editors

Azizur Rahman Siddiqui is professor and head, Department of Geography, University of Allahabad. Having finished his BSc, MSc, MPhil, and PhD degrees from Aligarh Muslim University, Aligarh, Prof Siddiqui first joined Chaudhary Charan Singh University, Meerut, in the year 1999 as lecturer and later the University of Allahabad in 2001. Prof Siddiqui has completed NNRMS training course and three international training programmes sponsored by the United Nations. He has also obtained postgraduate diploma in geo-information science with specialisation in geo-hazards from IIRS, Dehradun, and ITC, the Netherlands. His specialisation is in arid zone research, urban environment issues, and application of remote sensing and GIS in land degradation studies. From 2016, he has served as the secretary of Indian Institute of Geomorphologists (IGI). Prof Siddiqui has visited France, Russia, China, Belgium, the Netherlands, Switzerland, Italy, Austria, Greece, and Nepal for academic purposes and is a recipient of an excellence award in the year 2018 by the University of Allahabad.

Avijit Sahay is currently a postdoctoral scholar in the Department of Humanities and Social Sciences, Indian Institute of Technology Bombay. He did his PhD on riverbank erosion in Majuli Island of Assam in the Department of Geography, University of Allahabad, Allahabad. His works on Majuli Island have been presented in international conferences and published in national and international journals. He is also associate editor of *the Oriental Anthropologist* – an international journal of the science of man and now published by SAGE Publications.

Introduction: Resilience, Adaptation, and Migration – Exploring the Range of Human Response to Climatic Change

1

Avijit Sahay and Azizur Rahman Siddiqui

Introduction

Climate change and extreme natural events have become a major problem across the globe. The capacity of societies to adapt to such changes in the environment is called resilience, but sometimes, extreme natural events make settlements inhospitable, unproductive, and uninhabitable, which lead to abandonment of settlement and therefore, cause mass migration of populations (Hugo, 2010; Adamo & de Sherbinin, 2011; McLeman, 2011; Wrathall, 2012; Ferris, 2020). Such environment-induced migration is most pronounced in developing regions of the world as high disaster risk and a general poverty of the population have rendered societies increasingly vulnerable to climate change. The problem has got exacerbated with the recent multiple earthquake events in Himalayan region, extreme cyclones in the tropics, and the continuing riverbank erosion of Brahmaputra in India and Bangladesh.

Resilience and adaptation are inter-related concepts used in climate science and disaster studies to describe the capacity of a society to adjust to changing environment. Relationships between the two terms have been subject of extensive academic research (Adger, 2006; Folke, 2006; Berkes, 2007; Kuenzer et al., 2020), and they have been explored to understand the linkages between vulnerability and adaptive capacity in global change science (Gallopín, 2006; Vogel et al., 2007; Lei et al., 2013). Other works have focused on the nexus of vulnerability, resilience, and adaptation to integrate sustainability studies within a coupled human-environment system (Turner II, 2010; Endfield, 2012).

While linkages between resilience and adaptation are well established, there is an understanding that migration is the result of low adaptation capacity (Klepp, 2017; Kaczan & Orgill-Meyer, 2020). Such simplistic explanation of migration fails to interpret that in certain situations, in situ adaptation to climatic and environmental change becomes prohibitive and thus, the only adaptive strategy available to affected communities is migration. This is especially true in the case of slow-onset hazards and displacement of population in the context of climate change. Slow-onset disasters refer to environmental degradation processes such as droughts and desertification, increased salinisation, rising sea levels, or thawing of permafrost. Each of these is related to climate change and causes progressive degradation of land to the limit that it causes the carrying capacity of a place to diminish and thus, human habitation becomes challenging.

A. Sahay (✉)
Department of Humanities and Social Sciences, IIT Bombay, Mumbai, Maharashtra, India
e-mail: avijitsahay@iitb.ac.in

A. R. Siddiqui
Department of Geography, University of Allahabad, Prayagraj, Uttar Pradesh, India

A. R. Siddiqui, A. Sahay (eds.), *Climate Change, Disaster and Adaptations*, Sustainable Development Goals Series, https://doi.org/10.1007/978-3-030-91010-5_1

Thus, in many parts of the world, climate change and natural hazards have become a primary cause of internal migration. This is especially true of developing regions, where the triple factors of disaster risk, vulnerability of the society to hazards in general, and poverty have caused millions of people to become internally displaced persons (IDPs). Coupled to this is the growing awareness that climate change is going to cause unprecedented displacement of people from fragile ecosystems and low-lying areas of the world in the coming decades of this century. Considering all these factors in a holistic manner, it becomes clear that the ongoing and the predicted displacement of people induced by environmental change is one of the major challenges of this century and beyond.

The present book tries to explore human adaptation to climatic and environmental change as well as migration as a part of response continuum within the resilience capacity of a society. The major aim here is to develop a conceptual framework to understand environment-induced displacement.

Social-Ecological Systems and Social-Ecological Regime Shifts

The term social-ecological system is an amalgamation of social systems and ecological systems. An ecological system is an interdependent system of organisms (Anderies et al., 2004), while social system comprises relationships between individuals in a society. A social-ecological system is, therefore, a concept that suggests that human-natural systems are composed of a set of relationships that maintain them in relatively stable states (Wrathall, 2012). In other words, social-ecological system (SES) can be defined as complex, integrated systems in which humans are a part of, not apart from nature. The stable state in such a system results from the capacity of the system to absorb and adapt to external forces without undergoing any fundamental change, and this capacity to absorb and adapt is called resilience (Wrathall, 2012). Thus, when external shocks affect a system, resilient relationships allow the system to return to its previous stable state (Adger, 2000; Walker et al., 2004; Gallopín, 2006; Wrathall, 2012).

However sometimes external forces are strong enough to abruptly change the stable state of a social-ecological system. Such a change in the stable state is called social-ecological regime shift. Regime shifts may happen because of climate change or sudden disruptions in the environment in the form of hazards and disasters. Some scholars argue that in the face of social-ecological regime shift, migration is the only adaptive strategy that can be employed by a society (Barnett & Jones, 2002), while others argue that migration happens in locations where adaptation is no longer possible (Oliver-Smith, 2002).

Slow-Onset Events and Sendai Framework

While much of attention in academia and policy-making is focused on sudden events that lead to environmental, ecological, or climate change, a strong argument could be made for the effects of slow-onset disasters and the multiple impacts that it has. A slow-onset event can be defined as one that does not emerge from any single or distinct source, but one that emerges gradually over time and because of the confluence of a number of different but inter-related events. Two major slow-onset events recognised now are desertification and riverbank erosion. A major characteristic of slow-onset events is that it renders the affected area not fit for further human habitation and as such, causes ex situ resettlement of population. The ever-increasing body of evidence on the adverse impacts of slow-onset events was finally recognised by the United Nations in the Sendai Framework for Disaster Risk Reduction. The Preamble of the Sendai Framework identifies the lessons learned and gaps identified from the Hyogo Framework (UNDRR, 2015) and mentions in Point 4 that:

> Recurring small-scale disasters and slow-onset disasters particularly affect communities, households and small and medium-sized enterprises, constituting a high percentage of all losses

Thus, there is an acceptance that slow-onset disasters have not been given the importance that they deserve and that these are significant events in their own right. The Framework further elaborates on the nature on scope of disasters in Point 15 (UNDRR, 2015) and states that the Framework will cover:

> …small-scale and large-scale, frequent and infrequent, sudden and slow-onset disasters caused by natural or manmade hazards, as well as related environmental, technological and biological hazards and risks

Climate Change as a Slow-Onset Event

While discussing slow-onset events and their place in hazard studies, it is important to note that in the next few decades of this century, slow-onset events are going to play an increasing role in human mobility. Global warming and its associated effects of climate change, sea level rise, and desertification are all slow-onset events and are likely to cause displacement of a large number of people living in vulnerable areas, especially the coastal regions. Thus, effectively dealing with slow-onset events is going to be one of the challenges of this century, and it is hoped that this book will provide the epistemological and methodological understanding of the event.

Present Initiative

The book deals with changes in environment and how local communities deal with these changes. Environmental change occurs as a result of natural hazard or as climate change. The proposed book examines human responses to droughts, floods, erosions, pandemics, and climate change with special reference to human mobility in places not suitable for re-habitation. In total, the book contains 19 chapters.

Chapter 1 contains an introduction to the topic as well as to the book. This chapter introduces the concepts of social-ecological systems and social-ecological regime shifts and also explores the impact of slow-onset disasters.

Chapter 2 examines the long-term trend of drought conditions over arid and semiarid districts of Rajasthan. The chapter uses data from MODIS satellite to find normalized difference vegetation index and enhanced vegetation index of drought-prone areas of the state and concludes that with certain exceptions, terminal drought intensity has increased, especially in the northern and southwestern parts of Rajasthan.

Chapter 3 assesses how climate change and human actions are exacerbating the already existing hazards in North Sikkim and how this is endangering the stability of glacial lakes in Sikkim.

Chapter 4 studies climate change-induced migration in Pacific island states with special reference to Fiji. The chapter identifies five hotspots in the Pacific that are likely to become major source of climate change-related migration in the future, viz. (a) urban areas; (b) urban atolls; (c) non-urban atolls; (d) coastal, delta, and riverine communities; and (e) communities prone to drought. The article explores how a dangerous mix of poverty, backwardness, and geographical location has made out migration inevitable in the Pacific Islands.

The Himalayan mountain system represents a very fragile climatic region that is susceptible to major changes in environment and ecology because of climate change. Chapter 5 explores these issues in the Kashmir Valley with the help of indicators like temperature, precipitation, and reduction in the size of glaciers to draw baseline results demonstrating the immediate effects of climate change.

A very thought-provoking question is asked in Chap. 6. The chapter argues that it is technological development that is the main cause of climate change, and this has led to an increase in incidences of extreme natural events like hazards and disasters. This chapter explores how the phase of technological innovation accelerated with the growth of human population has caused anthropogenic climate change which in turn has influenced the social, political, and economic drivers of human mobility in complex ways.

A gendered perception of climate change and resilience is explored in Chap. 7. The chapter reviews the gendered perceptions in climate sensitivity and resilience, whereby the authors have tried to propose a women-centric climatic hazard management. On the basis of this review, the article concludes that women are not as vulnerable as they may seem and are also willing to come forward to lead climate change mitigation challenges from the front.

Chapter 8 explores glacial lake outburst floods (GLOFs) in Lahaul and Spiti and the changes in the frequency of it because of climate change. GLOFs have immense potential of flooding in downstream areas, causing disastrous consequences due to release of large volumes of water in a very short interval of time. The study tries to assess the probable causes of GLOF and evaluate the impact in case such a situation arises.

Chapter 9 attempts to provide a comprehensive analysis of environment-induced displacement on the basis of disaster risk pattern and emergency response system in developing countries. In the chapter, even though climate change- and disaster-induced displacement (CCDID) happens in both developing and developed countries, it is the developing countries (or the Global South) that bear the brunt of it because of lack of proper preparedness and awareness among people. The study thus hopes to build a holistic understanding of disaster-induced displacement in developing countries.

A theoretical framework for understanding adaptation and resilience is presented in Chap. 10. With a particular focus on oceanic impacts of climate change, the chapter explores different types of adaptation methods including ecosystem-based responses to climate change.

The impact of climatic variability on agriculture is examined in Chap. 11. With the help of analytical hierarchy process, the chapter prepares a land suitability assessment of Nalanda District of Bihar. Based on different parameters studied like soil texture, soil pH, drainage density, physiography, etc., the chapter proposes several techniques that can be adopted by the farmers in the study area to increase productivity even during times of climatic variability.

In Chap. 12, a balanced approach to increase agricultural productivity in desert region of India is presented. The chapter argues that certain changes in the environment brought about by climate change like sporadic showers in May and reduction of atmospheric dust caused by slowing down of winds present a favourable scenario of agriculture. The chapter also shows how overuse of irrigation in arid and semiarid regions has caused large-scale damage to land, and thus, to exploit the above-mentioned change in the environment for better agriculture, proper planning and sustainable use of land is suggested.

The coronavirus or COVID-19 pandemic has upended our usual way of life in almost every corner of the world. A particularly devastating impact of it has been felt on the urban daily wage workers and labourers in India. Chapter 13 explores the role of internal displacement by daily wages based workers and labourers from hotspots of COVID-19 and how a health disaster soon became an economic disaster for large sections of the country's population.

A psychological analysis of the impact of the COVID-19 pandemic is given in Chap. 14. Using online survey to collect data and demographic characteristics and coping strategies of different groups of people to lockdown, the chapter explores psychological adaptions to our lifestyle during the pandemic and shows that perceived stress has a positive correlation in sociocultural adaptation and psychological functioning.

Taking a more historical perspective on human migration, chapter 15 titled "Human Migration: An Evolutionary Perspective" shows how climate change, availability of food, agricultural revolution, and pure human inquisitiveness made possible the migration of *Homo sapiens* to Europe and South Asia around 50,000 years ago and America around 15,000 years back.

Chapter 16 discusses the impact of climate change on tidal and coastal areas of West Bengal, and the different environmental problems that these cause. The paper looks at the internal migration of tidal communities because of the loss of livelihoods and settlement sites caused by potential erosion of sandy beaches, sand dunes, and mud banks. The study uses satellite-derived database like ALOS PALSAR DEM, LANDSAT, and Google Earth and repeat survey across the shore transects in the region at regular interval.

As also IMD cyclone data and INCOIS generated wave data is used in the present study to highlight the nature of change along the shoreline.

Chapter 17 examines heavy traffic congestion in market areas of major cities using the example of Azadpur Mandi for case study. The Mandi is the main supply chain of producing fresh vegetables and fruits in Delhi, but the chapter shows that its infrastructure is not sufficient to cater to the massive volume of various type of vehicles, such as heavy motor vehicles and light motor vehicles and slow motor vehicles, and as such has become a major cause of pollution as well as traffic congestion.

Chapter 18 explores natural disaster events in Sri Lanka. Using data from various disasters to have affected the country, the chapter shows these calamities have forced people to move away voluntarily from their original residences. However, the government and some NGOs have now prepared a plan for resettlement and are once again displacing these people to newer government-approved areas. This new displacement pattern has become a big conflict between communities, and the chapter argues that preference should be given to voluntary migration instead of forced government-led displacement.

Using temperature and precipitation data from 62 meteorological stations in South India, Chap. 19 investigates increase in climatic extremes caused by global environmental changes.

Conclusion

Migration due to environmental change is now being recognised as a major issue in climate and disaster studies. The United Nations Office for Disaster Risk Reduction also now recognises this as necessary to achieve the targets of Sendai Framework for Disaster Risk Reduction (SFDRR). However, there is a need to understand migration as a part of resilience and adaptation strategies to better coordinate the efforts to mitigate the expected impacts of climate change. It is hoped that the present book will contribute in developing a greater understanding on different adaptive strategies with migration being recognised as one of them.

References

Adamo, S. B., & de Sherbinin, A. (2011). The impact of climate change on the spatial distribution of populations and migration. In *Population distribution, urbanization, internal migration and development: An international perspective* (pp. 161–195). Department of Economic and Social Affairs Population Division.

Adger, W. N. (2000). Social and ecological resilience: Are they related? *Progress in Human Geography, 24*(3), 347–364.

Adger, W. N. (2006). Vulnerability. *Global Environmental Change, 16*(3), 268–281.

Anderies, J., Janssen, M., & Ostrom, E. (2004). A framework to analyze the robustness of social-ecological systems from an institutional perspective. *Ecology and Society, 9*(1), 18.

Barnett, J., & Jones, R. (2002). Forced migration: The influence of environmental security. *Global Change, 7*(4), 3.

Berkes, F. (2007). Understanding uncertainty and reducing vulnerability: Lessons from resilience thinking. *Natural Hazards, 41*(2), 283–295.

Endfield, G. H. (2012). The resilience and adaptive capacity of social-environmental systems in colonial Mexico. *Proceedings of the National Academy of Sciences, 109*(10), 3676–3681.

Ferris, E. (2020). Research on climate change and migration where are we and where are we going? *Migration Studies, 8*(4), 612–625.

Folke, C. (2006). Resilience: The emergence of a perspective for social–ecological systems analyses. *Global Environmental Change, 16*(3), 253–267.

Gallopín, G. C. (2006). Linkages between vulnerability, resilience, and adaptive capacity. *Global Environmental Change, 16*(3), 293–303.

Hugo, G. (2010). Climate change-induced mobility and the existing migration regime in Asia and the Pacific. Oxford: Hart Publishing.

Kaczan, D. J., & Orgill-Meyer, J. (2020). The impact of climate change on migration: A synthesis of recent empirical insights. *Climatic Change, 158*(3), 281–300.

Klepp, S. (2017). Climate change and migration. In Oxford research encyclopedia of climate science. Oxford: Oxford University Press.

Kuenzer, C., Heimhuber, V., Day, J., Varis, O., Bucx, T., Renaud, F., … Glamore, W. (2020). Profiling resilience and adaptation in mega deltas: A comparative assessment of the Mekong, Yellow, Yangtze, and Rhine deltas. *Ocean & Coastal Management, 198*, 105362.

Lei, Y., Yue, Y., Zhou, H., & Yin, W. (2013). Rethinking the relationships of vulnerability, resilience, and adaptation from a disaster risk perspective. *Natural Hazards, 70*(1), 609–627.

McLeman, R. A. (2011). Climate change, migration and critical international security considerations. IOM, International Organization for Migration.

Oliver-Smith, A. (2002). Theorizing disasters: Nature, power, and culture. Catastrophe & culture. School of American Research Press, Santa Fe, 23–47.

Turner, B. L., II. (2010). Vulnerability and resilience: Coalescing or paralleling approaches for sustainability science? *Global Environmental Change, 20*(4), 570–576.

UNDRR. (2015, March). Sendai framework for disaster risk reduction 2015–2030. In *Proceedings of the 3rd United Nations world conference on DRR, Sendai, Japan* (pp. 14–18).

Vogel, C., Moser, S. C., Kasperson, R. E., & Dabelko, G. (2007). Linking vulnerability, adaptation, and resilience science to practice: Pathways, players, and partnerships. *Global Environmental Change, 17*, 349–364.

Walker, B., Holling, C. S., Carpenter, S., & Kinzig, A. (2004). Resilience, adaptability and transformability in social–ecological systems. *Ecology and Society, 9*(2), 5.

Warner, K. (2010). Global environmental change and migration: Governance challenges. *Global Environmental Change, 20*(3), 402–413.

Wrathall, D. J. (2012). Migration amidst social-ecological regime shift: The search for stability in Garifuna villages of northern Honduras. *Human Ecology, 40*(4), 583–596.

Check for updates

2 Spatio-temporal Variation of Drought Events over the Eastern Rajasthan (India): A Geo-spatial Approach

Arnab Kundu, Brijmohan Bairwa, Rashmi Sharma, N. R. Patel, K. K. Chattoraj, and Azizur Rahman Siddiqui

Introduction

Drought is considered as a devastating natural hazard that deteriorates the socioeconomic progress of society (Shaheen & Baig, 2011). It has long-term impacts which seriously affect crop production, animals, physical situation, and the economy as a whole. Since the past 30 years, several areas of the world have experienced water crisis, and drought caused distinct effect on regional economies (Ghulam et al., 2008). Drought can be categorized in agricultural, hydrological, meteorological and socio-economic perspective. Meteorological dry season is a circumstance of precipitation deviation from long-term average rainfall of a region. Farming dry season happens while soil dampness and precipitation are insufficient during the growing season. Hydrological dry spell occurs as a consequence of meteorological dry spell that causes water level decrease in reservoirs, lakes, streams, waterways, and groundwater level (Rathore, 2004). Throughout the recent decades, numerous studies across the world have concentrated on dry spell moderation technique. The initial phase in detailing a readiness arranging and relief measure is dry season checking. By observing dry season over a significant stretch of time (over 10 years), early dry spell cautioning frameworks can be created to guarantee worldwide food security. Contingent upon the degree of assurance, the degree of results might be diminished by diminishing the greatness of the water deficiency and by getting better the public mindfulness (social factor). Dry season is described as a multidimensional wonder (seriousness, span, extent). Consideration has been paid so far to disentangle these measurements to arrive at a pragmatic system to evaluate the seriousness of dry season.

Semi-arid and arid regions of Rajasthan (India) have become recurrent victim of tragic droughts, resulting in massive economic losses and degradation of natural resources of the area. Droughts are considered as serious issue of the society and believed to be an action of by legendry adoration of the Rain God; occasionally animal and human sacrifices were employed to satisfy the Rain God for saving people from the miseries of drought (Narain et al., 2006).

A. Kundu (✉) · K. K. Chattoraj
Department of Geo-Informatics,
P.R.M.S. Mahavidyalaya, Bankura University,
Bankura, West Bengal, India

B. Bairwa · R. Sharma
School of Earth Sciences, Banasthali Vidyapith,
Tonk, Rajasthan, India

N. R. Patel
Department of Agriculture and Soil, Indian Institute
of Remote Sensing, Indian Space Research
Organization, Dehradun, Uttarakhand, India

A. R. Siddiqui
Department of Geography, University of Allahabad,
Prayagraj, Uttar Pradesh, India

A. R. Siddiqui, A. Sahay (eds.), *Climate Change, Disaster and Adaptations*, Sustainable Development Goals Series, https://doi.org/10.1007/978-3-030-91010-5_2

In comparison to the ancillary and statistical measurements, remote sensing offers more reliable approach for studying drought facts over vast geographical regions. Drought detection through satellite-based data has been prevalently acknowledged in current years for its advantages like, synoptic view, replication of data acquirement, and reliability. Among various indices, NDVI has been acknowledged worldwide for detecting agricultural drought in diverse regions with changing environmental situations (Nicholson & Farrar, 1994; Kogan, 1995; Seiler et al., 2000; Anyamba et al., 2001; Ji and Peters, 2003; Dutta et al., 2013, 2015; Kundu et al., 2016, Kundu et al., 2020). Satellite-based NDVI is a convenient index for estimating and observing environmental issues, for example, crop health simulation, yield evaluation, land degradation, dryland issues, etc. (Aboelghar et al. 2010; Kundu & Dutta, 2011; Kundu et al., 2014a, 2014b, Kundu et al., 2015, Kundu et al., 2017; Kundu, 2018; Patel et al., 2019).

For monitoring agricultural drought, vegetation health and soil moisture are being considered as major indicators. The impact of precipitation and soil dampness on crops may be proficiently reflected by indices, for instance, NDVI and EVI, which play a significant role for drought occurrences. The MODIS-derived satellite images were employed to get spatial statistics on the status of vegetation. In light of the drowsy beginning of the disaster, drought allows a compromising time among the main indication and where individuals will be affected. Consequently, this can be applied as beginning pointer of drought event and could be utilized as a trouble-free remote sensing tool to map drought situations for crops (Shahabfar & Eitzinger, 2011). The term was EVI initially instituted by Huete et al., (1994). EVI can be aligned with stress and varieties identified with dry season.

So as to assess and research meteorological dry season, the SPI was utilized and planned by Mckee et al. (1993). Theoretically, SPI is corresponding to a Z score, yet isn't only the fluctuation from the mean and isolated by the standard deviation. It is a probability record where the likelihood of experiential precipitation is changed into a list for any time scale (3, 6, 12, 24, and 4-year time scales). A careful clarification of the SPI was given by Guttman (1998), Hayes et al. (1999), and Lloyd-Hughes and Saunders (2002). In several studies, SPI was also employed for monitoring drought in arid and semi-arid areas (Patel et al., 2007; Dutta, 2010; Dutta et al., 2013, 2015; Liu et al., 2016; Venkataraman et al., 2016; Okpara et al., 2017; Kundu, 2018; Kamble et al., 2019; Kundu et al., 2020).

In spite of the fact that the SPEI is recently uncovered, it has been applied in random examinations that have explored dry season irregularity (Potop, 2011; Paulo et al., 2012; Spinoni et al., 2013; Choudhury et al., 2021), dry season change (Allen et al., 2011), dry season gear (Vicente-Serrano et al., 2011), atmosphere adjustment (Wolf & Abatzoglou, 2011; Yu et al., 2013), and location of dry spell consequences for hydrological (Lorenzo-Lacruz et al., 2010; Wolf, 2012), rural (Potop et al., 2012), and biological frameworks (Vicente-Serrano et al. 2010a, 2010b; Toromani & Pasho, 2011; Vicente-Serrano et al., 2012; Lévesque et al., 2013).

Palmer (1965) used the PDSI based on water balance hypothesis and taking into account precipitation, soil wetness, runoff, and potential evapotranspiration (PET). This drought index includes actual component and henceforth can be applied to observe long-standing advancement of drought. Thus, this index has been widely utilized in drought monitoring over different regions of the world (Diaz, 1983; Dai et al., 2004; Sheffield & Wood, 2008; Dai, 2011; Sheffield et al., 2012; Yu et al., 2019; Ogunrinde et al., 2020).

MK test is one of the most widely utilized nonparametric tests for distinguishing time arrangement and pattern examination (Mann, 1945; Kendall, 1955). Kundu et al. (2014a, 2014b) analyzed several climate models to assess precipitation pattern and forecast using MK test and Sen's slope over eastern part of India. Trends of diverse climate aspects, viz., rainfall, temperature, wind speed, and relative humidity, were considered for studying the climatic variables in the eastern part of Rajasthan (Kundu et al., 2015).

Kumar et al. (2016) carried out the yearly pattern assessment for water yield utilizing MK test. Furthermore, flood occurrances were observed through long-term India Meteorological Department (IMD) data, Climatic Research Unit (CRU), and Community Climate System Model 4 (CCSM4) utilizing MK test over the Vaigai basin (India) by Nagalapalli et al. (2019). In their study, Bhatt et al. (2020) used MK test for long-term climate variability in response to NDVI as well as crop dynamics. Hence, this chapter aims to examine the long-term spatiotemporal trends of drought monitoring in eastern Rajasthan (India) during 2001–2020 through remote sensing and ancillary datasets.

Study Site

The eastern Rajasthan (study region) geographically located from 23°N to 28°15'N latitude and 72°15′E to 78°16′E longitude, principally comprises 21 districts with a total area of 146,800 km^2 (Fig. 2.1). This area has distinct agro-ecological environment, quite different from the arid western part of Rajasthan. The Arabian part of monsoon hits the eastern slant of Aravalli, and the eastern part gets orographic precipitation, while the western part stays dry. There is a distinct difference in diurnal and seasonal temperature. Generally, the normal temperature in winter ranges somewhere in the range of 22 and 25 °C, and in summer the normal temperature goes from 30 to 38 °C. Nonetheless, this region gets better precipitation (yearly normal 60–100 cm) than the western part, where dry season is predominant. In the eastern part of Rajasthan, rainfall begins in June and highest precipitation occurs in the months of August and September. This region gets practically 90% of the total annual precipitation during monsoon. There is an inordinate impact of beginning and span of storm upon the presence of downpour took care of kharif harvests of this locale. Since the eastern portion of Rajasthan gets more precipitation and depleted by various waterways and their feeders, it has been ideal for farming. There are primarily two yield seasons, for example, kharif and rabi, in the study area. Kharif crops (jowar, bajra, sorghum, and maize) are dependant upon monsoonal precipitation for water, while rabi crops (wheat, grain, mustard, gram) are grown in dry winter season. Kharif crops are planted in June to July and reaped during the months of September to October. Rabi crops are planted during October to November and reaped during March to April. Dry spells and frequent droughts in this region are generally connected with El Nino events which adversely affect the agro-based economy to a large extent (Parida & Oinam, 2015).

Materials and Methods

Datasets

NDVI and EVI

MODIS vegetation files, made on 16-day extends and at various spatial objectives, give consistent spatial and temporal assessments of vegetation greenness, leaf area, chlorophyll, and covering structure. Vegetation indices (MOD13Q1) are atmospherically-corrected level 3 product (version 6) at 250 m spatial resolution. These two indices even more satisfactorily depict the overall extent of vegetation states and cycles.

Rainfall

In order to reveal the causes behind positive or negative NDVI trend, long-term trend of rainfall was analyzed. It is well documented in previous researches that NDVI and rainfall are firmly associated with each other (Dutta et al., 2013). Estimating the inter-relationship between NDVI and rainfall requires long-term reliable rainfall data with good spatial resolution. Unlike the station based sparsely on distributed rainfall data, the satellite-based merged rainfall estimates provide long-term spatial data with better resolution and accuracy. The Climate Hazards Group InfraRed Precipitation with Station (CHIRPS) datasets with 0.05-degree resolution were collected for the period 2001–2020.

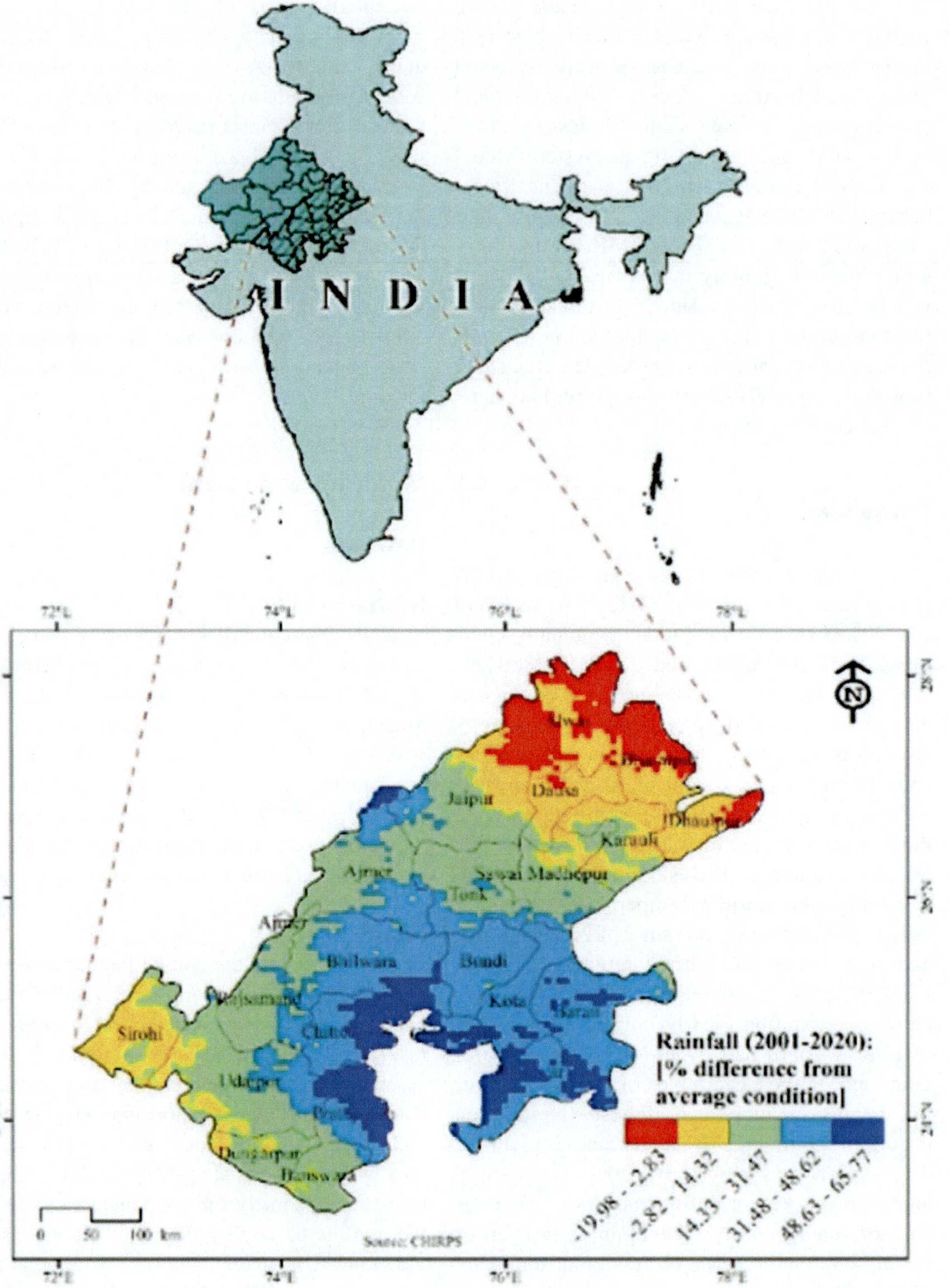

Fig. 2.1 Location map of the study area

Soil Moisture

The Climate Forecast System (CFS) gridded soil moisture data were used in this study for JJAS months since 2001–2020. The spatial resolution of data is $1^{\circ} \times 1^{\circ}$. The type of soil moisture is 0–10 cm underground liquid volumetric (non-frozen).

Methodology

Remote Sensing-Based Drought Indices

Drought indices require data on precipitation, soil wetness and several components associated with vegetation health. To observe and track drought condition in the investigation region, several remote sensing-based datasets and indices have been used. The satellite-based agricultural drought indices, viz., NDVI and EVI, and meteorological indices, viz., SPI, SPEI, and PDSI, were computed through satellite-derived datasets for this study.

NDVI

Rouse et al., (1973) proposed NDVI as a robust indicator for monitoring vegetation health and greenness. It is represented as (Eq. 2.1):

$$\mathrm{NDVI} = \left(\mathrm{NIR} - \mathrm{RED}\right) / \left(\mathrm{NIR} + \mathrm{RED}\right) \quad (2.1)$$

where NIR and RED are reflectance in near-infrared and red regions of electro-magnetic spectrum. The principle explanations for selecting these two bands are as follows: they are generally influenced by the concentration of chlorophyll in green vegetation and the difference among soil and vegetation is at a most extreme in red and NIR bands (Gutman, 1991; Kogan & Sullivan, 1993; Kogan, 1995, 1997). The NDVI esteem fluctuates somewhere in the range of −1 and +1, rock and exposed soils have a comparative reflectance in both red and NIR bands and result in NDVI almost zero, and vegetated lands have values which range from 0.1 to 0.7, while qualities lesser than 0.1 demonstrate no vegetation. Mists, water, and snow yield negative qualities because of bigger visible reflectance than of NIR reflectance.

EVI

For estimating the drought's impacts on vegetation while additionally measuring the degree of the dryness, MODIS satellite (MOD13Q1) data is made out of EVI measures (Iglesias et al., 2007). The EVI is an enhancement of the broadly utilized NDVI that likewise utilizes the contrast among blue and red reflectance to represent atmospheric aggravation (Solano et al., 2010) (Eq. 2.2):

$$\mathrm{EVI} = G \times \frac{\rho\mathrm{NIR} - \rho\mathrm{red}}{\rho\mathrm{NIR} + C1 \times \rho\mathrm{red} - C2 \times \rho\mathrm{blue} + L} \quad (2.2)$$

where G (gain factor) = 2.5; ρNIR, near-infrared reflectance; ρred, red reflectance; $C1 = 6$ (coefficient for aerosol resistance); $C2 = 7.5$ (coefficient for aerosol resistance); and $L = 1$ (canopy background factor) (Huete et al., 1994, 1997).

SPI

The SPI was developed by Mckee et al. (1993) and is a well-accepted index for evaluating meteorological drought studies. Besides, it is helpful for responding to the irregular dry or wet weather conditions (Moreira et al., 2012; Adhyani et al., 2017). This index helps to compute the amount to which rainfall in a given phase has diverged from the past (Sirdaş & Sen, 2003). The potency of SPI is its flexibility and capacity to calculate rainfall shortage distinctively associated to probability (McKee et al., 1993; Svoboda et al., 2012). The SPI of a place was calculated by separating the differentiation of periodic rainfall at a location and its long-term periodic mean by the standard deviation of the periodic rainfall (Eq. 2.3).

$$g\left(x\right) = \frac{1}{\beta^{\alpha}\Gamma\left(\alpha\right)} x^{\alpha-1} e^{-\frac{x}{\beta}} \quad \text{for}\, x > 0 \quad (2.3)$$

where α is the shape, β is the scale parameter ($\beta > 0$), x is the precipitation ($x > 0$), and τ (α) is the gamma function.

There are seven classes of SPI as per Mckee et al. (1993). The negative value of SPI denotes

Table 2.1 Classes of SPI by Mckee et al. (1993)

SPI values	Intensity
>2.0	Extremely wet
1.5 to 1.99	Very wet
1.0 to 1.49	Moderately wet
−0.99 to 0.99	Near normal
−1.0 to −1.49	Moderately dry
−1.5 to −1.99	Severely dry
< −2	Extremely dry

drought condition whereas ends when the value turn into positive (Table 2.1).

SPEI

The calculation of SPEI requires the potential evapotranspiration (PET) and precipitation information. The accompanying advances were utilized to register the SPEI esteems. The initial step is to gauge PET utilizing Hargreaves model. The distinction (D) between the precipitation (P) and PET for the month i was processed as appeared in Eq. 2.4.

$$D_i = P_i - \mathrm{PET}_\mathrm{i} \quad (2.4)$$

The determined D_i should be possible on numerous timesteps. The distinction in a given month j and year i relies upon the picked timesteps, k, for example, the collected contrast for 1 month in a specific year.

PDSI

The PDSI was proposed by Palmer (1965) and is an extensively used drought indicator to measure the long-term drought conditions, for an area at a certain time. It has been calculated by monthly rainfall and temperature along with info on the water-holding limit of soils. It takes into consideration the moisture received (rainfall) along with the moisture stored in soil, representing the probable loss of moisture owing to temperature effects.

MK Trend Analysis

The MK test, formulated by Mann (1945) and Kendall (1975), statistic S is given in Eq. 2.5:

$$S = \sum_{i=1}^{n-1} \sum_{j=i+1}^{n} \mathrm{Sgn}\left(x_j - x_k\right) \quad (2.5)$$

Where,

$$\mathrm{Sgn}\left(\mathrm{x_j} \;\; \mathrm{x_k}\right) = \begin{cases} 1 \text{ if}\left(\mathrm{x_j} \;\; \mathrm{x_k}\right) > 0 \\ 0 \text{ if}\left(\mathrm{x_j} \;\; \mathrm{x_k}\right) = 0 \\ \;\; 1 \text{ if}\left(\mathrm{x_j} \;\; \mathrm{x_k}\right) < 0 \end{cases}$$

where S is MK measurement and sgn is the signum work. The utilization of pattern test is never really time series x_i that is positioned from $i = 1, 2, \ldots n - 1$ and x_j and which is positioned from $j = i + 1, 2, \ldots n$. For $n < 10$, at that point estimation of $|S|$ is contrasted straightforwardly with the hypothetical dissemination of S inferred by Mann and Kendall.

The mean of S is $E(S) = 0$ (if $n \geq 10$) and the variance σ^2 (Eq. 2.6).

$$\mathrm{Var}\left(S\right)\left(\sigma^2\right) = \frac{1}{18}\left[\begin{array}{l} n\left(n-1\right)\left(2n+5\right) \\ -\sum_{j=1}^{p} t\left(t_j - 1\right)\left(2t_j + 5\right) \end{array}\right] \quad (2.6)$$

where p is the number of the tied groups in the dataset and t_j is the number of data points in the jth tied group. The statistic S is almost normal distributed presented that the following Z-transformation (Eq. 2.7).

$$Z_{\mathrm{MK}} = \begin{cases} \dfrac{S-1}{\sigma 2} & \text{If } S > 0 \\ & \text{If } S = 0 \\ \dfrac{S+1}{\sigma 2} & \text{If } S < 0 \end{cases} \quad (2.7)$$

A positive or negative value of Z_{MK} determines that the data trend to increase or decrease with time, respectively. Sen's estimator employed to predict the magnitude of trend. As per Sen (1968), Eq. 2.8 represents the slope (T_i) of data couples.

$$T_i = \frac{x_j - x_k}{j - k} \quad \text{for } i = 1, 2, \ldots, N \quad (2.8)$$

Here, x_j and x_k are data values at time j and k ($j > k$), respectively. The median of these N values of T_i is denoted as Sen's estimator of slope (Eq. 2.9).

$$Q_i = \begin{cases} T_{N+1/2}, & \text{Nisodd} \\ \frac{1}{2}\left(T_{\frac{N}{2}} + T_{\frac{N+2}{2}}\right), & \text{Niseven} \end{cases} \quad (2.9)$$

Finally, Q_{median} is registered by a two-sided test at 100 $(1 - \alpha)$ % certainty stretch, and afterward a genuine slope can be picked up by the nonparametric test. Positive estimation of Q_i shows an upward or expanding pattern, and a negative estimation of Q_i gives a descending or diminishing pattern in the time series.

Results and Discussion

NDVI

NDVI maps were prepared for JJAS months (kharif season) during 2002 and 2015. The study reveals 2002 as a severe drought year and 2015 as normal year (Fig. 2.2). During 2002, severe drought situations were clearly observed in southern and eastern districts over eastern Rajasthan (Fig. 2.2). The stressed conditions were gradually dominated from June to September months in 2002. Severe drought situations prevailed over almost north, south, southwestern, eastern, and western districts in the month of September. The vice versa situation was observed in 2015, even though few districts were affected by drought stress. The spatial variation of NDVI indicated vegetation health within the area. The range of NDVI was 0.11–0.50. This type of spatial variability mainly occurred for uneven distribution of monsoonal rainfall over that area. Similarly, the NDVI image of both the years showed significant temporal change within the study area.

EVI

Vegetation is the initial aspect which can be influenced by drought; thus, remote sensing indices have been generated for the estimate of drought based on brightness values of the land cover types. EVI is an useful index for studying canopy type, differences, architecture, and plant physiognomy. In this study, EVI fluctuates from 0.20 to 0.64 and has been classified into five classes (very low, low, moderate, high, and very high). For analysis, we considered drought (2002) and normal (2015) year according to NDVI for the months of JJAS (kharif season). The drought severity was observed noticeably in the months of August and September in 2002, and the normal condition was found in 2015 (Fig. 2.3).

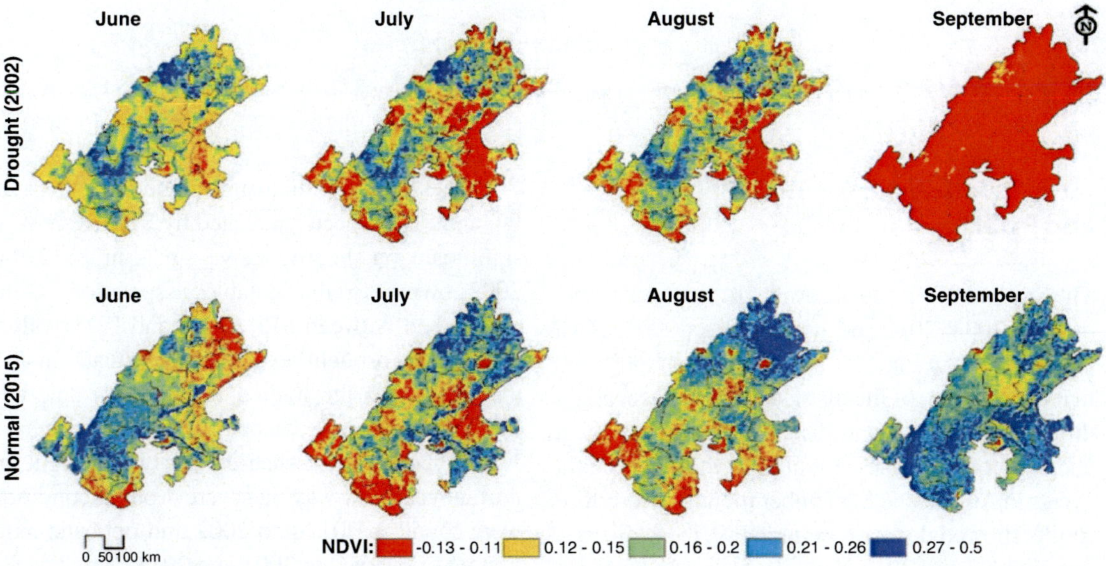

Fig. 2.2 Spatial pattern of NDVI for normal and drought years

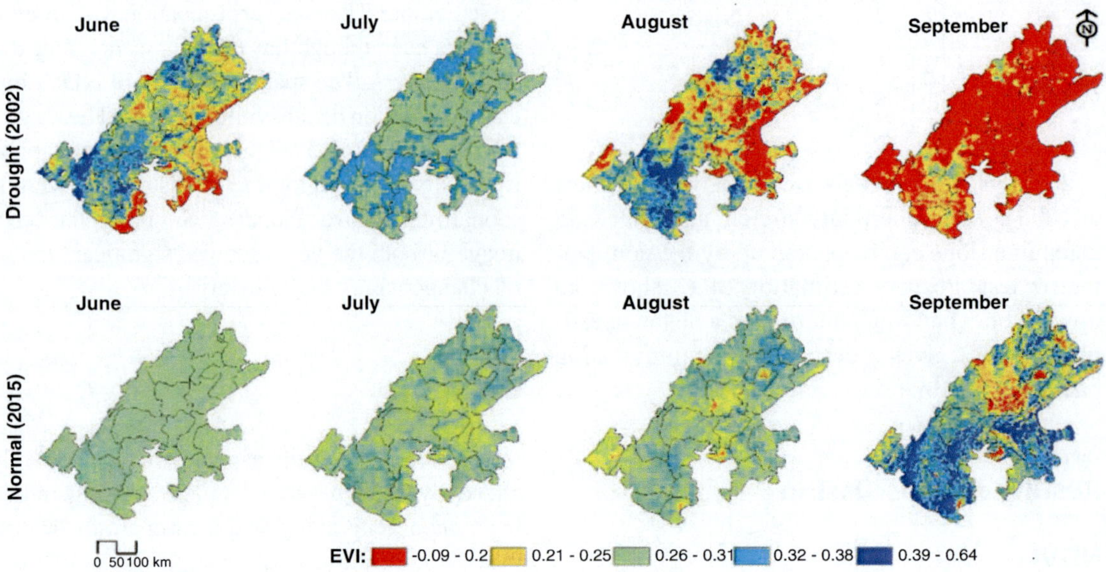

Fig. 2.3 Spatial pattern of EVI for normal and drought years

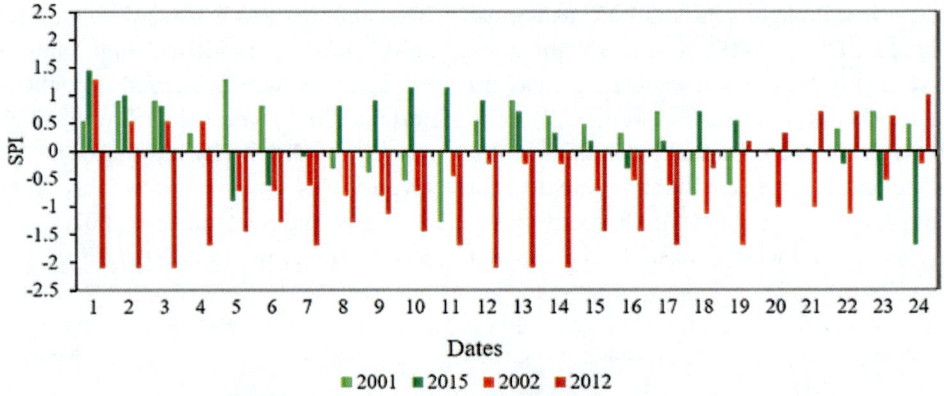

Fig. 2.4 SPI distribution for drought (2002, 2012) and normal (2001, 2015) years

Inter-Relationship of SPI, SPEI, and PDSI

The results of SPI values during JJAS period for the drought 2002, 2012 and normal years 2001, 2015 are presented in Fig. 2.4. SPI distribution indicated that the majority of the districts were influenced by drought during 2002 and 2012. The Baran, Jhalawar, and Chittorgarh districts suffered severe drought events in August and September months. Moreover, results showed decrease in rainfall in these districts during 2002 and 2012, that adversely affected vegetation cover soil moisture; and caused expansion of bare soil areas.

These results similarly displayed that the rainfall and PET aspects presented by SPEI were more significant on the vegetative stress in 2012 than 2002 owing to the notable association, which established between SPI, SPEI, and PDSI values. Besides, the drought severity stayed greatly in 2012 than 2002 over the studied districts. The date-wise SPEI values were calculated for the months of JJAS over the entire districts for two drought and normal years. By way of, severe drought conditions were found in 2012 and 2002 and opposite situation was observed in 2001 and 2015 (Fig. 2.5).

The study area is located in semi-arid region; thus, mostly the region is rain-fed, and vegeta-

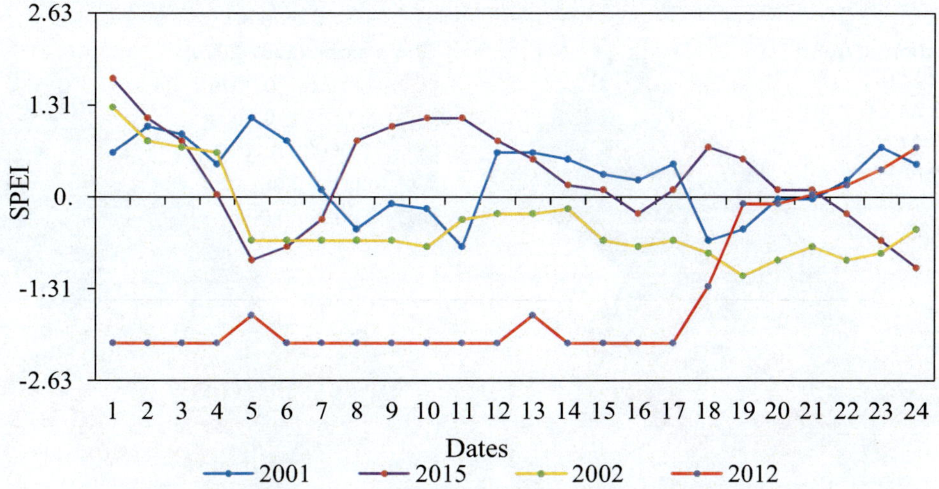

Fig. 2.5 Date-wise variation in SPEI for drought (2002, 2012) and normal (2001, 2015) years

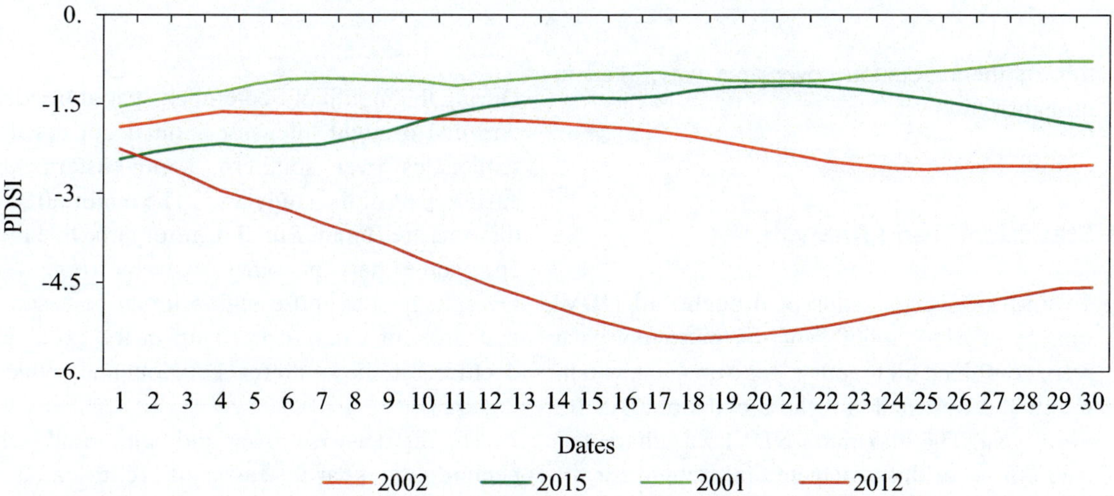

Fig. 2.6 Variation of PDSI in drought (2002, 2012) and normal (2001, 2015) years

tion growth is usually dependent on rainfall. The region has a distinct seasonal rainfall pattern and its geographical location made the area susceptible to vegetation stress. In order to recognize the relationship between vegetation and meteorological parameters, the SPI, SPEI, and PDSI were compared with NDVI and EVI. As Fig. 2.6 shows, there is an apparent variance among the results of 2001, 2002, 2012, and 2015, as the years 2002 and 2012 faced severe drought stress over the region.

During 2001 and 2015, the area showed ordinary vegetation and rainfall aspects. However, the vegetation experienced stress and loss of vegetation health in 2002 and 2012, respectively, owing to requiring rainfall. It was observed that, in the months of August and September, there was a huge rise in the parts that were classified under the first three classes of drought (extreme, severe, and moderate) in 2002 and 2012. For example, the part of these three drought classes was increased. Likewise, a significant deteriora-

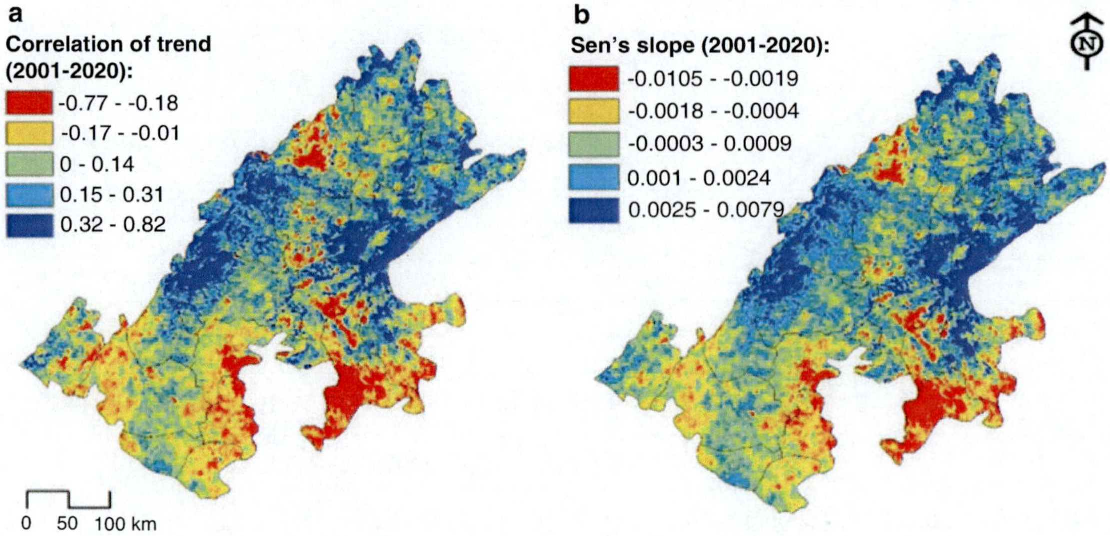

Fig. 2.7 Correlation of NDVI trend (Kendall Tau-b) (**a**) and slope of trend (Sen's slope) (**b**) for 2001–2020

tion of the vegetation cover area was noted in drought years.

Statistical Trend Analysis

To find long-term trends of drought, all NDVI images of JJAS months of the particular year were combined all together. Like this, a new output image was made for the months of JJAS for every year. The maximum NDVI was characterized utmost aridness detected in drought for the period of a specific month for a particular year, which specified terminal drought concentration. Trends in drought concentration were calculated for each pixel using climate engine tool. The results of linear regression were tested for mathematical significance using t-test, whereas those of MK test were tested with Z-statistic. With t-test, it was set up that, at scattered sites over north-western and south-western, the trends were significant for 82% confidence level. The correlation of NDVI trend (Kendall Tau-b) ranges from 0.10 to −0.50 and slope of trend (Sen's slope) ranges between 0.01 and –0.05. During the month of September, spatial trends in terminal drought intensity pointed out negative tendencies over southern, south-eastern, and eastern part of the study area. The major affected districts are Baran, Bundi, Chittorgarh, Jhalawar, and some part of other districts (Fig. 2.7). Overall, for the entire region it can be assumed that drought strength is rising in the north and declining in the south (except Kota and connecting area).

The district-wise jowar and bajra yield (tons/hectare) of kharif season is represented in Fig. 2.8 which indicates a noteworthy temporal variation for drought (2002) and normal (2015) years. By comparing NDVI, EVI, SPI, and SPEI values, it was examined that values have positive relationship with crop yield. Specifically, with increase in all indices, crop yield increase and decrease with indices values drop down for drought and normal years over the all districts. Dutta et al. (2013, 2015) attempted a similar study with satellite, meteorological indices and crop yield data, and they examined drought events in the eastern Rajasthan which amply matched with our study.

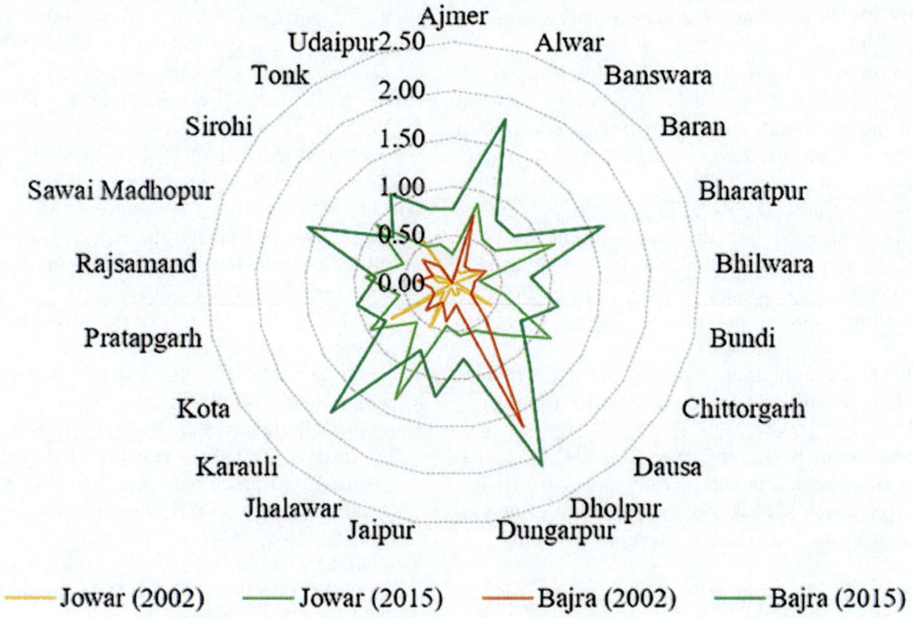

Fig. 2.8 Comparison of crop yield (jowar, bajra) for drought (2002) and normal (2015) years

Conclusion

In this study, we assessed agricultural and meteorological drought events in eastern Rajasthan (India) using satellite remote sensing and MK trend test for monitoring period of 2001–2020. First, we considered NDVI and EVI time series based on MODIS time series data for agricultural drought. Then, we integrated meteorological indices, viz., SPI, SPEI, and PDSI for establishing meteorological droughts over there. Furthermore, we computed all results through nonparametric MK trend test and Sen's slope approach. Finally, we investigated diverse characteristics, spatial spreading, and long-term trend of the droughts (agricultural and meteorological) in correspondence with crop (kharif) yield. It was discovered that a majority of the southern, south-eastern, and some portion of northern study region observed an increase drought amount with negative trends. With 82% confidence, it can be quantified that the terminal drought intensity increased at scattered sites over northern and south-western part of the study region. However, over eastern and western region, decreasing trends were recorded. In conclusion, results of this study may provide valuable information to combat repeated drought events, which can be beneficial for the decision-makers to diminish the devastating effects of drought by adopting suitable planning and management policy.

References

Aboelghar, M., Arafat, S., Saleh, A., Naeem, S., Shirbeny, M., & Belal, A. (2010). Retrieving leaf area index from SPOT4 satellite data. *The Egyptian Journal of Remote Sensing and Space Science, 13*, 121–127.

Adhyani, N. L., June, T., & Sopaheluwakan, A. (2017). Exposure to drought: Duration, severity and intensity (Java, Bali and Nusa Tenggara). In *IOP conference series: Earth and environmental science* (Vol. 58, no 1, p. 012040). IOP Publishing.

Allen, K. J., Ogden, J., Buckley, B. M., Cook, E. R., & Baker, P. J. (2011). The potential to reconstruct broadscale climate indices associated with south-east Australian droughts from Athrotaxis species, Tasmania. *Climate Dynamics, 37*, 1799–1821.

Anyamba, A., Tucker, C. J., & Eastman, J. R. (2001). NDVI anomaly patterns over Africa during the

1997/98 ENSO warm event. *International Journal of Remote Sensing, 22*, 1847–1859.

Bhatt, D., Kundu, A., Mall, R. K., & Prudhvi Raju, K. N. (2020). Dynamics of vegetation response to seasonal rainfall in the Gomati river basin (India) using earth observation data sets. *Journal of Scientific Research, 64*, 20–31.

Choudhury, A., Dutta, D., Bera, D., Kundu, A. (2021). Regional variation of drought parameters and long-term trends over India using standardized precipitation evapotranspiration index. *Journal of Environmental Management*, 296, https://doi.org/10.1016/j.jenvman.2021.113056

Dai, A. (2011). Drought under global warming: A review. *Wiley Interdisciplinary Reviews: Climate Change, 2*, 45–65.

Dai, A., Trenberth, K. E., & Qian, T. (2004). A global dataset of palmer drought severity index for 1870–2002: Relationship with soil moisture and effects of surface warming. *Journal of Hydrometeorology, 5*, 1117–1130.

Diaz, H. F. (1983). Drought in the United States. *Journal of Climate and Applied Meteorology, 22*, 3–16.

Dutta, D. (2010). A comparative approach for predicting agricultural drought using PS-n crop growth model, NOAA-AVHRR NDVI and ARIMA Model (M.Sc. Thesis). Faculty of Geo-Information and Earth Observation (ITC), University of Twente, Netherlands.

Dutta, D., Patel, N. R., & Kundu, A. (2011). Analyzing the performance of auto regressive integrated moving average (ARIMA) model for predicting agricultural productivity in eastern Rajasthan. *Research Journal of Agricultural Sciences, 2*, 555–559.

Dutta, D., Kundu, A., & Patel, N. R. (2013). Predicting agricultural drought in eastern Rajasthan of India using NDVI and standardized precipitation index. *Geocarto International, 28*, 192–209.

Dutta, D., Kundu, A., Patel, N. R., Saha, S. K., & Siddiqui, A. R. (2015). Assessment of agricultural drought in Rajasthan (India) using remote sensing derived Vegetation Condition Index (VCI) and Standardized Precipitation Index (SPI). *Egyptian Journal of Remote Sensing and Space Science, 18*, 53–63.

Ghulam, A., Li, Z.-L., Qin, Q., Yimit, H., & Wang, J., (2008). Estimating crop water stress with ETM+ NIR and SWIR data. *Agricultural and Forest Meteorology, 148*, 1679–1695.

Gutman, N. B. (1991). A sensitivity analysis of the Palmer Hydrologic Drought Index. *Water Resources Bulletin, 27*, 797–807.

Guttman, N. B. (1998). Comparing the palmer drought index and the standardized precipitation index. *Journal of the American Water Resources Association, 34*, 113–121.

Hayes, M. J., Svoboda, M. D., Wilhite, D. A., & Vanyarkho, O. V. (1999). Monitoring the 1996 drought using the standardized precipitation index. *Bulletin of the American Meteorological Society, 80*, 429–438.

Huete, A., Justice, C., & Liu, H. (1994). Development of vegetation and soil indices for MODIS-EOS. *Remote Sensing of Environment, 49*, 224–234.

Huete, A. R., Liu, H. Q., Batchily, K., & van Leeuwen, W. J. D. (1997). A comparison of vegetation indices over a global set of TM images for EOS-MODIS. *Remote Sensing of Environment, 59*, 440–451.

Iglesias, A., Garrote, L., Flores, F., & Moneo, M. (2007). Challenges to manage the risk of water scarcity and climate change in the Mediterranean. *Water Resources Management, 21*, 227–288.

Ji, L., & Peters, A. J. (2003). Assessing vegetation response to drought in the northern Great Plains using vegetation and drought indices. *Remote Sensing of Environment, 87*, 85–98.

Kamble, D. B., Gautam, S., Bisht, H., Rawat, S., & Kundu, A. (2019). Drought assessment for kharif rice using standardized precipitation index (SPI) and vegetation condition index (VCI). *Journal of Agrometeorology, 21*, 182–187.

Kendall, M. G. (1955). *Rank correlation methods*. Griffin.

Kendall, M. G. (1975). *Rank Correlation Methods*. 4th Edition, Charles Griffin, London.

Kogan, F. N. (1995). Application of vegetation index and brightness temperature for drought detection. *Advances in Space Research, 15*, 91–100.

Kogan, F. N. (1997). Global drought watch from space. *Bulletin of the American Meteorological Society, 78*, 621–636.

Kogan, F. N., & Sullivan, J. (1993). Development of a global drought watch system using NOAA/AVHRR data. *Advances in Space Research, 13*, 219–222.

Kumar, M., Denis, D. M., & Suryavanshi, S. (2016). Long-term climatic trend analysis of Giridih district, Jharkhand (India) using statistical approach. *Modeling Earth Systems and Environment, 2*, 1–10.

Kundu, A. (2018). An appraisal of drought dynamics in Bundelkhand Region (India) using geo-spatial techniques (Ph.D. Thesis, pp. 1–234). Faculty of Engineering and Technology, Sam Higginbottom University of Agriculture, Technology and Sciences (collaboration with Indian Institute of Remote Sensing, ISRO), India.

Kundu, A., & Dutta, D. (2011). Monitoring desertification risk through climate change and human interference using remote sensing and GIS techniques. *International Journal of Geomatics and Geosciences, 2*, 21–33.

Kundu, A., Dutta, D., Patel, N. R., Saha, S. K., & Siddiqui, A. R. (2014a). Identifying the process of environmental changes of Churu district, Rajasthan (India) using remote sensing indices. *Asian Journal of Geoinformatics, 14*, 14–22.

Kundu, A., Dwivedi, S., & Chandra, V. (2014b). Precipitation trend analysis over eastern region of India using cmip5 based climatic models. *International Archives of the Photogrammetry Spatial Information Science, 48*, 1437–1442.

Kundu, A., Patel, N. R., Saha, S. K., & Dutta, D. (2015). Monitoring the extent of desertification processes in western Rajasthan (India) using geo-information science. *Arabian Journal of Geosciences, 8*, 5727–5737.

Kundu, A., Dwivedi, S., & Dutta, D. (2016). Monitoring the vegetation health over India during contrasting monsoon years using satellite remote sensing indices. *Arabian Journal of Geosciences, 9*, 144. https://doi.org/10.1007/s12517-015-2185-9

Kundu, A., Patel, N. R., Saha, S. K., & Dutta, D. (2017). Desertification in western Rajasthan (India): Assessment using remote sensing derived rain-use efficiency and residual trend methods. *Natural Hazards, 86*, 297–313.

Kundu, A., Patel, N. R., Denis, D. M., & Dutta, D. (2020). An estimation of hydrometeorological drought stress over the central part of India using geo-information technology. *Journal of the Indian Society of Remote Sensing, 48*, 1–9.

Lévesque, M., Saurer, M., Siegwolf, R., Eilmann, B., Brang, P., Bugmann, H., & Rigling, A. (2013). Drought response of five conifer species under contrasting water availability suggests high vulnerability of Norway spruce and European larch. *Global Change Biology, 19*, 3184–3199.

Liu, Z., Wang, Y., Shao, M., Jia, X., & Li, X. (2016). Spatiotemporal analysis of multiscalar drought characteristics across the Loess Plateau of China. *Journal of Hydrology, 534*, 281–299.

Lloyd-Hughes, B., & Saunders, M. A. (2002). A drought climatology for Europe. *International Journal of Climatology, 22*, 1571–1592.

Lorenzo-Lacruz, J., Vicente-Serrano, S. M., López-Moreno, J. I., Beguería, S., García-Ruiz, J. M., & Cuadrat, J. M. (2010). The impact of droughts and water management on various hydrological systems in the headwaters of the Tagus River (Central Spain). *Journal of Hydrology, 386*, 13–26.

Mann, H. B. (1945). Nonparametric tests against trend. *Econometrica, 13*, 245–259.

McKee, T. B., Doesken, N. J. & Kleist, J. (1993). The relationship of drought frequency and duration to time scales. In: *Proceedings of the 8th conference on applied climatology* (vol. 17, pp. 179–183). Boston: American Meteorological Society.

Moreira, E. E., Mexia, J. T., & Pereira, L. S. (2012). Are drought occurrence and severity aggravating? A study on SPI drought class transitions using log-linear models and ANOVA-like inference. *Hydrology and Earth System Sciences, 16*, 3011–3028.

Nagalapalli, S., Kundu, A., Mall, R. K., Thattai, D., & Rangarajan, S. (2019). An appraisal of flood events using IMD, CRU, and CCSM4-derived meteorological data sets over the Vaigai river basin, Tamil Nadu (India). *Sustainable Water Resources Management, 5*, 1731–1744.

Narain, P., Rathore, L. S., Singh, R. S., & Rao, A. S. (2006). Drought assessment and management in arid Rajasthan. *Central Arid Zone Research Institute, Jodhpur and National Centre for Medium Range Weather Forecasting, Noida.*

Nicholson, S. E., & Farrar, T. J. (1994). The influence of soil type on the relationships between NDVI, rainfall, and soil moisture in semiarid Botswana: I. NDVI response to rainfall. *Remote Sensing of Environment, 50*, 107–120.

Ogunrinde, A. T., Oguntunde, P. G., Olasehinde, D. A., Fasinmirin, J. T., & Akinwumiju, A. S. (2020). Drought spatiotemporal characterization using self-calibrating Palmer Drought Severity Index in the northern region of Nigeria. *Results in Engineering, 5*, 100088.

Okpara, J. N., Afiesimama, E. A., Anuforom, A. C., Owino, A., & Ogunjobi, K. O. (2017). The applicability of standardized precipitation index: Drought characterization for early warning system and weather index insurance in West Africa. *Natural Hazards, 89*, 555–583.

Palmer, W. C. (1965). Meteorological drought. U.S. Research Paper No. 45. U.S. Weather Bureau, Washington, DC.

Parida, B. R., & Oinam, B. (2015). Unprecedented drought in North East India compared to Western India. *Current Science, 109*, 2121–2126.

Patel, N. R., Chopra, P., & Dadhwal, V. K. (2007). Analyzing spatial patterns of meteorological drought using standardized precipitation index. *Meteorological Applications, 14*, 329–336.

Patel, N. R., Mukund, A., & Parida, B. R. (2019). Satellite-derived vegetation temperature condition index to infer root zone soil moisture in semi-arid province of Rajasthan, India. *Geocarto International*, https://doi.org/10.1080/10106049.2019.1704074

Paulo, A. A., Rosa, R. D., & Pereira, L. S. (2012). Climate trends and behaviour of drought indices based on precipitation and evapotranspiration in Portugal. *Natural Hazards and Earth System Sciences, 12*, 1481–1491.

Potop, V. (2011). Evolution of drought severity and its impact on corn in the Republic of Moldova. *Theoretical and Applied Climatology, 105*, 469–483.

Potop, V., Možný, M., & Soukup, J. (2012). Drought at various time scales in the lowland regions and their impact on vegetable crops in the Czech Republic. *Agricultural and Forest Meteorology, 156*, 121–133.

Rathore, M. S. (2004). State level analysis of drought policies and impacts in Rajasthan, India. Colombo, Sri Lanka: IWMI. 40p. (*Working paper* 93: *Drought Series Paper* No. 6).

Rouse, J. W., Haas, R. H., Schell, J. A., & Deering, D. W. (1973). Monitoring vegetation systems in the Great Plains with ERTS. In 3rd ERTS *Symposium*, NASA SP-351 I, pp. 309–317.

Seiler, R. A., Kogan, F., & Wei, G. (2000). Wei Monitoring weather impact and crop yield from NOAA AVHRR data in Argentina. *Advances in Space Research, 26*, 1177–1185.

Sen, P. K. (1968). Estimates of the regression coefficient based on Kendall's tau. *Journal of the American Statistical Association, 63*, 1379–1389.

Shaheen, A., & Baig, M. A. (2011). Drought severity assessment in arid area of Thal Doab using remote sensing and GIS. *International Journal of Water Resources and Arid Environments, 1*, 92–101.

Shahabfar, A., & Eitzinger, J. (2011). Agricultural drought monitoring in semi-arid and arid areas using MODIS data. *Journal of Agricultural Science, 149*, 403–414.

Sheffield, J., & Wood, E. F. (2008). Projected changes in drought occurrence under future global warming from multi-model, multi-scenario, IPCC AR4 simulations. *Climate Dynamics, 31*, 79–105.

Sheffield, J., Wood, E. F., & Roderick, M. L. (2012). Little change in global drought over the past 60 years. *Nature, 491*, 435–438.

Sirdaş, S., & Sen, Z. (2003). Spatio-temporal drought analysis in the Trakya region, Turkey. *Hydrological Sciences Journal, 48*, 809–820.

Solano, R., Didan, K., Jacobson, A., & Huete, A. (2010). MODIS Vegetation Index. *User's Guide (MOD13 Series)*. Version 2.00. Tucson, AZ. Available online at: https://vip.arizona.edu/MODIS_UsersGuide.php (Accessed June 30, 2021).

Spinoni, J., Antofie, T., Barbosa, P., Bihari, Z., Lakatos, M., Szalai, S., Szentimrey, T., & Vogt, J. (2013). An overview of drought events in the Carpathian Region in 1961-2010. *Advances in Science and Research, 10*, 21–32.

Svoboda, M., Hayes, M., & Wood, D. (2012). *Standardized precipitation index user guide*. World Meteorological Organization.

Toromani, E., & Pasho, M. S. E. (2011). Growth responses to climate and drought in silver fir (Abies alba) along an altitudinal gradient in southern Kosovo. *Canadian Journal of Forest Research, 41*, 1795–1807.

Venkataraman, K., Tummuri, S., Medina, A., & Perry, J. (2016). 21st century drought outlook for major climate divisions of Texas based on CMIP5 multimodel ensemble: Implications for water resource management. *Journal of Hydrology, 534*, 300–316.

Vicente-Serrano, S. M., Begueria, S., & Lopez-Moreno, J. I. (2010a). A multi-scalar drought index sensitive to global warming: The standardized precipitation evapotranspiration index. *Journal of Climate, 23*, 1696–1718.

Vicente-Serrano, S. M., Lasanta, T., & Gracia, C. (2010b). Aridification determines changes in leaf activity in Pinus halepensis forests under semiarid Mediterranean climate conditions. *Agricultural and Forest Meteorology, 150*, 614–628.

Vicente-Serrano, S. M., López-Moreno, J. I., Gimeno, L., Nieto, R., Morán-Tejeda, E., Lorenzo-Lacruz, J., Beguería, S., & Azorin-Molina, C. (2011). A multi-scalar global evaluation of the impact of ENSO on droughts. *Journal of Geophysical Research, 116*, D20109. https://doi.org/10.1029/2011JD016039

Vicente-Serrano, S. M., Zouber, A., Lasanta, T., & Pueyo, Y. (2012). Dryness is accelerating degradation of vulnerable shrublands in semiarid Mediterranean environments. *Ecological Monographs, 82*, 407–428.

Wolf, J. W. (2012). *Evaluation of drought metrics in tracking streamflow in Idaho* (MS. Degree). College of Graduate Studies. University of Idaho. http://nimbus.cos.uidaho.edu/abatz/PDF/wolf_ms.pdf.

Wolf, J. F., & Abatzoglou, J. (2011). The suitability of drought metrics historically and under climate change scenarios. In *47th annual water resources conference, Albuquerque, NM*, 7–10 November 2011.

Yu, M., Li, G., Hayes, M. J., Svoboda, M., & Heim, R. R. (2013). Are droughts becoming more frequent or severe in China based on the standardized precipitation evapotranspiration index: 1951–2010? *International Journal of Climatology*. https://doi.org/10.1002/joc.3701

Yu, H., Zhang, Q., Xu, C. Y., Du, J., Sun, P., & Hu, P. (2019). Modified palmer drought severity index: Model improvement and application. *Environment International, 130*, 104951.

Evaluation of Hydro-geomorphic Responses to Climate Change in North Sikkim District, Sikkim, India

3

Kapil Ghosh and Sunil Kumar De

Introduction

Human societies have always been impacted by various geomorphic processes throughout history, including unusual streamflow, glacial dynamics, soil erosion, sedimentation, and sediment transport with environmental, social, and economic significance. The way of human response to these hydro-geomorphic processes has become great public concerns in recent days (Dey & Mishra, 2017; Piao et al., 2010; Yu et al., 2017; Zhang et al., 2008). These processes are greatly influenced by the changes of climatic elements, especially temperate and precipitation. Spatial and temporal changes of streamflow, surface runoff, sedimentation, soil moisture, and groundwater reserve are the important responses to changes of rainfall pattern and quantity (Ghosh & Ghosh, 2019). Moreover, precipitation is one of the important influencing factors in various geomorphic processes in the humid tropical region of the world (Ghosh et al., 2019). Therefore, any hydro-geomorphic process could be considered as a reliable evidence of climate change.

The menace of climate change create adverse impacts on river flow, glacial features, agricultural production, water availability, etc. (IPCC, 2007). Different researches have been conducted in Indian subcontinent to analyze the changing trend and intensity of temperature, and the studies established that there is a vivid consistency in terms of the rising temperature over India and the entire world for a long time (Pant & Kumar, 1997; Arora et al., 2005; Dash et al., 2007). Moreover, the previous study indicates that there is hardly any clear pattern about the amplification and reduction of mean rainfall over the country (Mooley & Parthasarthy, 1984; Thapliyal & Kulshreshtha, 1991; Lal, 2001; Kumar et al., 2010). But some previous research works focused that several portions of Asia have experienced the lesser occurrence of torrential rainfall as well as the amount of annual rainfall and the number of rainy days are also lowered (Khan et al., 2000; Shrestha et al., 2000; Mirza, 2002; Lal, 2003; Min et al., 2003; Goswami et al., 2006; Dash et al., 2007). The eastern Himalayan region extending from Arunachal Pradesh to Darjeeling of West Bengal mainly comprises Arunachal Pradesh, Sikkim, Bhutan, and extreme northern part of West Bengal state. Humid and hot summer, slight winter, and intense monsoon are major climatic characteristics of the eastern Himalayan region. This mountainous region witnesses climatic variability which induces several geomorphic and hydrological disasters and

K. Ghosh (✉)
Department of Geography, Diamond Harbour Women's University, Sarisha, West Bengal, India

S. Kumar De
Department of Geography, North Eastern Hill University (NEHU), Meghalaya, India

A. R. Siddiqui, A. Sahay (eds.), *Climate Change, Disaster and Adaptations*, Sustainable Development Goals Series, https://doi.org/10.1007/978-3-030-91010-5_3

promotes environmental alterations. It must be mentioned that climate has a great influence upon the water resources of any territory. The changing trend of temperature and rainfall pattern has the potentiality to trigger the frequency of water-induced catastrophes such as glacial lake outburst flood (GLOF), flash flood, landslide, etc. whereas on the other side, water availability and water utilization pattern have been remarkably changed due to climate change which undoubtedly cause obstruction in the aspect of electricity supply and operation of water resource-based programs (Shrestha & Aryal, 2011; Yamada, 1992). Since the previous three decades, the increasing rate of temperature has been varied within 0.15 °C and 0.60 °C/decade in the Himalayan territory (Shrestha et al., 2010), and as a consequence, the extension and quantity of glacial lakes are rising abruptly. Recently, retreating of few debris-covered glaciers because of the extension of glacial lakes has been notified in the Himalayan region (Sakai et al., 2000). The lake water exerts tremendous pressure on the end moraine as a result of swift melting glacial ice due to global warming and increasing frequency of rainfall and the entire mechanism acts as a triggering factor to cause GLOF at devastating scale. Climate change, especially the increasing trend of temperature, augments the occurrence of flash flood along with GLOF due to the melting of valley glacier and glacial lakes. These disastrous events would be intensified due to severe summer rainfall caused by global warming (Chalise, 1994; Shrestha, 1997; IPCC, 2001a, 2001b).

Various studies suggest that the mountain glaciers are shrinking in a significant number and their changes are linked with climatic variations. This will have a profound impact on snow accumulation, stream runoff patterns, and glacial lake which are potential sources of danger to people and property in the valleys below them (ICIMOD, 2011). Furthermore, the continuous shrinkage of ice coverage due to global climatic change has a devastating aftermath on fluvial discharge during dry spells over the Himalayan region. The low flows of the Himalayan river are lesser compared with high flows, and the notable decrease of ice coverage tremendously diminishes the discharge during dry periods. Climatic change in this mountainous zone accentuates the magnitude of rainfall which aggravates the probability of landslide occurrence throughout this area. The fact is noteworthy that the hydrological sphere of this region is also mostly impacted due to the acute climate change because this regime has a high dependency upon the water supply due to glacial melting during dry spell. It must be mentionable in this context that there is a strong correlation in between temperature and discharge of rivers, and the previous study proved that the discharge rate is slightly increased in terms of the rising of temperature (Chalise et al. 2006). The necessary remedial measures must be adopted to cope with the climatic change and reduce the magnitude of its deadly consequences in the eastern Himalayan region for promoting environmental and socioeconomic sustainability.

Location of the Study Area

The North Sikkim district (4296 km^2) is a part of the inner mountain ranges of the Himalayas and the largest among the four districts of Sikkim. To the north and east, it is bordered by the vast Tibetan Plateau, while the southern and south-eastern boundary lies along the west, south, and east districts of Sikkim (Fig. 3.1). Elevation ranges from 8598 m (Khangchendzonga) to 512 m. (near Dikchu) and maximum area (1499.9 km^2) of the district has steep (30°–45°) to very steep slope (45°–60°). Geologically, the area is characterized by young fold mountains and falls in earthquake zone V. Seismic activity takes place between the Main Boundary Thrust (MBT) and the Main Central Thrust (MCT) which are the two major structural elements in the Eastern Himalaya. The district experiences humid climate with seasonally intense rainfall. Mean annual rainfall (925–3389 mm) tends to increase from north-west to south-east. The northernmost part of the district is covered with dense snow, which feeds the major river systems. The district is drained by the river Teesta and its tributaries in six major watersheds, namely, the Teesta upper right bank (Lachen Chu), Teesta

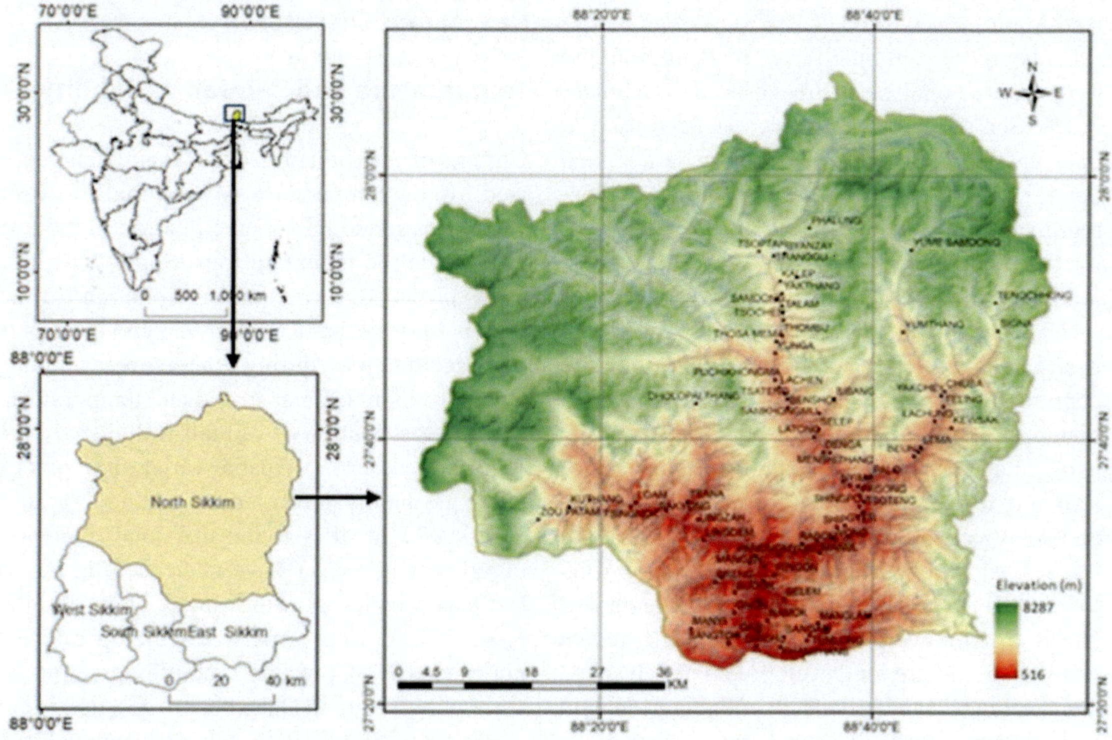

Fig. 3.1 Location map of North Sikkim district. Major drainage lines (light blue) are superimposed on the elevation map, which is based on SRTM DEM (30 m)

upper left bank (Lachung Chu), Yumthang Chu, Chhombo Chu, Zemu Chu, and Rangyong Chu. The major five predominant forest types in the district are East Himalayan Wet Temperate Forest, Oak Rhododendron Forest, East Himalayan Mixed Coniferous Forest, East Himalayan Sub-Alpine Birch Fir Forest, and Alpine Scrub Pasture. The soils are mostly loamy sand to silty clay loam in 30–100 cm thickness. The district accounts for only 7.13% of the state's population with a density of 10 persons per sq. km (Census 2011).

Material and Methods

The precipitation (1950–2002) and temperature (1950–2002) data used in this study are downloaded from the India Water Portal (http://www.indiawaterportal.org/met_data/). The changing nature of glaciers in the study area have been analyzed by using satellite imagery of LANDSAT TM 1990 and IRS P6-LISS III of 1997 and 2004. Inventory of glacial lakes have been prepared from Survey of India topographic maps and multidate satellite images including Google Earth images. Landslide locations are identified from the Survey of India topographic maps and satellite images. The Central Water Commission has a hydrological monitoring network of Teesta basin in Sikkim. Discharge data (1998–2002) of Lachen and Lachung stations and surface runoff, and sediment yield data (1992–2008) of Mangan station are collected and analyzed.

Long-term climate variability has been analyzed from monthly precipitation, maximum and minimum monthly temperature data. Monthly precipitation data of each year are grouped into pre-monsoon, monsoon, and post-monsoon along with annual precipitation in each year. Trend lines have been drawn to show the changing character of seasonal and annual precipitation during the considered period (1950–2002). Maximum temperature during summer and minimum temperature in each year has also been considered, and trend lines have been drawn. Mann-Kendall test (Mann, 1945; Kendall, 1975) of trend analysis is being used in time series precipitation and temperature analysis (Sreekesh & Debnath, 2016). Sen's slope (Sen, 1986) has been used to find out the rate of changes in time series data. Sen's slope is a widely used method of linear regression to estimate the true slope of an existing trend (as change per year) (Salmi et al., 2002; Salerno et al., 2014).

Glaciers and glacial melt lake have shown the best response to climate variation, and they are continually adjusting their size. The glacial boundaries were delineated from satellite images using image enhancement techniques. Data set of spatial extent of all glaciers for the year of 1990 and 2004 were prepared by digitizing the satellite images, and overlaying technique has been used to identify the areal changes. Glacial lake data of North Sikkim were extracted from the recent study of the Centre for Development of Advanced Computing (C-DAC), Pune, in collaboration with the Sikkim State Council of Science & Technology, Gangtok (Basnett et al., 2011). Annual rate of lake surface area changes in four phases (1965–1976, 1976–1989, 1989–1997, and during 1997–2010) were estimated. A scatter diagram was prepared to estimate the long-term (1965–2010) changes in total area of these glacial lakes. North Sikkim district comprises six major watersheds (Chhombo Chu, Zemu Chu, Yumthang Chu, Lachung Chu, Lachen Chu, and Rangyong Chu). These watershed boundaries were delineated form the SoI toposheets and old landslide scars were mapped. Frequency of old (mapped from toposheet) and new (mapped from satellite image) landslides were estimated. In order to assess the impact of temperature variability on river discharge during non-monsoon season, temperature and discharge data of Lachen and Lachung stations were analyzed. Polynomial (second-order) trend lines for both stations were drawn to show their relationship. Runoff and sediment yield relationship have been analyzed by considering annual runoff (Ha.M) and annual sediment yield (Ha.M) data from 1992 to 2008 of Mangan station (Teesta River). A scatterplot with a liner trend line was drawn to show the relationship among these two variables.

Result and Discussion

Temperature and Rainfall Variability

Climate of Sikkim Himalaya is mainly characterized by low temperature, high rainfall on windward slopes, and heavy precipitation in the form of snow at the mountain tops. South-west monsoon provided torrential precipitation in this region during June to September, and there is little precipitation during the winter months. However, there is great spatial and temporal variation in distribution of rainfall in the study area (Ghosh et al., 2019). Spatial variation in precipitation is mainly governed by altitude variation across Sikkim. It is found that total number of rainy days in a year have decreased during the last two decades (Bawa & Ingty, 2012; Rahman et al., 2012). Long-term average temperature do not exceed 20 °C, whereas monthly precipitation varies from 7 mm in December to 538 mm in July (Debnath et al., 2019). Analysis of annual average maximum and minimum temperature of north Sikkim show that there is a very little change in the maximum temperature, but the minimum temperature has increased by almost 2.5 °C during 1950 and 2012. Long-term (1950–2002) rainfall and temperature irregularities are specifically studied through Mann-Kendall trend analysis and Sen's slope analysis (Table 3.1).

Dendrochronology study in North Sikkim shows that 1978–1987 was the warmest (+0.25 °C) decade at Yumthang, North Sikkim, followed by cooler years in the early 1990s (Bhattacharyya & Chaudhary, 2006). Another study made by Yadava et al. (2015) shows that there is an alternative warming and cooling in decadal scale in Lachung and Lachen of North Sikkim since 1852, and they have identified the decade of 1996–2005 as the warmest period. Borgaonkar et al. (2018) found that the Sub-Himalayan West Bengal and Sikkim monsoon rainfall curve attained its peak in the 1990s.

From the temperature data of North Sikkim (1952–2012), it can be perceived that the mean minimum temperature is significantly increasing (Fig. 3.2) than the mean maximum temperature (Fig. 3.3). The trend line of the mean minimum

Table 3.1 Summary of long-term (1950–2002) trend analysis of seasonal precipitation and temperature of North Sikkim

Precipitation			Temperature		
Season	Kendall's tau	Sen's slope (mm/year)	Season	Kendall's tau	Sen's slope (mm/year)
Monsoon	−1.10	−2.894	Summer maximum	1.35	0.007
Pre-monsoon	2.63	2.164			
Post-monsoon	0.94	0.751	Winter minimum	3.36	0.027
Annual	0.10	0.396			

Data Source: (India Water Portal)

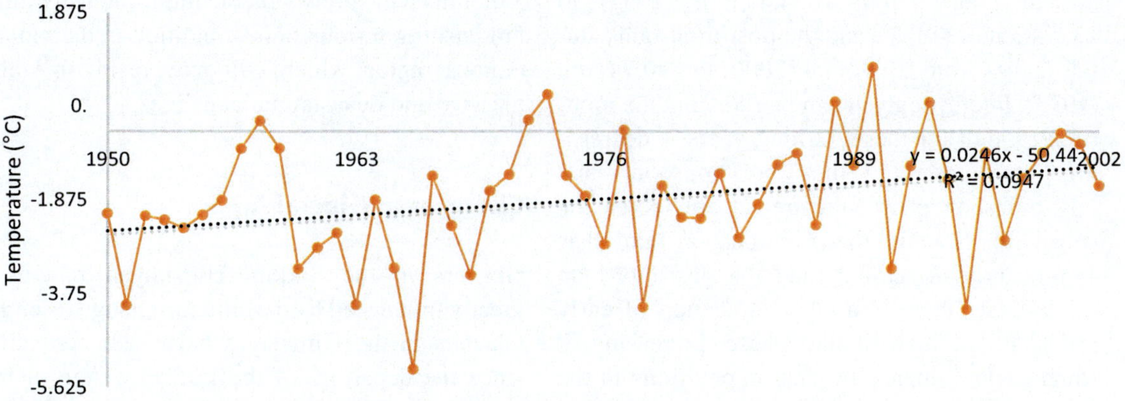

Fig. 3.2 Winter minimum temperature trend for the period 1952–2002 in North Sikkim

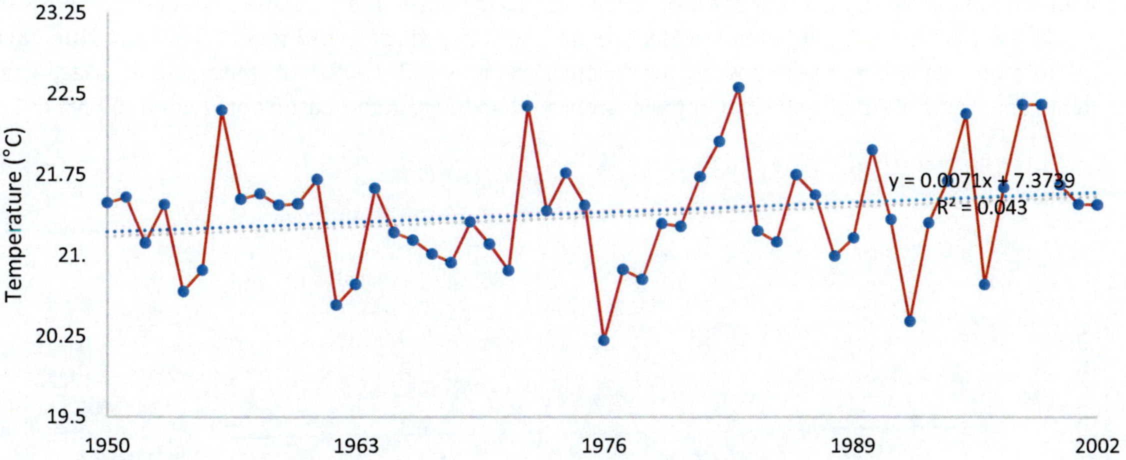

Fig. 3.3 Annual mean-maximum temperature trend for the period 1950–2002 in North Sikkim

temperature shows that it has increased about 1 °C in the past years whereas the mean maximum temperature is increased by 0.5 °C. So, it can be concluded that the winter season is getting warmer day by day as the minimum temperature is experienced by the winter months in this region. The summer season is also becoming warmer but not significant as the winter months ($R^2 = 0.0947$). The previous years data show a decreasing trend in temperature range where the difference between the maximum and minimum temperature is decreasing which will disrupt the heat balance of this region. Increasing of Winter temperature is becoming a serious issue for this region as it will trigger the glacier melting process, resulting in some major glacier outburst floods in the future.

To analyze the rainfall trend of 1950–2002, the monthly data has been divided into three time span, i.e., pre-monsoon, monsoon, and post-monsoon. The pre-monsoon rainfall ranges from 96.95 mm (1966) to 467.98 mm (1978), monsoon rainfall ranges from 1064.4 mm (1957) to 2309.83 mm (1952), and the post-monsoon rainfall ranges from 23.39 mm (2000) to 590.74 mm (1973). Interestingly, it can be seen that the monsoon rainfall for these 53 long years is decreasing, whereas in the pre-monsoon and post-monsoon phase, the trends are increasing (Fig. 3.4). Notably, the pre-monsoon rainfall is significantly increasing than the post-monsoon rainfall. So, there is a clear indication of early arrival of the high rainfall phase. Increasing of irregular and intense tropical depressions in the recent time may be responsible for this high rainfall in the pre-monsoon season. Increasing rainfall in the post-monsoon season may be associated with the high western disturbances. So, it can be stated that the total rainfall amount is decreasing in the monsoon season as it shows a decreasing trend. Intense heating of the Tibetan plateau due to extensive summer temperature in the recent times is expected to modify the whole monsoon mechanism which may be accountable for the shifting of the wet spells from monsoon to pre-monsoon seasons. The gradual shift of the high rainfall phase is creating a serious problem for the farmers for which the agricultural production is getting adversely affected. Lack of rainfall in monsoon time (June to September) can create water imbalance and difficulties in groundwater recharge. At the same time, high rainfall within 3 months will surely generate high-intensity rainfall ensuing serious flood condition in the mountainous region which will may result in some massive and devastating landslides.

Changes in Glacial Area

Glaciers in the Sikkim Himalayan region is greatly influenced by the climatic change recently. Glaciers in the Himalayas have been retreating since the departure of the Little Ice Age of the mid-nineteenth century (Fushimi & Ohata, 1980; Yamada, 1992; Karma et al., 2003) and during the last few decades, many glaciers in the Himalayas retreating rapidly. Karma et al. (2003) reveal that the glaciers in India and Bhutan Himalayas regions (87–100%) are retreating at a faster rate than those in the East Nepal region (57.3%).

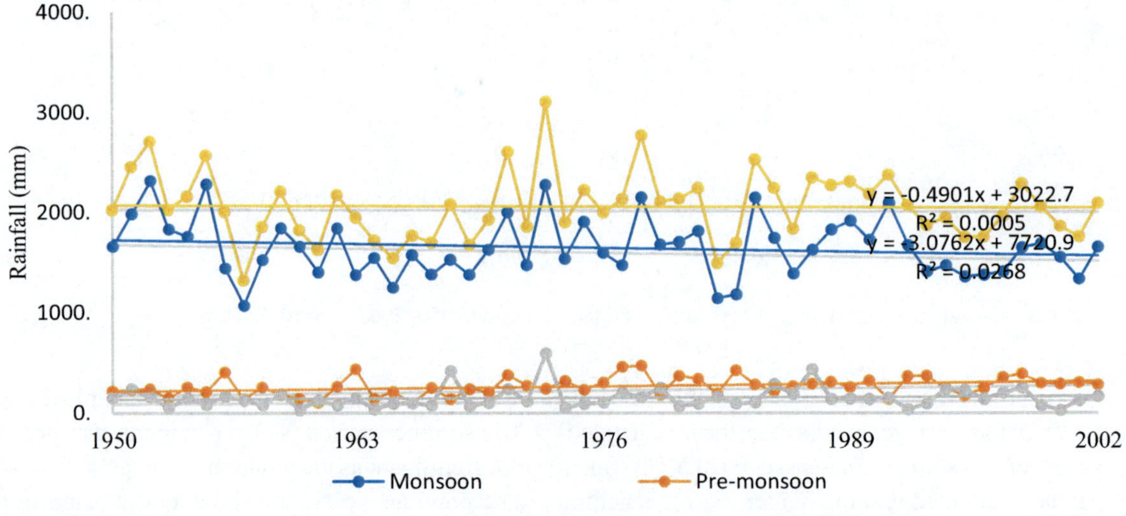

Fig. 3.4 Long-term (1950–2002) trend analysis of seasonal precipitation of North Sikkim district

Table 3.2 Changes of glaciers during 1990–2004 in Sikkim Himalayan region

Glacier area (km^2)	No. of glaciers 1990	Areal extent (km^2)		Loss in area between 1990 and 2004	
		1990	2004	km^2	Area (%)
<1	4	1.80	1.68	0.12	6.67
1–5	9	18.66	17.81	0.85	4.56
5–10	4	26.94	26.33	0.61	2.26
>10	3	154.73	153.94	0.79	0.51
Total	20	202.13	199.76	2.37	1.17

The Sikkim State Council of Science & Technology have monitored twenty glaciers in the Sikkim Himalayan region. The result shows that in 1990, the total area of the 20 glaciers was 202.13 (±4.68) km^2) and it has been reduced to 201.56 (±4.71) km^2 in 1997 and 199.76 (±4.63) km^2 in 2004. The change in glacier area also varies for different sizes of glaciers (Table 3.2). All glaciers are not retreating at a same rate. The glaciers above 10 km^2 have lost their area by 0.51%, whereas the smaller glaciers (<1 km^2) have lost their area by 6.67% during this period (1990–2004). This result indicates that smaller glaciers are more sensitive to temperature rise and they have significant response to climatic variations in this region. The study also reveals that during 1990–1997, 14 glaciers have been retreated and 6 glaciers have shown no advance or retreat. In 2004, it is observed that 18 glaciers (out of 20) have been retreated.

Changes in Glacial Lake Area

Several studies show that many glaciers in this region are leaving glacial lakes and this fact can be related with the intermediate effects of long-term climate change. A study being carried out by the Centre for Development of Advanced Computing (C-DAC), Pune, jointly with the Sikkim State Council of Science & Technology, Gangtok, have revealed that many glacial lakes in the Sikkim Himalayan region have grown over the years revealing the impact of climate change on glacial lakes. The status of these glaciers have become a measuring stick of climate change (SAC 2010). Changes in glacial lake area during 1965–2010 of ten glacial lakes of Teesta and Rangit basin in North Sikkim district have been analyzed. Excepting the glacial lake feeding to the river Shako Chu, all glacial lakes are located above 5000 m altitude.

The study reveals that excepting the Gurudongmar Chho-A, all other lakes have been increasing in size since 1965 and these changes can be analyzed from the view point of climatic change response. The maximum rate (0.046 km^2/year) of lake area expansion is found in Gurudongmar Chho "B" during 1976–1989 (Fig. 3.5). The significant increase in the Gurudongmar Chho "B" (four times) and "C" (two times) during 1965–2010 clearly indicate the glacier retreat or melt and accumulation of more melt water in the moraine dammed lakes. The change detection analysis shows that total lake surface area was 4.88 km^2 in 1965 and it has been increased to 8.92 km^2 in 2010. That study on South Lhonak glacial lake also reveals that the lake has been increased by 0.79 km^2 from 1965 to 2010, which is almost four times of its area (0.24 km^2) of 1965. During the last 13 years of study period (1997–2010), this lake has been expanded by 0.39 km^2 at the rate of 0.03 km^2 per year. A scatterplot (Fig. 3.6) of individual lake area in 1965 and 2010 shows that the individual lake surface area has been increased at a significant level.

Response to Landslide

The annual and monsoon rainfall shows a considerable spatial and temporal variability in North Sikkim district. The relationship between rainfall and landslide was observed by Ayalew (1999). Heavy precipitation can initiate mass wasting through hydrostatic pressure and serves to lubricate the slides once these are in motion. Tsereteli (in UNESCO/UNEP, 1988) also found close relationship exists between landslide and the amount of precipitation. Landslides are more likely to occur if high amounts of intense rainfall are preceded by a period of light but incessant precipitation which generally saturates the soil

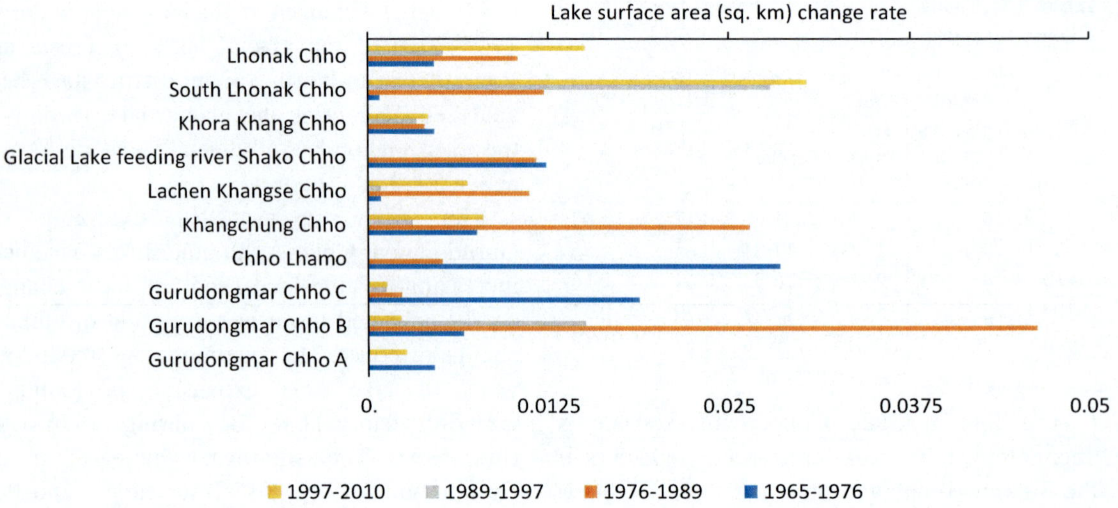

Fig. 3.5 Lake surface area change rate (km²/year) during 1965–2010

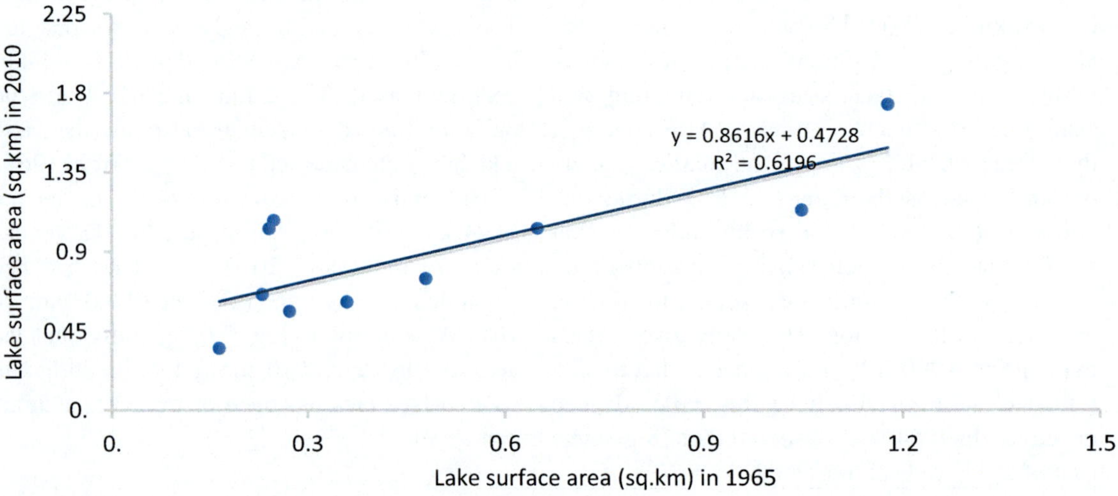

Fig. 3.6 Scatterplot of lake surface area of 1965 and 2010

(Ghosh et al., 2019). Most of the landslides in Sikkim Himalayan region are influenced by high rainfall intensity during monsoon period (Koley et al., 2019). Thus, spatiotemporal rainfall variability has a considerable role in influencing the landslide dynamics of the study area. Several studies show that individual rainfall events can cause slope failures in areas of limited extent or in large regions. The annual average rainfall (2010–2016) of North Sikkim hydro-meteorological division of India which encompasses the present study area (part of Sikkim Himalayas) is 3333.7 mm. During the monsoon period (June–September), the mean rainfall is 2136.9 mm, contributing 64.1% of annual rainfall (3333.7 mm). A study done by Sengupta et al. (2010) indicates that landslide at Lanta Khola in North Sikkim was initiated with normalized cumulative rainfall over a period of at least 15 days exceeds 250 mm. Koley et al. (2019) have identified 210 landslides which occurred during 2010–2016. It was found that out of these, 155 landslides were initiated by rainfall. They stated that 95.24% of landslide events occurred under the influence of 10-day antecedent rainfall prior to failure.

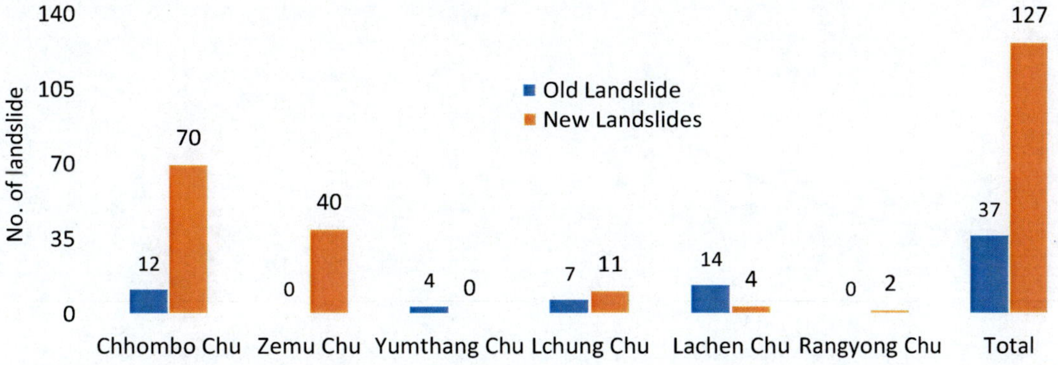

Fig. 3.7 Frequency of old and new landslides in various watersheds in the North District of Sikkim Himalaya

The status of landslide distribution (Fig. 3.7) in different watersheds in North Sikkim district show that number of new landslides (identified from satellite data of 2002) are comparatively larger than the old slides (identified from 1977 SoI toposheets of 1:50,000). The frequency of landslides in the last few years may be partially controlled by human interferences along with the geo-environmental factors of the region. But the climatic variability is controlling their frequency over different years.

Hydrological Impacts of Climate Warming

In addition to direct impact on glacial and glacial lake dynamics, climate change can have serious impacts on the hydrological regime of fluvial system as it is considerably dependent on snow melt water during dry season. In order to assess the impact of temperature on dry season flows at higher elevations, discharge data (1998–2002) from two hydrological stations (Lachen and Lachung) in North Sikkim were analyzed. Both stations indicate positive correlation between temperature and discharge (Fig. 3.8). Rising trend of the minimum temperature in this high-altitude region leads to glacial melting and contributes to the river flows. Although, due to the short data period a definite conclusion cannot be made at this stage. However, this positive correlation exists between temperature and river discharge at both stations during the dry season, supporting the postulation that climate warming in this high-altitude region can affect the hydrological regime of the Himalayan rivers.

Runoff and Sediment Yield

Surface runoff is the direct product of rainfall in a river basin. River discharge is directly controlled by the amount and frequency of precipitation in the river basin (Milliman & Farnsworth, 2013; Zhang et al., 2008) and amount of soil erosion (Anders et al., 2008), which turns to influence the sediment load (Yang & Lu, 2014). Global warming may result in a change in glacial melting (Piao et al., 2010) and thus, influences streamflows.

Metrological data analysis indicates that there are no significant changes in annual and post-monsoon precipitation, whereas pre-monsoon rainfall has increased significantly. Although there is insignificant change in annual rainfall, the total surface runoff and sediment discharge have increased significantly (Fig. 3.9), and that runoff could be the results of land use changes in the district. The inner-year distribution runoff and sediment load are studied on a seasonal basis. Hydraulic erosion is considered to be the main driver of the sediment yield which could be considered as the results of hydraulic erosion (Yang et al., 2004, 2005; Xu, 2009), which can be directly persuaded by the local rainfall intensity and duration. Scatterplot of sediment yield and surface runoff show that sediment load weakly depends on

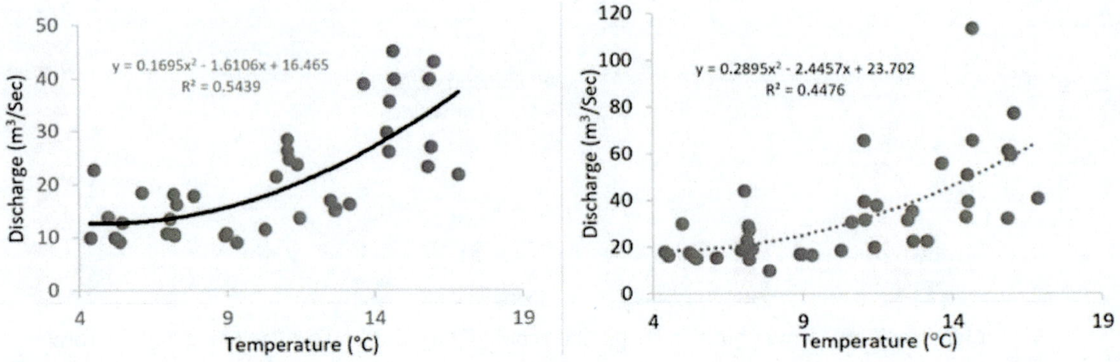

Fig. 3.8 Relationship between temperature and dry season discharge at Lachung River (left) and Lachen River (right)

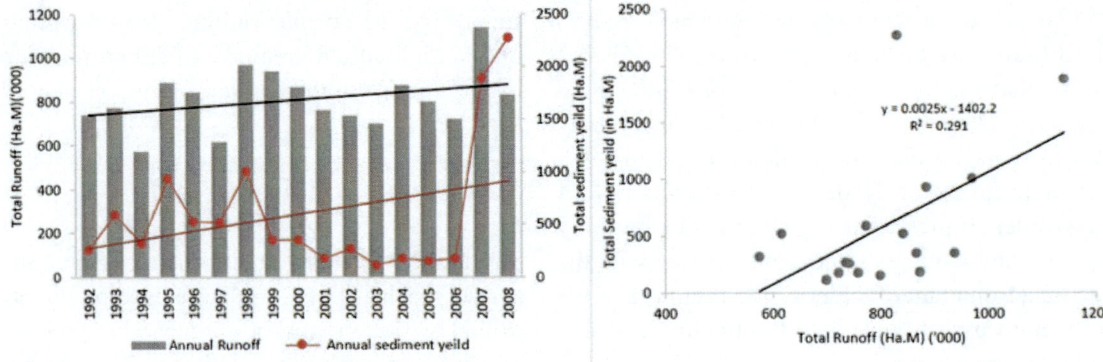

Fig. 3.9 Variation of runoff and total sediment load in the Teesta River (Mangan station). The straight solid lines represent the trend of variation during 1992–2008 (left). Scatterplots of annual surface runoff and annual sediment yield in the Teesta River at Mangan station (right)

surface runoff ($r^2 = 0.29$). Therefore, it can be said that the sediment load is more dependent on the water discharge compared to precipitation or precipitation intensity. Glacial melting and large numbers of landslide along the river side may contribute significant amount of sediment to the streamflow. However, the outcome of regression analysis may not imply that rainfall and resultant surface runoff is weakly or insignificantly correlated to sediment yield load in a physical sense, because of the limitation in number of meteorological stations (Zhang et al., 2019).

Conclusion

Long-term metrological data analysis has revealed the fact of climate change and its impact on hydro-geomorphic processes in the high-altitude region of Sikkim Himalaya. Time series analysis shows that there is no significant change in average temperature, but the average minimum temperature during winter has increased almost by 2.5 °C during 1950–2012. This significant change in winter temperature could have influence in glacial dynamics in high-altitude regions. However, there is great spatial and temporal variation in distribution of rainfall in the study area, as precipitation is mainly governed by altitudinal variation across the Sikkim. There are insignificant changes in annual precipitation, but pre-monsoon shows increasing trend at a significant level, whereas monsoon and post-monsoon seasons do not reflect any trend. In analysis of glacial and glacial lake inventories using existing records, toposheet, and satellite imageries of the present study area to understand the climate change impact on their dynamics, it is found that many glacial lakes have expanded over the last five decades. Therefore, it has become urgent to identify the rapidly growing glacial lakes and understand their growth mechanism. This scientific investigation, inventories, prioritization, and monitoring could help to develop management strategies of future disastrous GLOF events. Landslides in the Sikkim Himalayan region are greatly controlled by intensity and amount of precipitation. Increasing trend in pre-monsoon rainfall may have provided extra time to soil and pours rocks to become saturated which increases the landslide vulnerability in this region. It is found that the number of rainy days has decreased in the last two decades, but annual rainfall has not been changed significantly, which reveals the fact of increasing trend in rainfall intensity. Intense rainfall during short time span could produce more surface runoff and result in sediment load. However, runoff and erosion are controlled by land use and land cover pattern along with inherent characteristics of the land surface. Therefore, it cannot be concluded that climate change is the only responsible factor in runoff and sediment yield dynamics. In addition, it would be very uncertain to conclude the exact influence of temperature change on surface runoff, river water discharge, and sediment discharge based on limited data set. Future work will be attempted to incorporate the other factors of hydro-geomorphic process along with precise metrological data of considerable metrological stations for better understanding of the degree of impact of climate change on surface changes.

References

Anders, A. M., Roe, G. H., Montgomery, D. R., & Hallet, B. (2008). Influence of precipitation phase on the form of mountain ranges. *Geology, 36*(6), 479–482.

Arora, M., Goel, N. K., & Singh, P. (2005). Evaluation of temperature trends over India. *Hydrological Sciences Journal, 50*(1), 81–93.

Ayalew, L. (1999). The effect of seasonal rainfall on landslides in the highlands of Ethiopia. *Bulletin of Engineering Geology and the Environment, 58*, 9–19.

Bawa, K. S., & Ingty, T. (2012). Climate change in Sikkim: A synthesis. In M. L. Arrawatia & S. Tambe (Eds.), *Climate change in Sikkim patterns, impacts and initiatives* (pp. 413–424). Information and Public Relations Department, Government of Sikkim.

Bhattacharyya, A., & Chaudhary, V. (2006). Late-summer temperature reconstruction of the eastern Himalayan region based on tree–ring data of Abies densa. *Arctic, Antarctic, and Alpine Research, 35*, 196–202.

Borgaonkar, H. P., Gandhi, N., Ram, S., & Krishnan, R. (2018). Tree-ring reconstruction of late summer temperatures in northern Sikkim (eastern Himalayas). *Palaeogeography Palaeoclimatology Palaeoecology, 504*, 125–135.

Basnett, S., Kulkarni, A., Shresta, D.G. (2011). Glacier Studies in Sikkim Himalaya. https://doi.org/10.13140/RG.2.2.24023.88487

Chalise, S. R. (1994). Mountain environment and climate change in the Hindu Kush–Himalaya. In M. Beniston (Ed.), *Mountain environments in changing climate* (pp. 382–404). Routledge.

Chalise, S.R., Shrestha, M.L., Bajracharya, O.R., Shrestha, A.B. (2006). Climate change impacts on glacial lakes and glacierized basins in Nepal and implications for water resources. In Climate Variability and Change. *Hydrological Impacts; IAHS Publications: Hamburg, Germany*, *308*, 460–465.

Dash, S. K., Jenamani, R. K., Kalsi, S. R., & Panda, S. K. (2007). Some evidence of climate change in twentieth-century India. *Climatic Change, 85*, 299–321.

Debnath, M., Sharma, M. C., & Syiemlieh, H. J. (2019). Glacier dynamics in Changme Khangpu Basin, Sikkim Himalaya, India, between 1975 and 2016. *Geosciences, 9*(259), 1–21.

Dey, P., & Mishra, A. (2017). Separating the impacts of climate change and human activities on streamflow: A review of methodologies and critical assumptions. *Journal of Hydrology, 548*, 278–290.

Debnath, M., Syiemlieh, H.J., Sharma, M.C., Kumar, R., Chowdhury, A., Lal, U. (2018.) Glacial lake dynamics and lake surface temperature assessment along the Kangchengayo-Pauhunri Massif, Sikkim Himalaya, 1988–2014. *Remote Sens. Appl. Soc. Environ., 9*, 26–41.

Fushimi, H., & Ohata, T. (1980). Fluctuations of glaciers from 1970 to 1978 in the Khumbu Himal. *Journal of the Japanese Society of Snow and Ice (Special Issue), 41*(4), 67–70.

Ghosh, G. K., & Ghosh, K. (2019). Spatio–temporal analysis of rainfall pattern in North District, Sikkim, India. *International Advanced Research Journal in Science, Engineering and Technology, 6*(5), 88–92.

Ghosh, K., Ghosh, G. K., & De, S. K. (2019). *Application of frequency ration model for landslide hazard zonation of North Sikkim District, Sikkim, India.* Geographical Review of India 81 (in press).

Goswami, B. N., Venugopal, V., Sengupta, D., Madhusoodanam, M. S., & Xavier, P. K. (2006). Increasing trends of extreme rain events over India in a warming environment. *Science, 314*, 1442–1445.

ICIMOD. (2011). *Glacial lakes and glacial lake outburst floods in Nepal*. Report.

IPCC. (2001a). In J. J. McCarthy, O. F. Canziani, N. A. Leary, D. J. Dokken, & K. S. White (Eds.), *Climate change 2001: Impacts, adaptation and vulnerability. Contribution of working group II to the third assessment report of the intergovernmental panel on climate change*. Cambridge University Press.

IPCC. (2001b). In J. T. Houghton, Y. Ding, D. J. Griggs, M. Noguer, P. J. van der Linden, & D. Xiaosu (Eds.), *Climate change 2001: The scientific basis. Contribution of working group i to the third assessment report of the intergovernmental panel on climate change*. Cambridge University Press.

IPCC. (2007). Climate change 2007: Synthesis report.

Karma, T., Ageta, Y., Naito, N., Iwata, S., & Yabuki, H. (2003). Glacier distribution in the Himalayas and glacial shrinkage from 1963 to 1993 in the Himalayas. *Bulletin of Glaciological Research, 20*, 29–40.

Kendall, M. G. (1975). *Rank correlation methods*. Griffin.

Koley, B., Nath, A., Saraswati, S., Bandyopadhyay, K., & Ray, B. C. (2019). Assessment of rainfall thresholds for rain–induced landslide activity in North Sikkim road corridor in Sikkim Himalaya, India. *Journal of Geography, Environment and Earth Science International, 19*(3), 1–14.

Kumar, V., Jain, S.K., Singh, Y. (2010). Analysis of long-term rainfall trends in India. *Hydrol. Sci. J. 55*(4), 484–496.

Khan, T., Singh, O., Rahman, Md. (2000). Recent Sea Level and Sea Surface Temperature Trends Along the Bangladesh Coast in Relation to the Frequency of Intense Cyclones. *Marine Geodesy, 23*(2), 103–116.

Lal, M. (2001). Climatic change – Implications for India's water resources. *Journal of Indian Water Resource Society, 21*, 101–119.

Lal, M. (2003). Global climate change: India's monsoon and its variability. *Journal of Environmental Studies and Policy, 6*, 1–34.

Milliman, J. D., & Farnsworth, K. L. (2013). *River discharge to the Coastal Ocean: A global synthesis*. Cambridge University Press.

Min, S. K., Kwon, W. T., Parkand, E. H., & Choi, Y. (2003). Spatial and temporal comparisons of droughts over Korea with East Asia. *International Journal of Climatology, 23*, 223–233.

Mirza, M. Q. (2002). Global warming and changes in the probability of occurrence of floods in Bangladesh and implications. *Global Environmental Change, 12*, 127–138.

Mooley, D. A., & Parthasarthy, B. (1984). Fluctuations of all India summer monsoon rainfall during 1871–1978. *Climatic Change, 6*, 287–301.

Mann, H.B. (1945). Nonparametric Tests Against Trend. *Econometrica, 13*, 245–259.

Pant, G. B., & Kumar, K. R. (1997). *Climates of South Asia*. Wiley.

Piao, S., Ciais, P., Huang, Y., Shen, Z., Peng, S., & Li, J. (2010). The impacts of climate change on water resources and agriculture in China. *Nature, 467*(7311), 43.

Rahman, H., Karuppaiyan, R., Senapati, P. C., Ngachan, S. V., & Kumar, A. (2012). An analysis of past three decade weather phenomenon at the mid–hills of Sikkim and strategies for mitigating possible impact of climate change on agriculture. In M. L. Arrawatia & S. Tambe (Eds.), *Climate change in Sikkim patterns, impacts and initiatives, information and public relations department* (pp. 19–48). Government of Sikkim.

Sakai, A., Takeuchi, N., Fujita, K., & Nakawo, M. (2000). *Role of supraglacial ponds in the ablation process of a debris–covered glacier in the Nepal Himalayas* (pp. 119–132). IAHS Publication.

Sen, P. B. (1986). Estimates of the regression coefficient based on Kendall's tau. *Journal of the American Statistical Association, 63*(324), 1379–1389.

Sengupta, A., Gupta, S., & Anbarasu, K. (2010). Rainfall thresholds for the initiation of landslide at Lanta Khola in North Sikkim, India. *Natural Hazards, 52*, 31–42.

Shrestha, M. L. (1997). Development of climate scenarios with reference to Nepal. In *Proceedings of the workshop on climate change in Nepal (US Country Studies Program, Kathmandu, Nepal)* (pp. 16–32).

Shrestha, A. B., & Aryal, R. (2011). Climate change in Nepal and its impact on Himalayan glaciers. *Regional Environmental Change, 11*(Suppl 1), 65–77.

Shrestha, A. B., Wake, C. P., Dibb, J. E., & Mayewski, P. A. (2000). Precipitation fluctuations in the Nepal Himalaya and its vicinity and relationship with some large-scale climatological parameters. *International Journal of Climatology, 20*, 317–327.

Space Applications Centre (2010). National Wetland Atlas – Sikkim, under the "National Wetland Inventory and Assessment" project sponsored by Ministry of Environment and Forests, Govt. of India, Atlas: SAC/RESA/AFEG/NWIA/ATLAS/13/2010.

Salerno, F., Gambelli, S., Viviano, G., Thakuri, S., Guyennon, N., Dagata, C., Diolaiuti, G., Smiraglia C., Stefani, F., Bocchiola, D., Tartari, G. (2014). High alpine ponds shift upwards as average temperatures increase: A case study of the Ortles–Cevedale mountain group (Southern Alps, Italy) over the last 50years. *Global and Planetary Change, 120*, 81–91.

Salmi, T., Maatta,A., Pia, A., Tuija , R., Amnell, T. (2002). Detecting Trends of Annual Values of Atmospheric Pollutants by the Mann-Kendall Test and Sen's Solpe Estimates the Excel Template Application MAKESENS. *Publications on Air Quality*, 31.

Sreekesh, S., Debnath, M. (2016), Spatio-Temporal Variability of Rainfall and Temperature in Northeast India. In: Raju N. (eds) Geostatistical and Geospatial Approaches for the Characterization of Natural Resources in the Environment. *Springer, Cham*, https://doi.org/10.1007/978-3-319-18663-4_135

Shrestha, A., Cameron, W., Jack, D., Paul, M. (2010). Precipitation fluctuations in the Nepal Himalaya and its vicinity and relationship with some large scale climatological parameters. *International Journal of Climatology, 20*, 317 - 327.

Thapliyal, V., & Kulshreshtha, S. M. (1991). Climate changes and trends over India. *Mausam, 42*, 333–338.

UNESCO/UNEP. (1988). *Landslides and mudflows* (Vol. 1, p. 236). Washington State Department of Health Olympia.

Xu, J. (2009). Some rainfall–related thresholds for erosion and sediment yield in the upper Yangtze River basin. *Environmental Geology, 56*(6), 1183–1192.

Yadava, A. K., Yadav, R. R., Misra, K. G., & Singh, J. (2015). Tree ring evidence of late summer warming in Sikkim, Northeast India. *Quaternary International, 371*, 175–180.

Yamada, T. (1992). Report for the first research expedition to IMJA Glacier Lake, 25 March to 12 April 1992. WECS Report No. 3/4/120892/1/1, Seq. No. 412. Kathmandu: WECS/JICA.

Yang, X., & Lu, X. (2014). Estimate of cumulative sediment trapping by multiple reservoirs in large river basins: An example of the Yangtze River basin. *Geomorphology, 227*, 49–59.

Yu, X., Xie, X., & Meng, S. (2017). Modeling the responses of water and sediment discharge to climate change in the upper Yellow River Basin, China. *Journal of Hydrologic Engineering, 22*(12), 05017026.

Yang, S., Shi, Z., Zhao H., Li, P., Dai, S., Gao, A. (2004). Research Note: effects of human activities on the Yangtze River suspended sediment flux into the estuary in the last century. Hydrol. *Earth Syst. Sci. 8*(6), 1210–1216.

Yang, S., Zhang, J., Zhu, J., Smith, J., Dai, S., Gao, A., Li, P. (2005). Impact of dams on Yangtze River sediment supply to the sea and delta intertidal wetland response. *J. Geophys. Res. Earth Surf,* 110(F3). https://doi.org/10.1029/2004JF000271

Zhang, Q., Xu, C., Zhang, Z., Chen, Y., Liu, C., & Lin, H. (2008). Spatial and temporal variability of precipitation maxima during 1960–2005 in the Yangtze River basin and possible association with large–scale circulation. *Journal of Hydrology, 353*(3–4), 215–227.

Zhang, X., Yan, H., Yue, Y., & Xu, Q. (2019). Quantifying natural and anthropogenic impacts on runoff and sediment load: An investigation on the middle and lower reaches of the Jinsha River basin. *Journal of Hydrology: Regional Studies, 25*, 1–19.

4

Disasters, Climate Change, and Migration Concerns in Fiji and the Pacific: An Overview

Mohammad Afsar Alam and Mumtaz Alam

Introduction

Disasters, migration, and climate change are complex and interlinked environmental and societal issues. They have recently been given more attention primarily from the perspective of climate change or the so-called environmental migration (Foresight 2011; Piguet 2013; UNDP 2010).

Though economic and social factors can affect migration, the consequences of climate change may be the deciding factor for a person or family to relocate. For instance, river deltas and synclinal coastal areas can become less preferred for settlement due to limited income and food security choices. Habitability can be harmed as a result of reduced precipitation or elevated disease trajectories.

In world estimates for the number of refugees leaving as a consequence of climate change, according to the International Organisation for Migration (IOM), by 2050, the population could vary from 25 million to 1 billion, with 200 million being the most often mentioned number.

According to a review of the literature, five flashpoints in the Pacific are prone to become foundation zones for climate-induced migrants, urban atolls, urban areas, coastal, delta, non-urban atolls, and riverine populations, and drought-prone peoples are among them. Coastal areas are particularly prone to predicted climate change because ocean-based livelihoods are being destroyed as a result of amplified marine disturbances. Floods are a major threat to river deltas and are likely to escalate as a result of climate change. Also, various drought-stricken regions could accelerate the pace of migration in the Pacific. Highlands, atolls, and coastal regions of Papua New Guinea are examples of such migrations.

The Intergovernmental Panel on Climate Change (IPCC) stated as "small island states - especially the atoll nations of the Pacific are among the most vulnerable to climate change, seasonal-to-inter-annual climate variability, and sea level rise from which the most vulnerable of these island states are the Marshall Islands, Kiribati, Tuvalu, Tonga, the Federated States of Micronesia, and the Cook Islands".

The Pacific Climate Change Science Programme (PCCSP) recently released a report on the influence of climate-induced migration in this region (ABM and CSIRO, 2011). According to this study, the Pacific region would see an uptick in average temperature, the number of hot days, heavy precipitation, coastal acidification, and an increase in sea level. By 2080–2099, the Pacific's sea level increase is projected to fall in accordance with the global range of 0.18–0.59 m.

M. A. Alam (✉) · M. Alam
Department of Social Sciences, Fiji National University, Suva, Fiji
e-mail: mohammad.alam@fnu.ac.fj; mumtaz.alam@fnu.ac.fj

A. R. Siddiqui, A. Sahay (eds.), *Climate Change, Disaster and Adaptations*, Sustainable Development Goals Series, https://doi.org/10.1007/978-3-030-91010-5_4

Droughts are expected to occur less frequently but of greater intensity. The average annual rainfall is predicted to rise. Cyclones and tropical storms are also supposed to become less frequent however more intense.

Sea level rise, coastal flooding, saltwater infiltration, and more severe and prolonged droughts are the worse consequences of climate change for Pacific Islanders. As a result, the agent-based model predicts that by 2055, migration in Tuvalu and Kiribati would skyrocket under a mild climate change scenario (RCP 6) and population rise. If migration continues to grow, countries like Fiji, New Zealand, and Australia will have an increased number of migrants in the country. It is a matter of great concern because countries like Fiji already have limited resources. In case of any extra pressure, it may cause a lot of problems and will adversely affect the economy.

Natural disaster and climate change result in 80% migration within the same region and country and 20% international (Books, 2014). When migration occurs due to climate change, there is a direct effect on urbanization, brain drain, and rural-urban drift (Ekstrom, 2010). Recent migration has adversely affected urban areas. Asia's urban population increased from 24% of the total in 1970 to 42% in 2007, which is expected to hit 50% in the next decade, with a strong coastal concentration (United Nations, 2018). The countries of the Pacific are vast islands of volcanic origin or low-lying atoll and are similarly highly vulnerable (Books, 2014).

Many Pacific and Asian countries rely mostly on agricultural activities. Because of the change in the weather pattern, the activities are affected, and farmers and agriculturalists migrate for a better farming area or better opportunity. Fiji is said to be the theatre of climate change. In Fiji, places like Ba, Rakiraki, Labasa, or Vanua Levu are worst affected by climate change. People migrate overseas from these regions. From many Asian countries, too, because of the environmental problem, people are forced to move away from their original place (Bacolod, 2014).

Climate change has played and will continue to play a part in future migration, according to a rising consensus. An effort has been made in this report to address an outline of disaster-induced displacement due to climate change in Fiji and the Pacific.

As per Bronen (2010) "in recent decades, there has been an enormous relocation around Asia and Pacific which is a direct result of Climate Change and the debacle it has brought, due to which there are an incredible change in the portability, monetary and social changes". There is an extraordinary development that climate change will assume an indispensable job in movement later on also. A transitory reaction to environmental change will be dictated by past examples of migration whether ecological or financial, constrained, perpetual, or impermanent through longer-settled systems (Cloud, 2002).

A study by Oxford University shows that environmental change directly affects migration since individuals look for thriving, achievement, well-being, and future. Migration and climate change are interchangeable, and it tends to be overseen if strategy and practice can be increasingly successful in nations and worldwide limits. As per Clouds (2014), "relocation because of environmental change is overseen adequately, humanitarian emergencies will be minimized, clashes dodged, and nations can profit if arrangements and guideline can be made in connection with movement in light of climate change".

At the point when we consider climate change and migration, it takes a gander at the number of inhabitants in the nation through and from, the economy of the country, the financial status of the family after and before movement. Climate change influences relocation since individuals in the region may have endured fiascos, for example, flooding, intensive tropical cyclone, dry spell, and different perils which directly affected their vegetation and animals. This impact of climate change brings about 80% inter-regional and within the same country and 20% international (Books, 2014).

There are various reasons and numerous ways climate change has caused migration. Many Pacific and Asian nations depend for the most part on rural exercises, and in light of the adjustment in the climate accomplices, the activities are influenced, and ranchers and agriculturalists relocate for a superior cultivating zone or better chance. In Fiji, places like Ba, Rakiraki, and Labasa or Vanua Levu are flood-prone, and because of environmental change, individuals move abroad to be protected from such calamities and change in the weather pattern. Policies and regulation must be made as relocation extends, to adjust the population growth and save resources. Besides, environmental change influences resources in that locale and because of this, individuals decide to move away from such regions. There are many regions in Asia where due to ecological issue, individuals are compelled to move away from their original place (Bacolod, 2014).

There is no doubt about the role that climate change has played and will play in future relocation; accordingly, this study has endeavoured to discuss the disaster-induced migration regarding climate change in Fiji and the Pacific, whereby it will take a gander at the idea of climate change and migration, in Fiji Islands and the Pacific.

Climate-Induced Migration

As per the report of Clouds (1992), "At whatever point there is an ecological issue on account of environmental change, migration will undoubtedly occur". Therefore, we can say that environmental change and movement are exchangeable. Movement as a result of beachfront flooding, dry season, starvation, bushfire, volcanic emission, and sea level rise (Asian Development Bank). The best example was shown in the movie *Moana*. Environmental change and mankind have an immediate impact since man contributes to climate change.

Man burns trash, swears off the utilization of 3R's (reduce, reuse and recycle), rehearses deforestation, to name not many, and in light of such numbness of man, climate change and its effect return such a way that men are compelled to move away as a result of climate change and its consequences. Along these lines, environmental change sway that prompts automatic or constrained movement, or relocation, is probably going to provoke an increasingly massive reordering of relocation designs (Richardson, 1992). Progressive environmental change is likely to lead to a lack of long-term planning and inadequate adjustment, and rehabilitation measures will deteriorate the adaptive capacity of the victim groups. It could, thusly, upshot in edges of (maximum) tolerance being met, leading to a shift over time from direct changes to indirect changes in migration patterns. Likewise, concerning environmental hazards, indirect changes may result from a significant increase in the repeat or effect of risks or the perceptions of those perils (Bronen, 2010). A research report shows that "migration obviously and widely utilized as a significant piece of adjustment to react to massive changes in financial, ecological, and political settings across Asia and the Pacific" (Geey, 1999).

It is imperative to know that migration is not only for the destitute but for anybody who does not have the means to sustain against it. The inadequate resources and absence of any concrete policy made by the government can make anybody helpless to fight against climate change, and thus migration is bound to happen. So far as migration and climate change are concerned, in many ways, climate change leads to migration. Sea level rise, coastal flooding, bushfire, change in weather patterns, drought, and famine are the chief reasons for movements (Suphachalasai, 2014).

Migration has its disadvantages and advantages. For instance, when a family migrates from one region to another or from one country to another, it will take a lot of time to adjust. A farmer migration from a rural setting to a modern environment may have difficulty in finding jobs that are suitable for him; a modern family migration into a region or country may find it difficult to adjust as well. On the brighter side, when families migrate because of climate change and they tend to look for jobs that they like or dislike, it will have a diffract positive impact on the economy, because the sector flow will expand.

Climate change can cause two sorts of migration, voluntary and forced migration. Willful movement can either be transitory or perpetual. Right off the bat, willful relocation can diminish populace pressure, therefore expanding environmental sustainability in climate change-influenced territories. Furthermore, movement will improve income production by remittances. Remittances can help people fulfil their basic needs while also adding to the local economy. Thirdly, relocation can bring about aptitude move between the traveller accepting nation and the sending nation, increasing access of information technology to the communities. It is especially the situation when permanent migrants get back incidentally, or when impermanent transients come back, carrying with new abilities and skills. This will support the government and the economy also.

Forced migration happens when people are forced by certain factors, like rising sea level. The percentage shares of voluntary and forced migrations are 20% and 80%, respectively (Fig. 4.1). They are often called refugees. The term "environmental refugees" was first coined in 1985 as a report title for the United Nations Environment Programme. People are forced away from their homes and villages because the effects of climate change were greater and unbearable. This causes a lot of problem in the migration countries, such as limited resources and scarcity of land. It also affects the health of many people.

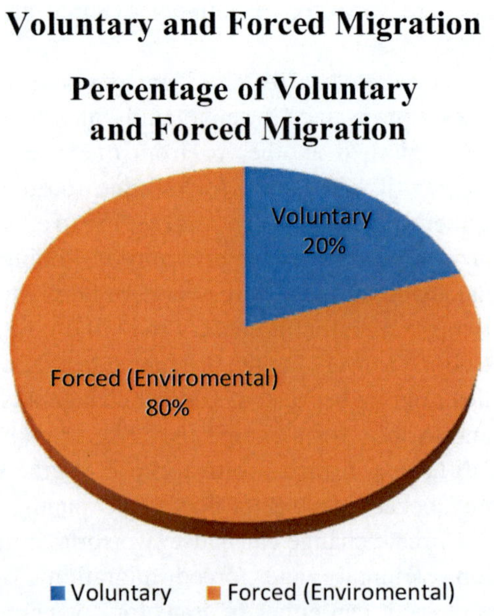

Fig. 4.1 Voluntary and forced migration

Figure 4.1 shows that there are two kinds of migration, i.e. forced migration and voluntary migration. Forced migration is where people migrate due to the impacts of climate change. Many are forced externally and internally. They are forced because of sea level rise and coastal flooding. Countries like Kiribati and Tuvalu have seen worse scenario of sea level rise and coastal flooding (Farbotko and Lazrus, 2012). This kind of migration can either be permanent or temporary. Voluntary migration is when people volunteer to move away from their original place. They do so because of the effects of climate change. Voluntary migration includes climate events that only temporarily predictably disrupt livelihoods, and the help of the government can solve it.

Migration in the Pacific

PCCM which stands for Pacific Climate Change and Migration is a European Union (EU) funded and implement through the Economic and Social Commission for Asia and the Pacific (ESCAP) with the partnership of the United Nations Development Program (UNDP), the International Labour Organization (ILO), and UNU (United Nations University). The UNU was accountable for the fieldwork, data analysis, and research design for the scheme.

According to the UN report, Pacific inhabitants have been migrating in reaction to both ecological and community change, as well as a way for families and individuals to carry out educational and economic opportunities. The heft of momentum migration from Pacific Island nations has seen departed from one's home nation, nearly half a million Pacific People, or roughly one-fourth of the population, residing abroad (Asian Development Bank).

According to the report of Late (2003) "climate change will result in increased migration as islanders are affected by Sea level rise, coastal erosion, saltwater intrusion, intensive cyclone, weather pattern and more frequent and worst droughts" (Fig. 4.2). Therefore, in the typical cli-

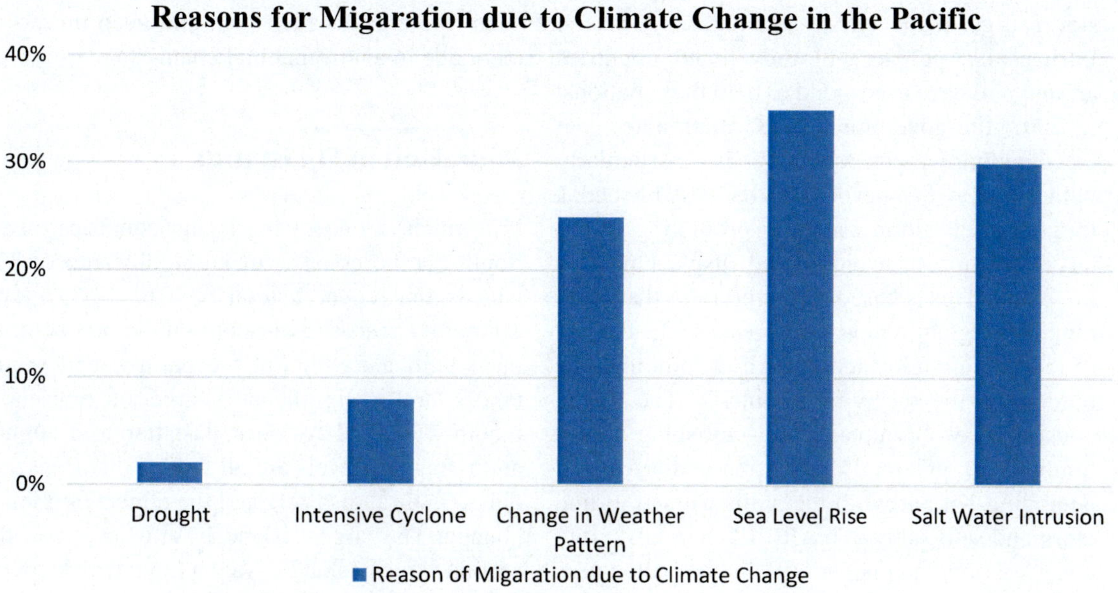

Fig. 4.2 Reasons for migration due to climate change in the Pacific

mate change and the growing population scenario, it has been projected that by 2055 migration will intensely increase in Kiribati and Tuvalu. It is a matter of grim concern because if the movement continues to grow, countries like Fiji, Australia, and New Zealand will have an increased number of migrants in the country. Say, for instance, in a country like Fiji, where there are already a limited number of resources and if the migrant's population will increase, it will exert extra pressure on the available resources and the economy as well.

Figure 4.2 shows the reasons why people migrate. Few of the many effects of climate change is shown above, and each of the impact either moves people permanently or temporarily from their original place.

The above figure depicts that the sea level rise is the biggest threat in the Pacific, and it causes immense damage. Places like Tuvalu, Kiribati, and the Kingdom of Tonga have more effects of these impacts. Saltwater intrusion also contributes to migration as it is linked with sea level rise. When the weather pattern changes, intensive cyclone and droughts are caused, and this also helps to migration.

If sea level rise or flooding worsens, more than 30% of households think relocation would be a possible solution. Internal migration is not an option since Nauru is a tiny island with massive phosphate extracting disruption. Sixteen per cent of Nauruans believe that migration would be inevitable if agricultural productivity declines.

In a journal article by Lejune (2006), she expressed that "there was a wide assortment of understandings of the existence, seriousness and impacts of climate change. Nonetheless, most participants were worried about the effect of environmental change and movement on island culture and sovereignty". A minor, however, firmly held faith in the islands is that "God will secure islanders, identified with God's pledge with Noah" (which means relocation is viewed as superfluous). Be that as it may, there is a variety of opinions on the scope to which islanders can adapt to climate change, comprising the capacity to move. If sea level keeps on rising, movement will be a must except if the administration accompanies an activity to control it.

As per Kuruppu (2009), some countries like Kiribati, Tuvalu, the Kingdom of Tonga, and other atoll nations often need inadequate surface and groundwater and paucity of agricultural land. They are mainly reliant on limited fishing and are vulnerable to saltwater intrusion. The entire coastal population are exposed to king tides, storm surge, sea level rise, and natural disasters. Intrinsically, people will move, and countries around will have to respond to help these nations. In 2008, the government of Kiribati stated that "recognizing their country is particularly vulnerable to Sea level rise, had established a long-term relocation vision for promoting voluntary migration to avoid forced displacement in the future. This is based on the premise that planning for migration now, will preserve the dignity of those being relocated as well as minimize the burden on the receiving countries. The vision consists of two components, establishing opportunities with potential destination countries, and equipping Kiribati with the skills needed to find work and settle abroad". Albeit, when taken into consideration, the uneducated and unskilled people will be left behind without any jobs vacant for them, and this will create a problem for the standing government.

According to the United Nations report (Busby & Purvis, 2004), "Tuvalu, Tonga, Samoa and Vanuatu are examples where migration (because of climate change) is working as a positive development strategy, and reducing climate change vulnerability". In Australia and New Zealand, all four countries have a substantial number of overseas migrants because migration is voluntary and offers a tool for producing remittances and income diversification. Despite a high fertility rate, Tuvaluan has only risen marginally in course of time, i.e. 0.27% per annum. Tuvalu's population growth is constrained by its high per capita access to foreign migration (United Nations, 2013). Kiribati, on the other hand, is vulnerable to climate change and has a fast-growing population with a 1.55% annual growth rate because of excessive fertility and reduced foreign migration (United Nations, 2013).

For the country to be prepared to relocate, a couple of notes are to be thought about. Right off the bat, relocation should be encouraged properly. Also, preparing could assist islanders with enhancing content in international jobs. Thirdly, find a way to ensure the manageable growth of country capitals, and lastly, more research is needed in other Small Island Developing States. In Table 4.1, names of individual ventures have been referenced which help and keep up movement due to environmental change.

Migration in Fiji Islands

Fiji which is one of the island countries of the South Pacific consists of an archipelago of 332 islands that cover a total area of 18,333 km^2. Tourism, sugar, and agricultural exports contribute significantly to Fiji's economy, and remittances are a significant source of household income. Tropical cyclones, flash floods, droughts, and rising sea levels are all threats that Fiji faces due to natural disasters and the effects of climate change. The largest island is Viti Levu, and the second-largest island is Vanua Levu which covers about 10,429 km^2 and 5556 km^2, respectively. More than 90% Fijian population are coastal dwellers, which is the hub of all activities such as infrastructure, agricultural production, and social centres. Climate change is expected to have a broad spectrum of consequences on Fiji's coastal resources.

Despite strong internal resistance, Small Island Developing States (SIDS), which involve all Pacific Island Countries (PICs) including Fiji, have been listed as some of the most vulnerable to climate change (Barnett and Campbell (2010).

Firstly, one of the most important consequences of global change is sea level increase. The world's focus has been attracted to the strong expected rates of potential sea level increase. Countries living in low-lying regions, as well as small islands, are particularly worried that their land areas will be reduced due to inundation and coastal flooding and that, at worst, a substantial proportion of their people will be compelled to move to other countries. As a consequence, as the implications of climate change become more apparent, this topic has garnered increased international interest (Mimura, 2013). Secondly, heightened coral bleaching can become more frequent as sea surface temperatures grow. This, combined with the lag in reef development, can result in a decrease in sediment supply, which is needed to sustain shoreline stability. Coral bleaching can also have a negative impact on

Table 4.1 Projects help and maintain migration caused by climate change in the Pacific

Sections	Project	Slow-onset events
Environmental Migration, Social Vulnerability and Adaptation (EMSVA)	Migration, Environment and Climate Change: Evidence for Policy (MECLEP) Resilience academy: Livelihood resilience in Kiribati Turning research into action loss and damage in vulnerable countries initiative	Changing rainfall and drought Riverbank erosion and drought, coastal erosion, salinization, and changing rainfall
Vulnerability Assessment, Risk Management & Adaptive Planning (VARMAP) garschagen@ehs.unu.edu	Preparing for Extreme And Rare events in coastaL regions (PEARL) The Political Economy of Urbanization and Climate Risk in Vietnam (PEUR)	Sea level rise Drought
Environmental Vulnerability and Ecosystem Services (EVES)	Catalysing action towards sustainability of deltaic systems with an integrated modelling framework for risk assessment (DELTAS)	Salinity intrusion Glacier melting

Source: Taken from Climate Change – The Pacific Islands Response

marine ecological diversity and ecosystems. Third, shifts in storminess trends, such as the increasing severity of tropical cyclones, may lead to an increase in the frequency of coastal inundation and erosion events. Reduced reef defence can intensify these processes (Korovulavula, 2016).

Viti Levu is currently suffering human-caused repercussions for its beachfront zone. Fast population development rates have intensified urban change incidents, erosion of catchments, pollution, and systematic exploitation of organic and physical beachfront properties, exposing large regions of the coast to disintegration and inundation events. As a consequence, coastal frameworks have a limited capacity to respond to environmental change, sea level rise, and human exercises, and coastal communities and infrastructure are vulnerable to increased exposure to natural disasters such as storm waves, high tides, and tsunamis, not to mention rising sea levels. For example, migration from lower to higher zones has already been considered. As a consequence of climate change, several cities have been evacuated.

The Suva Peninsula is the biggest and most populated urban area on Viti Levu. Some of the impacts of rising sea level on the Rewa Delta and Suva Peninsula are as follows:

1. Increasing water tables in low-lying areas
2. Reduced feasibility of in-ground septic systems and inundation of sewer syphoning frameworks
3. Exceeding of downtown Suva shore protection at more severe wave events under a 25 cm SLR situation
4. Mangrove shoreward withdrawal in Rewa Delta
5. Heightened sedimentation and flood vulnerability in the Rewa Delta drains

It was because of the above reasons the inhabitants of Suva wanted to move away from the peninsula.

Climate and the environment have a huge effect on human wellbeing. Climate change has the potential to intensify today's health issues, including mortality from severe weather conditions, cardiovascular and respiratory illnesses, infectious diseases, and starvation, thus weakening water and food sources, infrastructure, educational services, and social security systems (World Health Organization, 2015). Fiji too is facing a lot of medical problems by changing climatic conditions. However, people migrate/migrated to some other destinations. Due to climate change (temperature, rainfall, and humidity), the incidence of malaria, filariasis, and ciguatera, dengue fever, diarrheal infections, and nutrition-related ailment are serious, and people tend to move away from these regions and places (Naicker, 2012; Ali, 2013). Dengue fever is one of many infectious tropical diseases, including typhoid fever, leptospirosis, zika, and chikungunya, that are common to the Pacific Islands including Fiji and are on the rise, contributing to the already large toll of noncommunicable diseases (Moceituba & Tsang, 2015).

Dengue fever scourge possibilities are more significant in beachfront regions and decline inland. Pestilences are more likely to occur in western and northern compared to the eastern Viti Levu. The varying patterns of weather have intensified Fiji's weakness in virus-related sickness flare-ups. Fiji recorded a dry season that prompted incident of the diarrheal illness in 2011, combatted a post-flood leptospirosis flare-up in 2012, and subdued a dengue flare-up in 2013. In Fiji, the summer season also increases the risk of leptospirosis (a zoonotic disease caused by heavy rainfall) due to higher rainfall. Fiji has recorded 160 leptospirosis cases nationwide and more than 10 deaths in 2021. The health ministry meanwhile recorded 335 dengue cases and 53 typhoid cases, including 1 death from each (Xinhua News, 2021).

Relocation could be the last option, but it is also one of the most effective adaptation strategies for many coastal Fijian villagers currently experiencing similar challenges like Vunidogoloa (Vunidogoloa is located in the province of Cakaudrove, SavuSavu). As a result, people continue to move away from these areas as the risks become more and more severe (Charan et al., 2017). Vunidogoloa people were the nation's first population relocated as a result of climate change, which exposed it to flooding as high tides combined with heavy rainfall, increasing the river level. The village consists of 26 houses in Natewa Bay with 140 inhabitants; sea level rise, flooding, and erosion had significant impacts on the livelihood of the community. Designed migration by the Ministry of Economy in Fiji is a relatively recent solution to the consequences of environmental change, and it is only utilized if everything else fails. Migration is a complicated phase that can be stressful for many who are interested. There are a few houses, non-substantial viewpoints linked to migration, which can threaten identities, as well as numerous psychological, socioeconomic, cognitive, and cultural losses (Dutt, 2013; Benge & Neef, 2020) (Table 4.2).

Table 4.2 Village migrations due to climate change in Fiji

Successful migration of villages	Villages partially relocated	Planned relocations in initial stages
Vunidogoloa village in Cakaudrove Province	Vunisavisavi village in Cakaudrove Province	Narikoso village in Ono, Kadavu
Tukuraki village in Ba	Nagasauva village in Udu point, Cakaudrove	Waciwaci District School in Lakeba, Lau
Denimanu village on Yadua Island		

Source: Climate Change – The Fiji Islands Response, Fiji

The table above depicts displacement as a result of climate change in Fiji's various provinces. Sea level rise, coastal erosion, and saltwater infiltration were also significant causes for migration. Since the arguments were too compelling, there were no other choices except to move or transfer. According to the above table, three villages successfully migrated, two villages partly migrated, while two other villages are also preparing to move. These villagers are mainly from Fiji's second-largest island, Vanua Levu, in the provinces of Cakaudrove, Yadua Island, Kadavu, and Lau, with one from Ba province (see Fig. 4.3).

In Fiji, three towns have recently been entirely relocated, two have been partly relocated, and two more are in the early stages of resettlement, and the Fijian Green Growth Framework proposes that about 40 more settlements be relocated (DeLacy, 2014). The correct assessment of these societies' social dynamics is important. As a result, the Fijian government has developed a Relocation Manual, which includes a step-by-step structure of techniques for handling migration work in Fiji. The rules are proposed to be a supportive and concise guide for populations searching for assistance with climate-instigated migration with coordinating government ministries and, external organizations providing support for the movement procedure. According to

Fig. 4.3 First migration of Vunidogoloa village due to sea level rise

Fig. 4.4 Migration due to climate change in Fiji

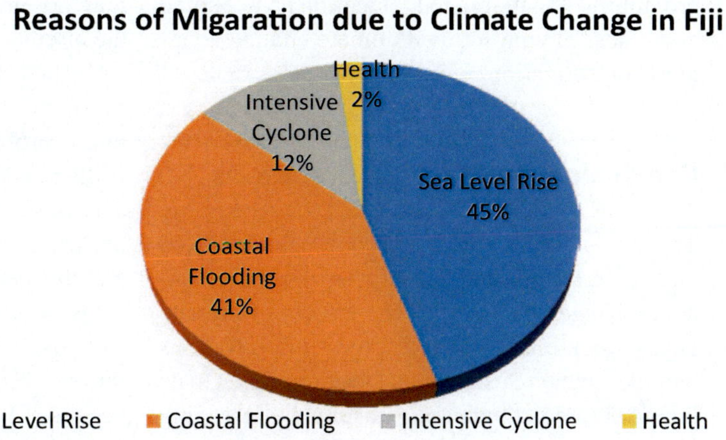

the available data on migration due to climate change in Fiji, the largest number of migrations from Fiji is driven by sea level rise (45%). The second-highest proportion of migrants is due to coastal floods (41%), followed by the intense cyclone (12%), and only 2% migrate for health reasons (Fig. 4.4). As a result, it is important to consider how serious the consequences are.

Global sea level increase will be dramatic by the end of the century, according to Fiji's National Climate Change Policy. Fiji has seen an annual increase in sea level of 6 mm (0.2 inches) since 1993, which is higher than the global average. The drastic rise in sea levels, as well as the salt-water interruption triggered by the intensified ferocity of seaside hurricanes, has left parts of the island nation barren. Rising sea levels, as well as higher temperatures and stronger El Nino cycles, render the island more susceptible to harmful food and water-borne illnesses. The future of Fiji depends on mitigating the consequences of climate change (COP 23, Fiji 2017–18) (Table 4.3).

The table above illustrates the probability of sea level rise and the year in which it would be most serious. This generation and government must take steps to tackle sea level rise. To help deter coastal floods and sea level rise, the government should build sea walls (Fig. 4.5). Table 4.4 indicates that the Fijian government has already taken the initiative to build sea walls for villages that could be forced to evacuate due to coastal floods and sea level increase.

Adapting to climate change increases people's standards of living, as well as their potential survival and livelihoods in the Pacific. Managing the effects of climate change on vulnerable people through adaptation and seeking to minimize possible climate change through mitigation would remain priorities. Mitigation and adaptation to climate change would reduce the need for climate-related migration. As a result, any actions promoting climate-related relocation do not come at the cost of conventional climate change adaptation or mitigation.

Conclusion and Recommendations

This study shows how individuals relocate from their unique spot because of the impacts of climate change. To facilitate the effects of climate change on climate transients, the United Nations includes in its Sustainable Development Goals the need to do alleviation measures by lessening greenhouse gas emissions and adapting of urban areas and infrastructures making them more resilient to changes. Mechanical advancement, vitality effectiveness, and the usage of sustainable power sources, among different measures, will play a key role later on for environmental protection, financial development, and social development.

Table 4.3 A summary of the increase in sea level

Year	2025	2050	2100
Scenario			
(Mid-range, best guess	11 cm	23 cm	50 cm
(High-range, worst case)	21 cm	43 cm	103 cm

Source: Climate Change – The Fiji Islands Response

Fiji has vowed to move to 100% clean energy by 2030 and has embraced a reforestation policy that is planned to store carbon from newly planted trees. Fiji has conducted efforts to track and dispatch a rapid solution to climate-related threats in cooperation with the Global Environment Facility, the Green Climate Fund, and a few United Nations offices.

Unfortunately, the Fiji government lacks the budgetary means to properly finance mitigation and climate change impact-related studies and appraisal assessments, as well as programmes that resolve existing climate change effects, as identified herein. As a result, Fiji will need assistance from the convention's financial system in the short and medium term, supplemented by other donor assistance. However, Fiji is doing whatever to fight against climate change so that migration can be minimized. Fiji is a small country; if people start to migrate, then the economy may not grow; and it will be a great problem on the circular flow.

In the other Pacific countries, migration is a must, because climate change has brought so many problems that they cannot survive. The skilled and the young have privileges to migrate and may adjust well to the new country. Still, the unskilled and uneducated people are left behind, and this causes the government to come up with ways to support the remaining people.

Climate change is a global challenge, but it is much more challenging in Fiji and the Pacific because of poverty, backwardness, and above all their locations in the midst of the Ocean. Therefore, migration is apparent and bound to happen, both inter-regional international. Whenever there is migration, there will be advantages and disadvantages. There are many ways that the government can do to combat climate change to help in migration. One of the best ways that the government can help is by creating awareness and educating. According to Climate Change – The Fiji Islands Response, "Education

Fig. 4.5 Construction of the sea wall

Table 4.4 Names of villages where the government has helped in building the sea walls

Sea walls completed	Plans for further sea walls
Kiuva village in Tailevu	Naivakacau in Tailevu – to prevent saltwater intrusion
Kumi village in Tailevu	Vabea village in Ono, Kadavu
Wainibokasi riverbank protection in Tailevu	
Riverbank protection in Buretu village in Tailevu	
Matainoco seawall in Tailevu	
Daku village in Tailevu – drainage improvement and saltwater intrusion protection works	

Source: Climate Change – The Fiji Island Response

initiatives and enhanced public awareness and action are recognized as key elements of a national strategy for addressing climate change and Sea level rise (SLR) issues. The focus of future activities will be directed at strengthening the coverage or inclusion of climate change in the following areas". Fiji, on the other hand, is committed to adopting community-level adaptation initiatives to mitigate the impact of climate change. It is also participating in a Canadian government-funded adaptation programme. The government will use this to implement strategies for environmental protection, such as reducing natural resource depletion and protecting Fiji's biodiversity, encouraging and endorsing sustainable waste management, combating the impact of climate change, enacting the Sustainable Bill, and eventually, public awareness and education.

Also, building sea walls could help prevent coastal flooding from sea level rise. The Fiji government has already taken the initiative to build sea walls to the village that can migrate because of coastal flooding and sea level rise.

References

Ali, M. (2013). Dengue: Factors driving the emerging epidemic. *Clinical Epidemiology, 5*, 461–463. https://doi.org/10.2147/CLEP.S54834

Australian Bureau of Meteorology (ABM) and Commonwealth Scientific and Industrial Research Organisation (CSIRO). (2011). *Climate change in the pacific: Scientific assessment and new research* (Vol. 1). Regional Overview.

Bacolod, E. D. (2014). Fiji 2020 agriculture sector policy agenda. In *A report prepared by the Ministry of Agriculture with the Technical Assistance from the Food and Agriculture Organization of the United Nations*.

Barnett, J., & Campbell, J. (2010). *Climate change and small island states: Power, knowledge and the South Pacific*. Earthscan.

Benge, L., & Neef, A. (2020). Planned relocation as a contentious strategy of climate change adaptation in Fiji. In *Climate-induced disasters in the Asia-Pacific region: Response, recovery, adaptation*. Emerald Publishing Limited.

Books, M. M. (2014). *Asian Development Bank*. Asian Development Bank.

Bronen, R. (2010). Forced migration of Alaskan indigenous communities due to climate change. In *Environment, forced migration and social vulnerability* (pp. 87–98). Springer: Berlin Work Press.

Busby, J., & Purvis, N. (2004). *The security implications of climate change for the UN system. Environmental change and security project report*. United Nations Press.

Charan, D., Kaur, M., & Singh, P. (2017). Customary land and climate change induced relocation—A case study of Vunidogoloa Village, Vanua Levu, Fiji. In *Climate change adaptation in Pacific countries* (pp. 19–33). Springer.

Cloud, D. (2002). *Traversing laser locating system in migration*. York University Press.

Clouds, D. (1992). *A regional geography*. Cambridge University Press.

Clouds, D. (2014). *The Leodicario show*. Dicapario Press.

COP23. (2017–18). https://cop23.com.fj/fiji-and-the-pacific/how-fiji-is-affected-by-climate-change/.

DeLacy, T. (2014). *Green growth and travelism: Concept, policy and practice for sustainable tourism*. Fijian Green Growth Framework.

Dutt, L. (2013). *Relocation of the islands*. The Fiji Times.

Ekstrom, M. (2010). A framework to diagnose barriers to climate change adaptation. *Proceedings of the National Academy of Sciences, 107*(51), 125–129.

Farbotko, C., & Lazrus, H. (2012). The first climate refugees? Contesting global narratives of climate change in Tuvalu. *Global Environmental Change, 22*(2), 382–390.

Geey, V. (1999). *Environment and planning*. Nelson Press.

Korovulavula, I. (2016). Climate change impacts in Ra and Kadavu provinces. *SPC Pacific Community Report*.

Kuruppu, N. (2009). Adapting water resources to climate change in Kiribati: The importance of cultural values and meanings. *Environmental Science & Policy, 12*(7), 799–809.

Late, L. R. (2003). *Quaternary cold events Recorded in Southern Ocean sediments in the Pacific*. Pacific Sector.

Lejune, F. (2006). Pacicifa. *Learning about Tuvalu and Kiribati, 1*(2), 26–30.

Mimura, N. (2013). Sea-level rise caused by climate change and its implications for society. *Proceedings of the Japan Academy. Series B, Physical and Biological Sciences, 89*(7), 281–301. https://doi.org/10.2183/pjab.89.281

Moceituba, A., & Tsang, M. (2015). Averting climate change's health effects in Fiji. *World Health Organization. Bulletin of the World Health Organization, 93*(11), 746.

Naicker, M. (2012). Climate change and health in Fiji: Environmental epidemiology of infectious diseases and potential for climate-based early warning systems. *Fiji Journal of Public Health, 1*, 7–13.

Richardson, B. C. (1992). The Caribbean in the wider world. In *A regional geography* (pp. 58–69). Cambridge University Press.

Suphachalasai, A. (2014). *Assessing the costs of climate change and adaptation in South Asia*. Asian Development Bank.

United Nations. (2013). *International migration report 2013*. UN.

United Nations. (2018). *2018 revision of world urbanization prospects*. UN.

World Health Organization. (2015). *Climate and health country profile 2015: Fiji* (No. WHO/FWC/PHE/EPE/15.35). World Health Organization.

Xinhua News. (2021). *More than 10 Fijians die from leptospirosis this year*. http://www.xinhuanet.com/english/asiapacific/2021-03/03/c_139780593.htm.

5

Assessing the Microclimatic Environmental Indicators of Climate Change of a Temperate Valley in the Western Himalayan Region

Majid Farooq, Humayun Rashid, Gowhar Meraj, Shruti Kanga, and Suraj Kumar Singh

Introduction

Climate change implies some substantial change for the long-term "average weather" in a specific area or the entire world (Welbergen et al., 2008). It requires changes in atmospheric fluctuations or average conditions over decades to millions of years. These shifts may be attributed to complex Earth cycles, natural factors like sunlight intensity fluctuations, and human behaviors in recent years (Houghton et al., 1996). The use of metrics and measures to demonstrate relationships between natural resources and financial information demonstrates significant concern. A measure of performance that aggregates information in an active type can be described as an indicator (Rogers et al., 1997). Because of their geographic vastness and climatic environments, the research community has to develop an extensive knowledge of climate change causes in numerous parts of the world. There is significant interest in the usage of metrics in order to assess the social and atmospheric status and impacts at the national or geographic level (Rogers, 1999). One hundred thirty metrics for domestic evaluations of the social, economic, environmental, and structural facets of sustainable growth have been established by the Sustainable Development Division of the United Nations (UNSD, 1996). Rogers (1999) reviewed several approaches used to create and incorporate indicators and integrate them into measures for a study of the historical (starting in the 1960s) and current environmental quality indices implementations. They propose the use, in the World Bank development diamonds, of new indices (cost of remediations, environmental elasticity) and environmental diamonds.

In the past, several metrics and indices were used for the measurement of water quality. To analyze water problems in Africa, Falkenmark (1989) used a per capita quantity of water available. In a report by Gleick (1990), a non-wounded index of storage ratio, demand ratio, usage of hydroelectricity, groundwater overdraw, and flux variability was compiled in 18 water supply regions in the USA to measure the climatic insta-

M. Farooq · G. Meraj
Centre for Climate Change & Water Research (C3WR), Suresh Gyan Vihar University, Jaipur, Rajasthan, India

Department of Ecology, Environment and Remote Sensing, Government of Jammu and Kashmir, Kashmir, India

H. Rashid · S. Kanga
Centre for Climate Change & Water Research (C3WR), Suresh Gyan Vihar University, Jaipur, Rajasthan, India

S. K. Singh (✉)
Centre for Sustainable Development, Suresh Gyan Vihar University, Jaipur, Rajasthan, India
e-mail: suraj.kumar@mygyanvihar.com

A. R. Siddiqui, A. Sahay (eds.), *Climate Change, Disaster and Adaptations*, Sustainable Development Goals Series, https://doi.org/10.1007/978-3-030-91010-5_5

bility of water systems as indices of regional variability. Raskin et al. (1997) use hydrological and socioeconomic metrics to include a weighted and aggregated risk index, where no stress, medium stress, stress, and stress were named each country/region. A vast number of metrics are also provided by the US Environmental Protection Agency (U.S. EPA, 1997) to measure the environmental quality of watersheds. The states of the water bays and their resistance to potential issues have also been calculated with indicators (U.S. EPA, 1997). The metrics need to be taken from a trustworthy and technically appropriate data source to draw a reasonable inference regarding a system. In order to recognize system changes, the metrics should also be reasonably adaptive to provide an essential framework for decision-making (July 2009).

GIS will be used as an instrument for monitoring climate change as temperature increases worldwide. The validity of the different hypotheses on climate change can be determined by GIS. GIS systems may also forecast future environmental situations reliably based on existing scenarios, which can be used in policy planning (Ghosh, 2009). The shifts in the world's environment reflect changes in land use, glaciers, temperature, and precipitation changes and changes in socioeconomic trends of people living in the valleys (Beniston, 2003; Cruz et al., 2007). Continuing climate change over the next decades is possible, with significant cascading effects on river streams, drainage reloads, environmental hazards, and biodiversity; nature, structure, and role of ecosystems; and livelihoods for humans (Ma et al., 2009). The Himalayan valleys provide scientists and researchers with the most excellent chance to study climate change because the extremes of environmental conditions in the immediate surrounding area hardly affect these areas. The findings were seen here to relate to the changing climatic conditions in the Kashmir Valley. Climate change, such as temperature, precipitation, and melting glaciers, has shown a warming pattern over the last 100 years and especially over the past 30 years (Muslim, 2012). This report aims at defining and evaluating a few climate change factors in the Kashmir Valley.

Study Area

A distinctive geographical character is present in the Kashmir Valley (Fig. 5.1), also referred to as the paradise on Earth. Located in northwest Himalayan folds, the valley has mountain ranges of 15,440 km^2 on almost every foot, with snow-capping lofty peaks. On the northeast slope, at about 2.770 m to the south, the mountain range is up to 5550 m. The only river outlet is Baramulla-George, where the flat grassy banks of the Jhelum River flood across the southern plains, headlong hurrying down its rugged course. The oval valley is full of dense alluvium deposits, which have bleached right at the bottom of the surrounding areas. Jhelum and its affluents have drained it, widespread among the Lidder, Indus, Pohru, Sandran, Bringi, Vishav, and Sukhnag. The valley has a length of around 130 km and a diameter of 40 km. The Kashmir Valley is classified into four physiographical divisions of the Jhelum valley plain, Karewas Valleys, Side Valleys, and the Greater Himalayan Range based on stratigraphy and altitude. The Kashmir Valley has a continental climate with exact seasonal characteristics. In the Indian subcontinent, the genesis of Kashmir's atmosphere is implicit in the process of wetting. However, there is an altered subtropical atmosphere within the valley surrounding the Himalayas.

Materials and Methods

Climatic data for six meteorological stations of the valley was obtained and then averaged. The primary measures for this analysis were an annual average temperature and annual gross precipitation. The diagrams were drawn, and weather patterns were studied. To analyze the temporal variability between the indicators concerning glacial extents, the investigation was carried out using LANDSAT ETM+ satellite data covering the Kashmir Valley. The glacier boundary was delineated using archival topographic maps. Change analysis was done using the overlay analysis tool in ArcGIS, and the statistics were generated. An investigation was carried out on the

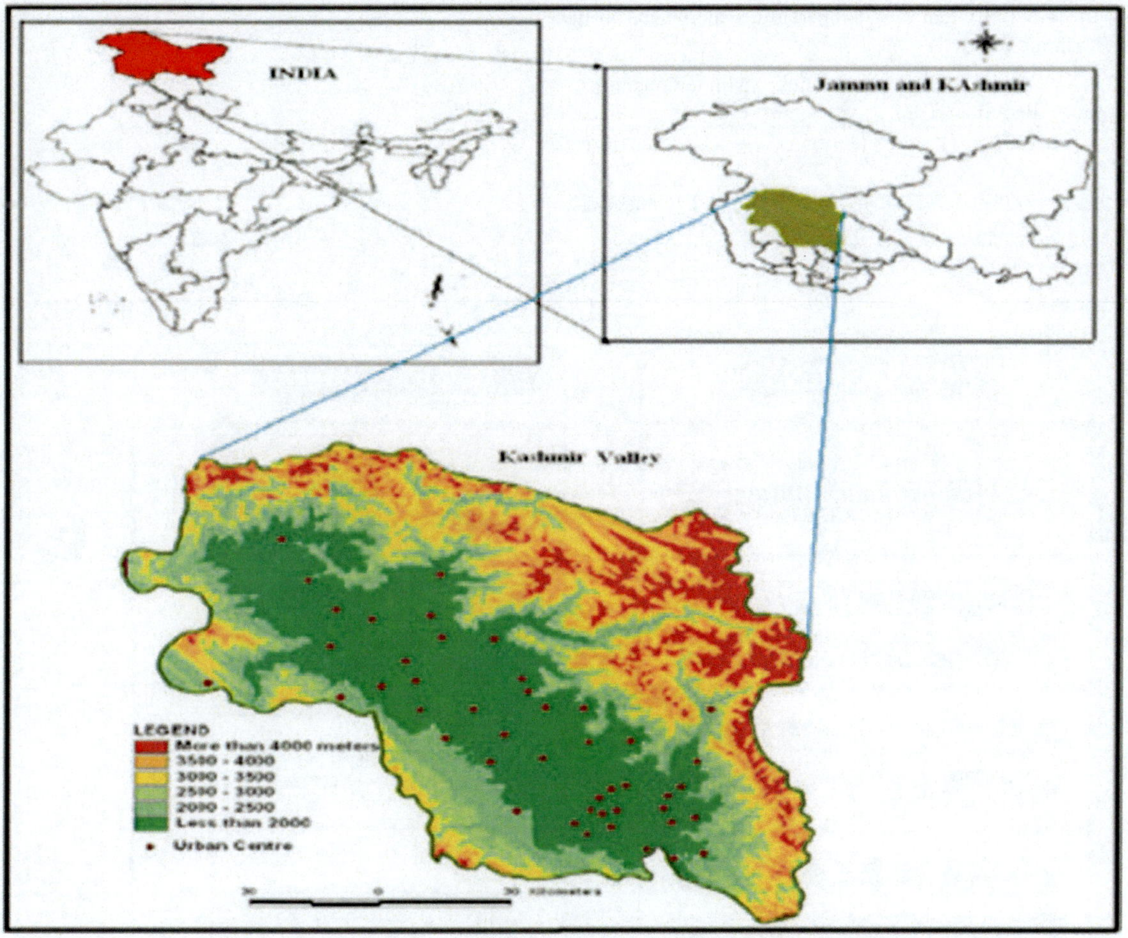

Fig. 5.1 Study area: Kashmir Valley. (Source: Extracted from Malik (2012))

changes in the extent of wetlands and lakes in the Kashmir Valley over the last 100 years using an archival topographical map and the LANDSAT ETM+ image of 2001. In the study of alpine vegetation, trend changes were visually interpreted to understand the effect of the climate change on the satellite images of LANDSAT MSS in 1976 and the 2011 IRS-1D LISS-3 results. In addition to this, field data were obtained and compared with previous floral data on temperate and alpine plant species presence. Visually associated with adjustments were developed to examine the effects of climate change on the structure of the LANDSAT MSS 1976 and IRS-1D LISS-3 lake satellite datasets. Flowering in the spring season was chosen as one indicator; however, the datasets used were chosen as blossoming is very dynamic and of concise duration. Landsat TM dated April 15, 1994, and aWIFS of March 25, 2012, about the Kashmir Valley were analyzed to study the phonological variations in the blossoming dates. Figure 5.2 shows the overall methodology employed in this study.

.

Results and Discussion

Climate

Temperature

The winter snowfall is influenced, and the comfort of summer afternoons is determined by temperature (Aono, 1998; Walther et al., 2002; Zhou et al., 2003). The temperature has been one of the most commonly used climate change markers

Table 5.1 Extremes of temperature experienced in the Kashmir Valley

S. no.	Period	Max temperature (°C)	Min temperature (°C)
1.	1901–1950	30.8 (July)	−2.3(January)
2.	1979–1996	32.4(July, 1979)	−5.4(January)

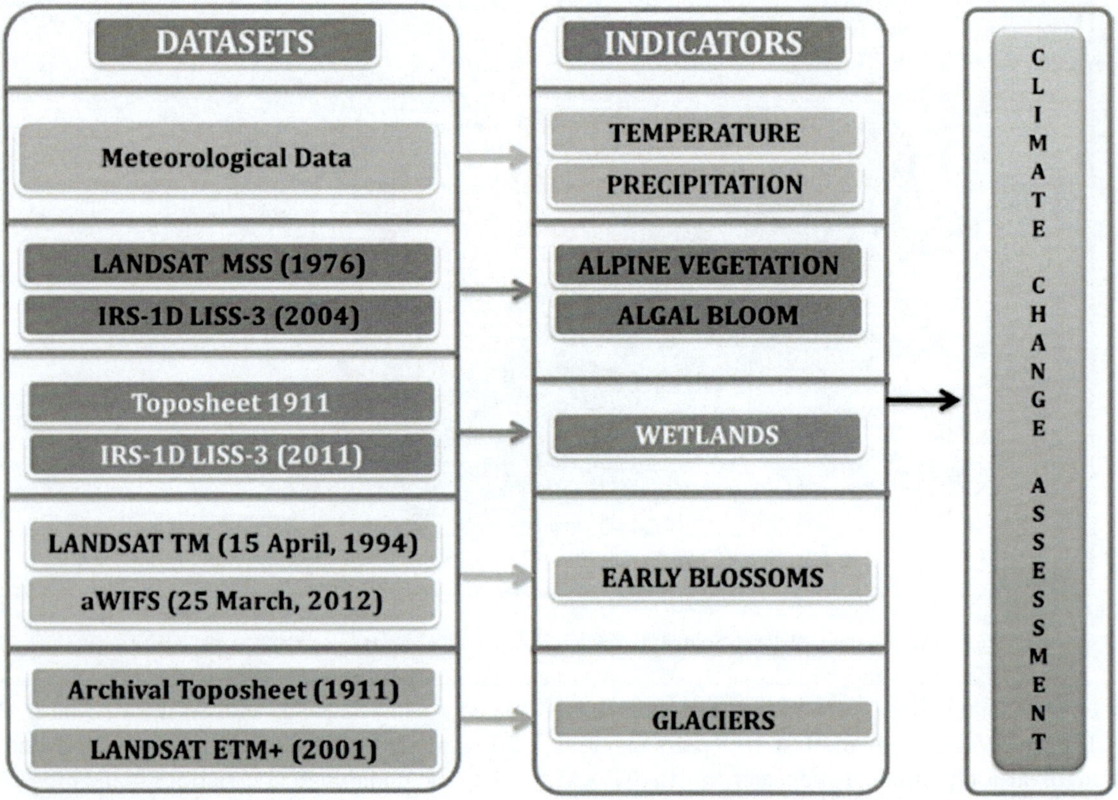

Fig. 5.2 Methodology workflow

(Livingstone & Dokulil, 2001; O'Reilly et al., 2003; Keller, 2007; Hampton et al., 2008) since 1900. It is possible to analyze the changing climate of the Kashmir Valley over the past century using a long-term instrumental record of average annual temperature. The maximum and minimum temperature variations from 1901 to 1950 and 1979 to 1996 are shown in Table 5.1. Figure 5.2 sweltering and freezing temperatures were evident from the data toward the end of the century. The annual average temperature for the Kashmir Valley shows considerable variability on annual and decadal measurements of temperature (Fig. 5.3).

The average annual temperature was much lesser in 1920 and 1957 compared to relatively warm years in 1946, 1970, and 1973. Extended cold periods were observed in the early 1900s and the late 1940s. The data (Fig. 5.3) shows a steady increase in temperature since 1998, clearly indicating a change in the temperature regime. Since current climate change proposals are focused on a rise of 2–3° for the broader population and habitats of the Himalayas, these temper-

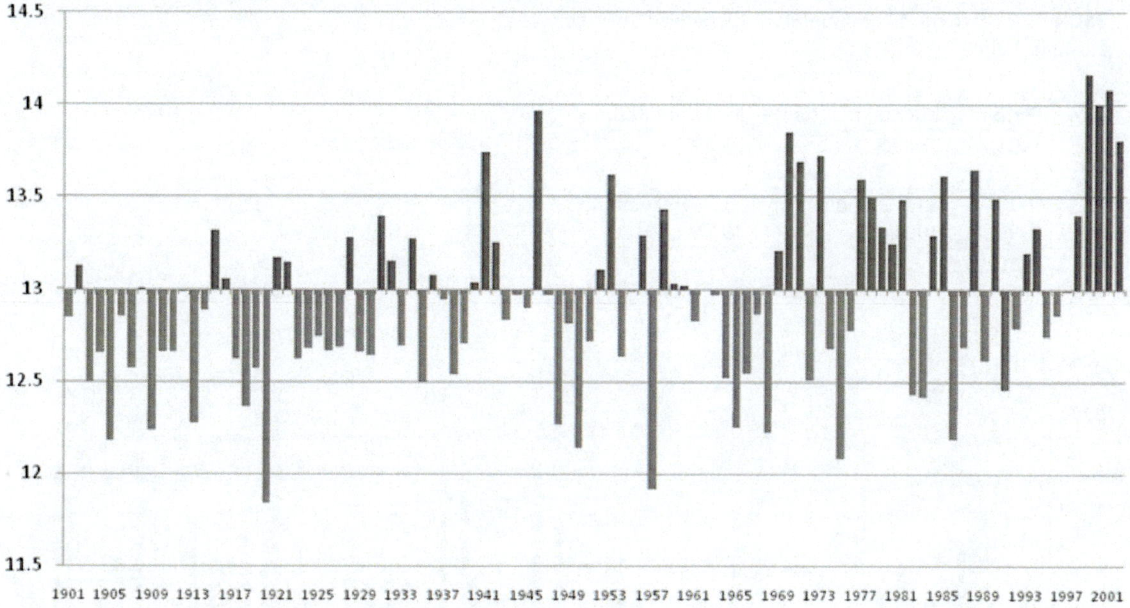

Fig. 5.3 Average annual temperature (°C) for Kashmir Valley from 1901 to 2002

atures can be disastrous (Anderson & Bowe, 2008; Hansen et al., 2008; Solomon et al., 2009). Analysis of meteorological data for the period 1901–2002 revealed a definite climate change in the region with overall maximum and minimum temperature showing an increase of about 1.43 °C (±0.84) and 0.5 °C (±0.3). The Himalayan region shows great variations in summer and winter temperatures. Winter mean and maximum temperature shows a significant increase, but mean and minimum summer temperatures show a consistent decline (Fowler & Archer, 2006; Romshoo & Rashid, 2010). This phenomenon is observed not just in the Kashmir Himalayas but also in adjoining mountain ranges like Karakoram range (Kumar & Pant, 1994; Yadav et al., 2004). Most of the region is undergoing warming trends. The annual mean temperature is increasing at a rate of 0.01 °C/year or more.

Precipitation

Precipitation is defined as liquid or solid condensation of water vapor falling from clouds or deposited from the air onto the ground. Precipitation can fall as rain, snow, sleet, hail, or freezing rain (IPCC, 2007, Groisman et al., 1999; Karl & Knight, 1998; Klein, 2003). The total annual precipitation in the Kashmir Valley from 1901 to 2002 was analyzed, and it has been found that there has been significant spatial variability in precipitation. Precipitation in the region increased from 909.23 mm to 1225.58 mm (±316.35). In recent decades, growing and diminishing precipitation patterns have been witnessed in the Greater Himalayas (Shrestha et al., 2000; Xu et al., 2007; Ma et al., 2009). According to Wake and Markham (2005), there has been a 2% increase in precipitation worldwide, and the Kashmir Valley shows a steady increase in precipitation since 1972. Drought impacts the water quality and quantity, forest health, and agricultural productivity in an area (Christensen et al., 2007; Daly et al., 1994; Russell et al., 1995; Murdoch et al., 2000; Schindler, 1997; Williams et al., 1996). The past trend and change projections suggest that temperatures will continue to rise and rainfall patterns will become more variable, with both localized increases and decreases (Fig. 5.4 and Table 5.2).

Glacier Recession

Glaciers are dynamic and fragile ice bodies on the landscape and are products of the climate and climatic changes. Change in climate is reflected

Table 5.2 Extremes of rainfall experienced in the Kashmir Valley

S. no.	Period	Max rainfall (cm)	Min rainfall (cm)
1.	1901–1950	181.55	65.89
2.	1979–1996	84.0 (Bandipur in 1994)	21.7 (Badgam in 1987)

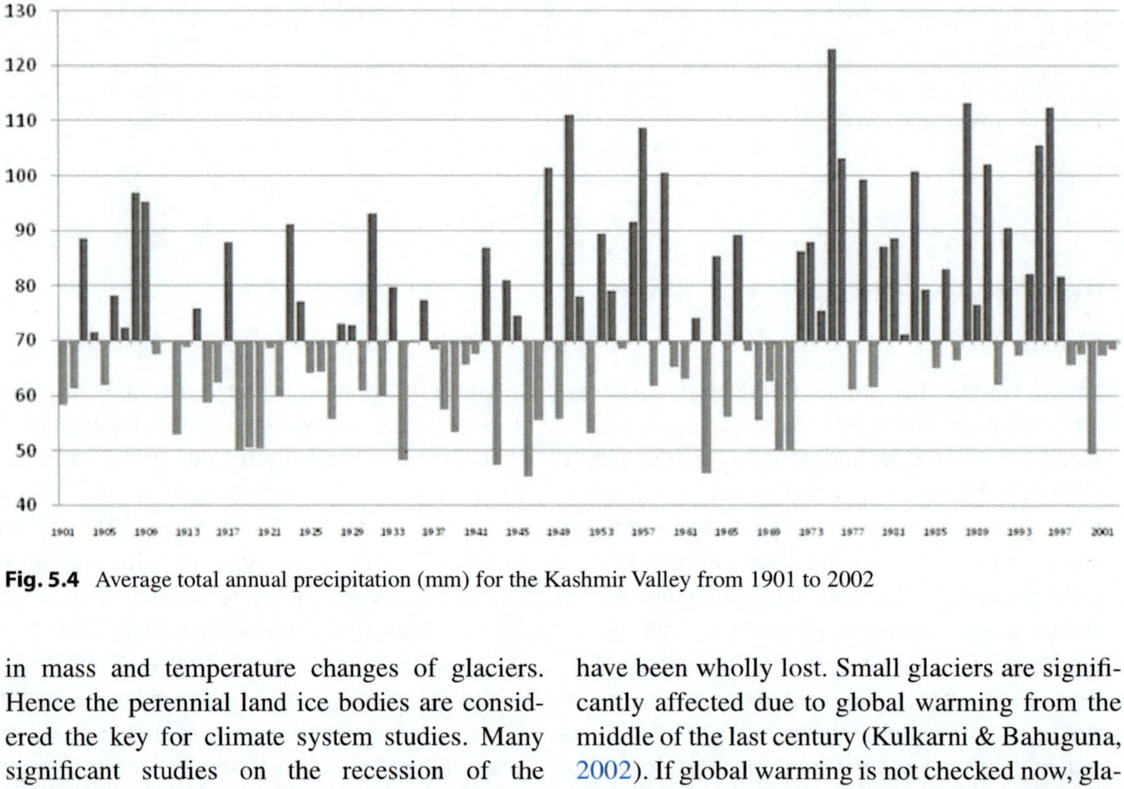

Fig. 5.4 Average total annual precipitation (mm) for the Kashmir Valley from 1901 to 2002

in mass and temperature changes of glaciers. Hence the perennial land ice bodies are considered the key for climate system studies. Many significant studies on the recession of the Himalayan glaciers have been made and are considered one of the prominent indicators of global climate change (Yao et al., 2004; Barnett et al., 2005; IPCC, 2007; Nogues-Bravo et al., 2007; Hasnain, 2008). Analysis of the areal extent of glaciers from 1911 to 2001 revealed that the extent of glaciers has reduced (Fig. 5.5) from 112 km^2 in 1911 to only 94.44 km^2 in 2001, showing a glacial retreat of 17.89 km^2. The observations made in this study suggest that small glaciers in Shepidddalau, Anantnag District, have been reduced from 0.5 to 0.3 km^2. Besides, more enormous glaciers in Bodapather, Tosamaidan, Anantnag District, are being fragmented into smaller glaciers (Fig. 5.6). Some of the glaciers have been wholly lost. Small glaciers are significantly affected due to global warming from the middle of the last century (Kulkarni & Bahuguna, 2002). If global warming is not checked now, glacial fragmentation and retreat are likely to increase in the future, which might profoundly affect the availability of water resources in this region (Table 5.3).

Wetland Loss

As predicted, the decrease or increase in precipitation will have important implications for all ecosystems, particularly wetlands, whose ecological character is very much dependent on its hydrological regime (Anderson et al., 2008; Winter, 2000). Climate change is known to have a significant impact on the hydrological regimes of the lakes and wetlands (Rashid et al., 2008). The study reveals the loss of nearly 50% of the

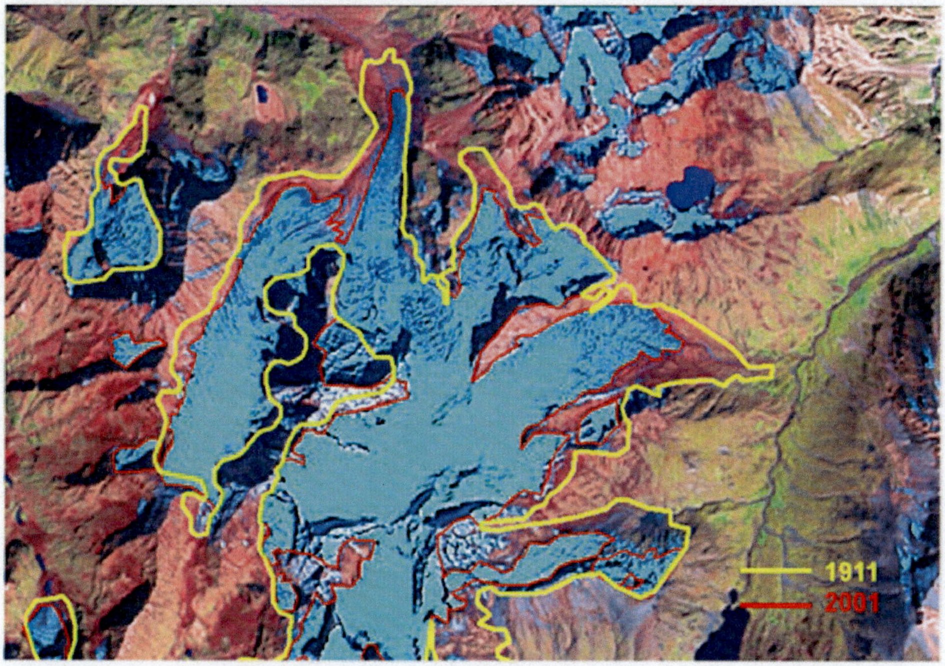

Fig. 5.5 ETM Plus image showing the change in the extent of Kolahoi glacier from 1911 to 2001

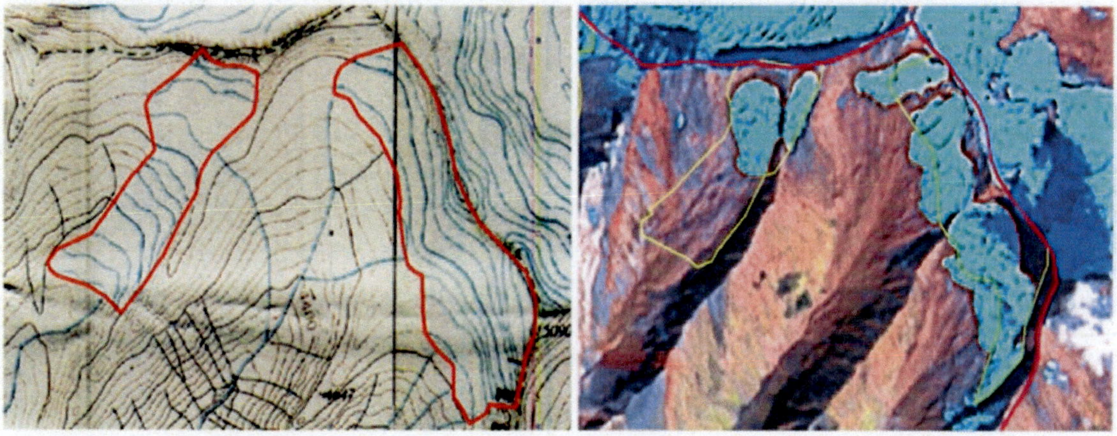

Fig. 5.6 Two glaciers on the archive map of 1911 having got fragmented now as per ETM plus image

wetlands over 100 years (Fig. 5.7). In 1911, the average size of marshy water sources was 356.85 km^2, but in 2011 it was lowered to 158.54 km^2. Climate change will severely affect the regions where precipitation is mainly in the form of snowfall. A rise in the temperature would potentially contribute to increased winter rainfall and decreased flow in spring and summer (Kwadijk & Middlekoop, 1994; Sealthun et al., 1998). The decline in precipitation and river discharge has adversely affected the water spread of the wetlands and lakes (Table 5.4).

Dynamics of Vegetation Patterns in Alpine Regions

The mountains are vulnerable to the adverse influence of climate change (Gairolal et al., 2008; Deshingkar, 1997; Ravindranath et al., 2006) and include the subalpine and the almond trees (the Himalayan arid and dry temperate wood and the

Table 5.3 Extent of glacial retreat in the Kashmir Valley from 1911 to 2001

S. no.	Year	Total extent km^2
1	1911	112. 33
2	2001	94.44
	Loss in spatial extent	17.89

Table 5.4 The extent of lakes and wetlands in the Kashmir Valley (1911–2011)

S. no.	Year	Category		Total
		Marshy	Water body	(km^2)
1	1911	271.70	85.15	356.85
2	2011	117.43	41.11	158.54
	Loss in spatial extent	154.27	44.04	

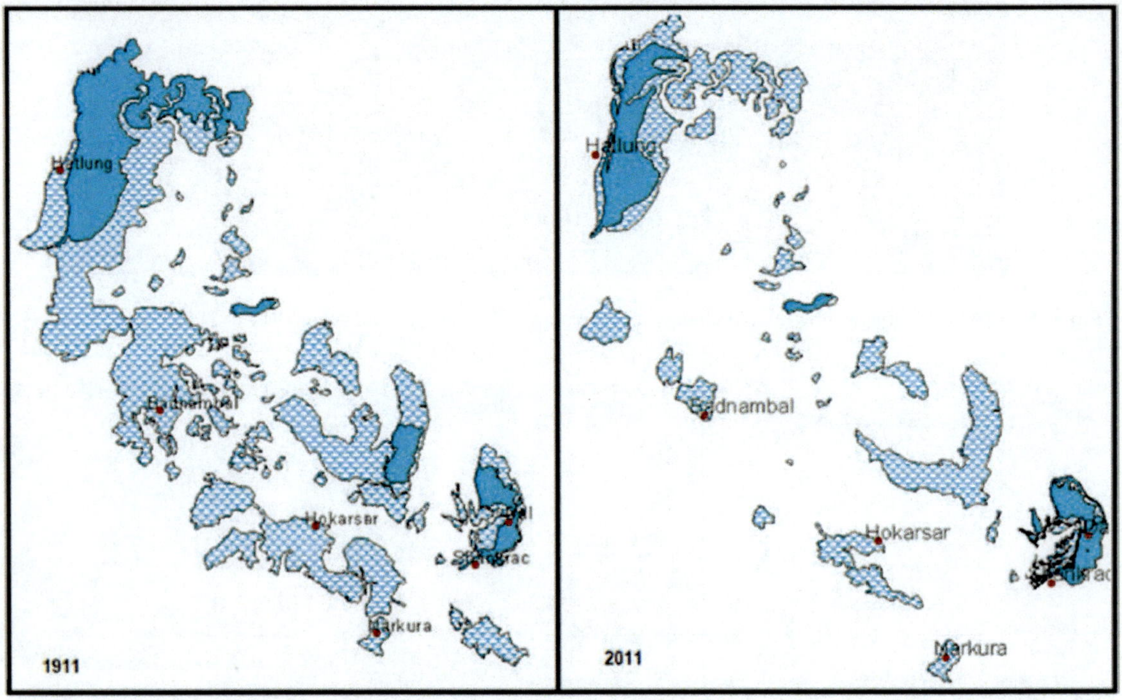

Fig. 5.7 Changes in the extent of lakes and wetlands from 1911 to 2011

Himalayan humid temperate forest). Field studies in the Pir Panjal area have shown that tree lines and forest precipitations into alpine pastures (Song et al., 2004), which respond to high temperatures, have migrated upward and have displaced adapted species in cold climatic conditions. The restoration of *Abies pindrow* has collapsed, and *Pinus wallichiana* has been replaced. This is a vital sign that the temperature in this area of *Pinus wallichiana* is optimal (Van Zonneveld et al., 2009) and not *Abies pindrow*. Grabherr et al. (1994) calculated that an increase of 0.50 °C in temperature per 100-m high would potentially result in a change of 8–10 m per decade in altitudinal lands. It has been estimated in montane environments to contribute to a change in isotherms of approximately 160 m or 150 km in height by 1 liter in average annual temperature. The Ecotone Alpine Lines is useful in tracking climate change, but studies are complicated by biogeography, the biological value of plants, site history, and anthropological impact (Fig. 5.8).

Under the influence of climate change, even the remote alpine regions have shown changes in species richness, about 10–12% (Neilson et al.,

2005; Gopalakrishnan et al., 2011). Alpine vegetation can be a sensitive "ecological indicator" for climate change (Pauli et al., 1996, 2003). Interim analysis of satellite images reveals that alpine areas, particularly rocky, stony waste, support a higher level of vegetation than in the past, which may have been attributed to elevated temperatures that make the area favorable conditions for vegetation growth. Below this highest alpine vegetation zone in the eastern Himalayas, geography, elevation (temperature), and precipitation are the main variables determining alpine plant distributions. Such geographically limited species are subjected to the isolation imposed by "sky islands" (Heald, 1967)

Weed Bloom in Aquatic Ecosystems

Although there is a shared awareness that lakes can be sentinels of climate change, they have not been extensively evaluated for their effectiveness (Schindler et al., 1996; Magnuson et al., 2000; Verburg et al., 2003). We identified algal bloom (Thackeray et al., 2008; Johnk, 2008; Huisman et al., 2004; Paerl & Huisman, 2008) as variables within a lake that act as indicators of the effects of climate change in the lake. Climate change would likely increase the amount of heavy spring rainfall over the century and contribute to the number of harmful blooms of algae (Scavia, 2012). In Wular Lake, having a total area of about 133 km^2, there has been tremendous growth, i.e., 22.59 km^2, in aquatic weeds over 40 years. The analysis of multi-date satellite images indicates the presence of algal blooms. Given sufficient resource availability, increasing temperatures generally accelerate growth and development rates of individual organisms, although changes in absolute abundances tend to be species-specific (Adrian et al., 2006; Reist, 2006; Blenckner, 2007). There is scientific evidence suggesting an increase in the water temperature of the Kashmir lakes over the last few decades (Solim & Wanganeo, 2008). The red algal bloom evident from satellite imagery of Wular and Dal lakes is also an indication of increased water temperature. Growth rates, abundance, and species composition can each be considered an indicator of climate change. An increase in temperatures has resulted in severe eutrophication because of too high macrophyte biomass (Solim & Wanganeo, 2008). At higher temperatures, bloom-forming cyanobacteria have a competitive advantage over other phytoplankton groups (Johnk et al., 2008; Wagner & Adrian, 2009) (Fig. 5.9).

Early Spring Bloom

Potential "bioindicators" of climate change are the change in the existence of plants and animals, such as flora, migrant birds, and insects' appearance in the spring (Hopp, 1974; Bradley et al., 1999; Memmott et al., 2007; Romanovskaja, 2007 Crepinsek et al., 2006). Native annual flowering plants are often measured, but these data can be problematic to interpret because of the year-to-year variation of the plants measured and their physical location (microclimate). For this

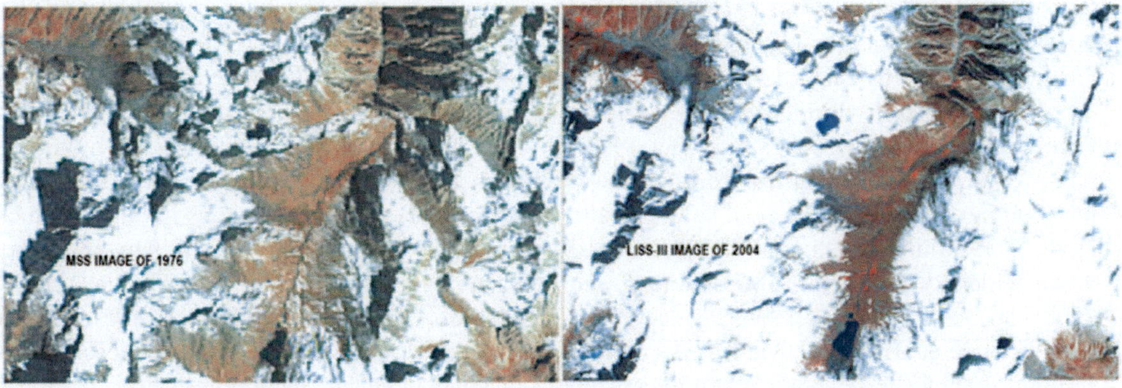

Fig. 5.8 Landsat MSS satellite image of 1976 and LISS-III image of 2004 depicting changes in ground cover in the alpine zone of Kashmir Valley

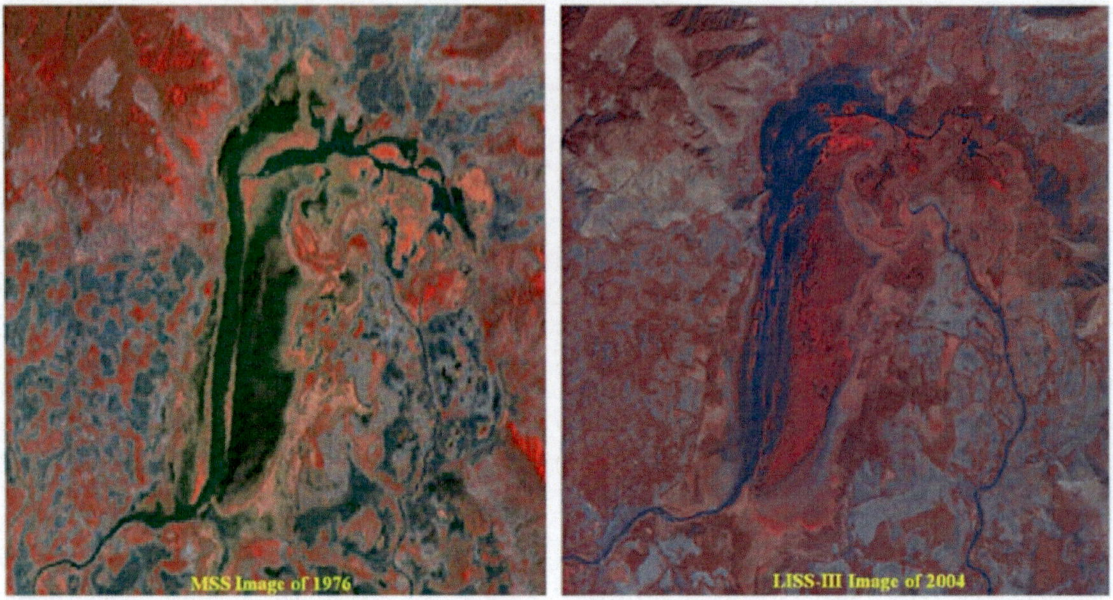

Fig. 5.9 Landsat MSS satellite image of 1976 and LISS-III image of 2004 depicting changes in the extent of aquatic weeds

reason, woody perennials have some advantages as indicator species compared to annuals (Lechowicz & Koike, 1995). Temporal satellite images used for observing the blooming patterns of Apple Orchards Multi-temporal satellite data indicate that there is a shift in the blossom dates of apples. The satellite data of March 25, 2012, showed the signs of phonological variation for blossoming, while as it was not there on even April 15 in 1994, clearly indicating that the apples were blooming earlier. An analysis of relation between full bloom dates and temperatures was carried out for the period 1974–2009. While there was no correlation between temperature and bloom during winter and spring, there was a negative correlation during autumn (August–October). Studies suggest that spring phenophases influenced by temperature changes are now coming earlier (Kramer et al., 2000; Keatley et al., 2002; Chmielewski et al., 2004; Lu et al., 2006; Miller-Rushing et al., 2007; Doi & Katano, 2008; Guedon & Legave, 2008; Primack et al., 2009; Vitasse et al., 2009). Other works on blossoming of apples have shown a significant relation between increasing spring temperatures and their influence on the pheno-phases (Chmielewski et al., 2004; Doi, 2007; Miller-Rushing et al., 2007; Doi & Katano, 2008), and in some circumstances, flowering is most sensitive to mean temperatures during the month before average blossoming date (Lu et al., 2006). The latest analyses in the New York State field of historical apple yields have found that colder temperatures between January 1 and the bud break are associated with lower yields (Wake & Markham, 2005). Kumar et al. (2012) identified the control gene that speeds up flowering time in response to temperature. Temperature alone could exert specific and precise control on the control gene (PIF4), which activates the flowering pathway in the presence of warm air temperature, but with cold air, the gene is unable to act. Finally, the cumulative impacts of rapid regional warming and decreased winter/early spring precipitation and associated extended dry spells in this season can be due to long-term temporal changes at apple phänological stages in Kashmir's Valley (Grab & Craparo, 2011). The temperature shift in the winter and early spring, which influences the dormancy and genetic evolution, is essential for the spring growth of plants. In tandem with milder winters, challenges to late spring frosts pose a

Fig. 5.10 LANDSAT MSS of April 15, 1994 and IRS aWIFS of March 2012 showing variation in blooming dates of apple orchards

risk for even frost hardy plants. If the crop varies in the atmosphere and spring temperatures differ further, the frost damage to flowers and fruit development will decrease. Results suggest an advance in spring phenology ranging from 10 to 15 days (Wolfe et al., 2005). The magnitude of this climate impact on phenology is similar to other analyses of ground observations for other plant species in mid- and northern latitudes (Bradley et al., 1999; Abu-Asab et al., 2001; Menzel et al., 2001). Our results are also qualitatively in agreement with the advancement in "green-up" based on satellite normalized difference vegetation index (NDVI) data (Fig. 5.10).

Conclusion

In conclusion, a review of proposed environmental indicators for climate change in the Kashmir Valley shows that data exist for many variables and show positive signs of climate change. Fluctuations in temperature and precipitation, loss of wetlands, dynamics of vegetation patterns in alpine regions, and growth of aquatic weeds in the lakes provide significant revelations related to climate change in this microclimatic setting. This study has shown that the changes on the glacier surface are due to climatic and topographic (local geomorphology) factors, which decreased overall glaciated area by 6% between 1962 and 2006. The increase in long-term temperatures during the critical period leading up to full bloom is, in part, a likely consequence of the reduced frequency and strength of frontal perturbations over time. More study is, however, required to determine the sensitivity and utility of these findings. Further attention is necessary to increase data quality and availability and develop new environmental monitoring and surveillance databases, especially for climate-sensitive morbidity.

References

Abu-Asab, M., Peterson, P. M., Shetler, S. G., & Orli, S. S. (2001). Earlier plant flowering in spring as a response to global warming in the Washington, D.C., area. *Biodiversity and Conservation, 10*, 597–612.

Adrian, R., Wilhelm, S., & Gerten, D. (2006). Life history traits of lake plankton species may govern their phenological response to climate warming. *Global Change Biology, 12*, 652–661.

Anderson, K., & Bows, A. (2008). Reframing the climate change challenge in light of post-2000 emissions trends. *Transactions of the Royal Academy-A: Mathematical, Physical and Engineering Sciences, 366*, 3863–3882.

Anderson, J., Arblaster, K., & Bartley, J. (2008). Climate change induced water stress and its impact on natural and managed ecosystems. In J. Anderson, K. Arblaster, & J. Bartley (Eds.), *Economic and scientific policy* (p. 108110). E.P. Publishers.

Aono, Y. (1998). Climate change in march temperature deduced from phenological record for flowering of cherry tree in Tokyo since the late 18th century. *Bulletin of Osaka Prefecture University, Series B, 50*, 11–19.

Barnett, T. P., Adam, J. C., & Lettenmaier, D. P. (2005). Potential impacts of a warming climate on water availability in a snow-dominated region. *Nature, 438*, 303–309.

Beniston, M. (2003). Climatic change in mountain regions: A review of possible impacts. *Climatic Change, 59*, 5–31.

Blenckner, T., Adrian, R., Livingstone, D. M., Jennings, E., Weyhenmeyer, G. A., George, D. G., Jankowski, T., Järvinen, M., Aonghusa, C. N., Noges, T., & Straile, D. (2007). Large-scale climatic signatures in lakes across Europe: A meta-analysis. *Global Change Biology, 13*, 1314–1326.

Bradley, N. L., Leopold, A. C., Ross, J., & Huffaker, W. (1999). Phenological changes reflect climate change in Wisconsin. *Proceedings National Academy of Sciences USA, 96*, 9701–9704.

Chmielewski, F.-M., Müller, A., & Bruns, E. (2004). Climate changes and trends in phenology of fruit trees and field crops in Germany, 1961–2000. *Agricultural and Forest Meteorology, 121*, 69–78.

Christensen, J. H., Hewitson, B., Busuioc, A., Chen, A., Gao, X., Held, I., Jones, R., Kolli, R. K., Kwon, W. T., Laprise, R., Magana Rueda, V., Mearns, L., Menéndez, C. G., Raisanen, J., Rinke, A., Sarr, A., & Whetton, P. (2007). Regional Climate Projections. In *Climate change 2007: The physical science basis. Contribution of working group I that the fourth assessment report of the Intergovernmental Panel on Climate Change, United Kingdom and USA*.

Crepinsek, Z., Kajfez-Bogataj, L., & Bergent, K. (2006). Modelling of weather variability effect on phyto phenology. *Ecological Modelling, 194*, 256–265.

Cruz, R., Harasawa, H., Lal, M., Wu, S., Anokhin, Y., Punsalmaa, B., Honda, Y., Jafari, M., Li, C., & Ninh, N. H. (2007). Asia. In M. Parry, O. F. Canziani, J. P. Palutikof, P. J. van der Linden, & C. E. Hanson (Eds.), *Climate change 2007: Impacts, adaptation and vulnerability. Contribution of working group II to the fourth assessment report of the intergovernmental panel on climate change* (pp. 469–506). Cambridge University Press.

Daly, C., Neilson, R. P., & Phillips, D. L. (1994). A statistical-topographic model for mapping climatological precipitation over mountainous terrain. *Journal of Applied Meteorology, 33*, 140–158.

Deshingkar, P. (1997). *Adapting to climate change in a forest based land use system – A case study of Himachal Pradesh, India*. Atmospheric Environmental Issues in Developing Countries, Stockholm Environment Institute.

Doi, H. (2007). Winter flowering phenology of Japanese apricot Prunus mume reflects climate change across Japan. *Climate Research, 34*, 99–104.

Doi, H., & Katano, I. (2008). Phenological timings of leaf budburst with climate change in Japan. *Agricultural and Forest Meteorology, 148*, 512–516.

Falkenmark, M. (1989). The massive water scarcity now threatening Africa—Why isn't it being addressed? *Ambio, 18*(2), 112–118.

Fowler, H. J., & Archer, D. R. (2006). Conflicting signals of climate change over the upper Indus basin. *Journal of Climate, 19*, 42764293.

Gairolal, S., Rawal, R. S., & Todaria, N. P. (2008). Forest vegetation patterns along an altitudinal gradient in sub-alpine zone of west Himalaya, India. *African Journal of Plant Science, 2*(6), 042–048.

Ghosh, P. (2009). Geomatics and impact of climate change with specific reference to mountain ecosystem. *Current Science, 97*(10), 1045–1046.

Gleick, P. H. (1990). Vulnerability of water systems. In P. E. Waggoner (Ed.), *Climate change and U.S. water resources*. Wiley.

Gopalakrishnan, R., Jayaraman, M., Bala, G., & Ravindranath, N. H. (2011). Climate change and Indian forests. *Current Science, 101*(3, 10), 348–355.

Grab, S., & Craparo, A. (2011). Advance of apple and pear tree full bloom dates in response to climate change in the southwestern cape, South Africa: 1973–2009. *Agricultural and Forest Meteorology, 151*, 406–413. https://doi.org/10.1016/j.agrformet.2010.11.001

Grabherr, G., Gottfriend, M., & Pauli, H. (1994). Climate effects on mountain plants. *Nature, 369*, 448.

Groisman, P. Y., Karl, T. R., Easterling, D. R., Knight, R. W., Jamason, P. F., Hennessy, K. J., Suppiah, R., Page, C. M., Wibig, J., Fortuniak, K., Razuvaev, V. N., Douglas, A., Forland, E., & Zhai, P. M. (1999). Changes in the probability of heavy precipitation: Important indicators of climatic change. *Climatic Change, 42*, 243–283.

Guedon, Y., & Legave, J. M. (2008). Analyzing the time-course variation of apple and pear tree dates of flowering stages in the global warming context. *Ecological Modelling, 219*, 189–199.

Hampton, S. E., Izmest'eva, L. R., Moore, M. V., Katz, S. L., Dennis, B., & Silow, E. A. (2008). Sixty years of environmental change in the world's largest freshwater Lake—Lake Baikal, Siberia. *Global Change Biology, 14*, 1947–1958.

Hansen, J., Sato, M., Kharecha, P., Beerling, D., Berner, R., Masson-Delmotte, V., Pagani, M., Raymo, M., Royer, D. L., & Zachos, J. C. (2008). Target atmospheric CO2: Where should humans aim? *The Open Atmospheric Science Journal, 2*, 217–231.

Hasnain, S. I. (2008). Impact of climate change on Himalayan glaciers and glacier lakes. In: *Proc. Taal 2007, The 12th lake conference* (pp. 1088–1091).

Heald, W. F. (1967). Sky Island. Van Nostrand, NY. Pauli, H., Gottfried, M., Reiter, K., Klettner, C., Grabherr, G., 2007. Signals of range expansions and contractions of vascular plants in the high Alps: Observations (1994–2004) at the GLORIA master site Schrankogel, Tyrol Austria. *Global Change Biology, 13*, 147–156.

Hopp, R. J. (1974). Plant phenology observation networks. In IN: Leith, H. (Ed.), *Phenology and seasonality modeling* (pp. 25–43). Springer-Verlag.

Houghton, J. T., Meira Filho, L. G., Callander, B. A., Harris, N., Kattenberg, A., & Maskell, K. (Eds.). (1996). *Climate change 1995: The science of climate change*. Cambridge University Press.

Huisman, J., Sharples, J., Stroom, J. M., Visser, P. M., Kardinaal, W. E. A., Verspagen, J. M. H., & Sommeijer, B. (2004). Changes in turbulent mixing shift competition for light between phytoplankton species. *Ecology, 85*, 2960–2970.

Intergovernmental Panel on Climate Change (IPCC). (2007). Climate change 2007: Impacts, adaptation and vulnerability. In *Contribution of working group II to the fourth assessment report of the intergovernmental panel on climate change*. Cambridge University Press.

Johnk, K., Huisman, J., Sharples, J., Sommeijer, B., Visser, P. M., & Stroom, J. M. (2008). Summer heatwaves promote blooms of harmful cyanobacteria. *Global Change Biology, 14*, 495–512.

Karl, T. R., & Knight, R. W. (1998). Secular trends of precipitation amount, frequency and intensity in the United States. *Bulletin of the American Meteorological Society, 79*(2), 231–241.

Keatley, M. R., Fletcher, T. D., Hudson, I. L., & Ades, P. K. (2002). Phenological studies in Australia: Potential application in historical and future climate analysis. *International Journal of Climatology: A Journal of the Royal Meteorological Society, 22*, 1769–1780.

Keller, W. (2007). Implications of climate warming for boreal shield lakes: A review and synthesis. *Environmental Reviews, 15*, 99–112.

Klein Tank, A. M. G., & Konnen, G. P. (2003). Trends in indices of daily temperature and precipitation extremes in Europe, 1946-99. *Journal of Climate, 16*, 3665–3680.

Kramer, K., Leinonen, I., & Loustau, D. (2000). The importance of phenology for the evaluation of impact of climate change on growth of boreal, temperate and Mediterranean forest ecosystems: An overview. *International Journal of Biometeorology, 44*, 67–75.

Kulkarni, A. V., & Bahuguna, I. M. (2002). Glacial retreat in the BaspaBasin, Himalayas, monitored with satellite stereo data. *Journal of Glaciology, 48*, 171–172.

Kumar, K., & Pant, G. B. (1994). Diurnal asymmetry of surface temperature trends over India. *Geophysical Research Letters, 21*, 677680.

Kumar, S. V., Lucyshyn, D., Jaeger, K. E., Alós, E., Alvey, E., Harberd, N. P., & Wigge, P. A. (2012). Transcription factor PIF4 controls the thermosensory activation of flowering. *Nature, 484*, 242–245. [PubMed].

Kwadijk, J., & Middlekoop, H. (1994). Estimation of impact of climate change on the peak discharge probability of the river Rhine. *Climate Change, 27*, 199224.

Lechowicz, M. J., & Koike, T. (1995). Phenology and seasonality of woody plants: An unappreciated element in global change research? *Canadian Journal of Botany, 73*, 147–148.

Livingstone, D. M., & Dokulil, M. (2001). Eighty years of spatially coherent Austrian lake surface temperatures and their relationship to regional air temperature and the North Atlantic Oscillation. *Limnology and Oceanography, 46*, 1220–1227.

Lu, P., Yu, Q., Liu, J., & Lee, X. (2006). Advance of tree-flowering dates in response to urban climate change. *Agricultural and Forest Meteorology, 138*, 120–131.

Ma, X., Xu, J. C., Luo, Y., Aggarwal, S. P., & Li, J. T. (2009). Response of hydrological processes to land cover and climate change in Kejie Watershed, Southwest China. *Hydrological Processes*. https://doi.org/10.1002/hyp.7233

Magnuson, J. J., Robertson, D. M., Benson, B. J., Wynne, R. H., Livingstone, D. M., Arai, T., Assel, R. A., Barry, R. G., Card, V., Kuusisto, E., & Granin, N. G. (2000). Historical trends in lake and river ice cover in the Northern Hemisphere. *Science, 289*, 1743–1746. and Errata 2001, Science 291: 254. [PubMed: 10976066].

Malik, M. I. (2012). Spatio-temporal analysis of urban dynamics in Kashmir valley (1901-2011) using geographical information system. *International Multidisciplinary Research Journal, 2*(8), 21–26.

Memmott, J., Craze, P. G., Waser, N. M., & Price, M. V. (2007). Global warming and the disruption of plant-pollinator interactions. *Ecology Letters, 10*, 710–717.

Menzel, A., Estrelly, N., & Fabian, P. (2001). Spatial and temporal variability of the phenological season in Germany from 1951–1996. *Global Change Biology, 7*, 657–666.

Miller-Rushing, A. J., Katsuki, T., Primack, R. B., Ishii, Y., Lee, S. D., & Higuchi, H. (2007). Impact of global warming on a group of related species and their hybrids: Cherry tree (Rosaceae) flowering at Mt. Takao, Japan. *American Journal of Botany, 94*, 1470–1478.

Murdoch, P. S., Baron, J. S., & Miller, T. L. (2000). Potential effects of climate change on surface-water quality in North America. *Journal of the American Water Resources Association, 36*, 347–366.

Muslim, M. (2012). Predicting climate change over Kashmir and assessing the Impact of climate change on Paddy Yield Ph.D. (Env.Sci.). Thesis submitted to Sher - e - Kashmir University of Agricultural Sciences and Technology Shalimar, pp.160–162.

Neilson, R. P., Pitelka, L. F., & Solomon, A. M. (2005). Forecasting regional to global plant migration in response to climate change. *Bios, 55*, 749–759.

Nogues-Bravo, D., Araujo, M. B., Errea, M. P., & Martinez-Rica, J. P. (2007). Exposure of global mountain systems to climate warming during the 21st century. *Global Environmental Change, 17*, 420–428.

O'Reilly, C. M., Alin, S. R., Plisnier, P.-D., Cohen, A. S., & McKee, B. A. (2003). Climate change decreases aquatic ecosystem productivity in Lake Tanganyika, Africa. *Nature, 424*, 766–768.

Paerl, H. W., & Huisman, J. (2008). Climate—blooms like it hot. *Science, 320*, 57–58.

Pauli, H., Gottfried, M., & Grabherr, G. (1996). Effects of climate change on mountain ecosystems upward shifting of alpine plants. *World Resource Review, 8*, 382–390.

Pauli, H., Gottfried, M., & Grabherr, G. (2003). Effects of climate change on the alpine and nival vegetation of the Alps. *Journal of Mountain Ecology, 7*(Suppl), 9–12.

Primack, R. B., Higuchi, H., & Miller-Rushing, A. J. (2009). The impact of climate change on cherry trees and other species in Japan. *Biological Conservation, 142*, 1943–1949.

Rashid, H., & Naseem, G. (2008). Quantification of loss in spatial extent and wetlands in the suburbs of Srinagar City during last century using geospatial approach. In M. Sengupta & R. Dalwani (Eds.), *Proceedings of Taal2007: The 12th world Lake conference, 2008* (pp. 653–658).

Raskin, P., Gleick, P. H., Kirshen, P., Pontius, R. G., Jr., & Strzepek, K. (1997). *Open file 85–166, U.S. Geological survey, U.S.* Department of the Interior.

Ravindranath, N. H., Joshi, N. V., Sukumar, R., & Saxena, A. (2006). Impact of climate change on forests in India. *Current Science, 90*(3), 354–361.

Reist, J. D., Wrona, F. J., Prowse, T. D., Power, M., Dempson, J. B., Beamish, R. J., King, J. R., Carmichael, T. J., & Sawatzky, C. D. (2006). General effects of climate change on Arctic fishes and fish populations. *Ambio, 35*, 370–380.

Rogers, P. R. (1999). Environmental indicators. *Journal of Water Resources Planning and Management, 125*(2), 74–75.

Rogers, P. R., Jalal, K. F., Lohani, B. N., Owens, G. M., Yu, C., Dufournaud, C. M., & Jun, B. (1997). *Measuring environmental quality in Asia*. Harvard University Press.

Romanovskaja, D., & Baksiene, E. (2007). Influence of thermal mode on seasonal phonological phenomena in Lithuania. *Ecologija, 53*, 15–20.

Romshoo, S. A., & Rashid, I. (2010). Potential and constraints of geospatial data for precise assessment of the impacts of climate change at landscape level. *International Journal of Geomatics and Geosciences, 1*(3), 386.

Russell, G. L., Miller, J. R., & Rind, D. (1995). A coupled atmosphere ocean model for transient climate change studies. *Atmosphere-Ocean, 33*(4), 683.

Scavia, D. (2012). Climate change likely to increase Lake Erie algae blooms and 'dead zones,' U-M ecologist says: http://www.ns.umich.edu/new/releases/20750-climatechange-likely-to-increase-lake-erie-algae-blooms-and-dead-zones-according-to-umichigan-ecologist.

Schindler, D. W. (1997). Widespread effects of climatic warming on freshwater ecosystems in North America. *Hydrological Processes, 11*, 225–251.

Schindler, D. W., Bayley, S. E., & Parker, B. R. (1996). The effects of climatic warming on the properties of boreal lakes and streams at the Experimental Lakes area, northwestern Ontario. *Limnology and Oceanography, 41*, 1004–1017.

Sealthun, N. R., Aittoniemi, P., & Bergsstorm, S. (1998). *Climate change impacts on runoff and hydropower in the Nordic countries* (TemaNord report No. 52, 170 pp.). Nordic Council of Ministers.

Shrestha, A. B., Wake, C. P., Dibb, J. E., & Mayyewski, P. A. (2000). Precipitation fluctuations in the Nepal Himalaya and its vicinity and relationship with some large-scale climatology parameters. *International Journal of Climatology, 20*, 317–327.

Solim, S., & Wanganeo, A. (2008). Excessive phosphorus loading to dal Lake, India: Implications for managing shallow Eutrophic Lakes in urbanized watersheds. *International Review of Hydrobiology, 93*, 148–166. https://doi.org/10.1002/iroh.200710934

Solomon, S., Plattner, G. K., Knutti, R., & Friedlingstein, P. (2009). Irreversible climate change due to carbon dioxide emissions. *Proceedings of the National Academy of Sciences of the United States of America, 106*, 1704–1709.

Song, M., Zhou, C., & Ouyang, H. (2004). Distributions of dominant tree species on the Tibetan plateau under current and future climate scenarios. *Mountain Research and Development, 24*, 166–173.

Thackeray, S. J., Jones, I. D., & Maberly, S. C. (2008). Long-term change in the phenology of spring phytoplankton: Species-specific responses to nutrient enrichment and climatic change. *Journal of Ecology, 96*, 523–535.

U.S. Environmental Protection Agency. (1997). *The index of watershed indicators* (Rep. No. EPA-841-R-97-010). Washington, DC: Office of Water.

United Nations Division for Sustainable Development. (1996). *Indicators of sustainable development framework and methodology*. United Nations Division for Sustainable Development.

Van Zonneveld, M., Jarvis, A., Dvorak, W., Lema, G., & Leibing, C. (2009). Climate change impact predictions on Pinus patula and Pinus tecunumanii populations in Mexico and Central America. *Forest Ecology and Management, 257*(7), 1566–1576.

Verburg, P., Hecky, R. E., & Kling, H. (2003). Ecological consequences of a century of warming in Lake Tanganyika. *Science, 301*, 505–507.

Vitasse, Y., Delzon, S., Dufrêne, E., Pontailler, J.-Y., Louvet, J.-M., Kremer, A., & Michalet, R. (2009). Leaf phenology sensitivity to temperature in European trees: Do within-species populations exhibit similar responses? *Agricultural and Forest Meteorology, 149*, 735–744.

Wagner, C., & Adrian, R. (2009). Cyanobacteria dominance: Quantifying the effects of climate change. *Limnology and Oceanography, 54*, 2460–2468.

Wake, C., & Markham, A. (2005). Indicators of climate change in the northeast 2005. *Clean Air-Cool Plane*.

Walther, G.-R., Post, E., Convey, P., Menzel, A., Parmesan, C., Beebee, T. J. C., Fromentin, J.-M., Hoegh-Guldberg, O., & Bairlein, F. (2002). Ecological responses to recent climate change. *Nature, 416*, 389–395.

Welbergen, J. A., Klose, S. M., Mark, W. N. N., & Eby, P. (2008). Climate change and the effect of temperature extremes in Australian flying flocks. In *Proceedings of the Royal Society B* (pp. 304–410). Royal Society Publishing Inc.

Williams, M. W., Losleben, M., Caine, N., & Greenland, D. (1996). Changes in climate and hydrochemical responses in a high elevation catchment in the Rocky Mountains, USA. *Limnology and Oceanography, 41*, 939–946.

Winter, T. C. (2000). The vulnerability of wetlands to climate change: A hydrological landscape perspective.

Journal of the American Water Resources Association, 36, 305311.

Wolfe, D. W., Schwartz, M. D., Lakso, A. N., Otsuki, Y., Pool, R. M., & Shaulis, N. J. (2005). Climate change and shifts in spring phenology of three horticultural woody perennials in northeastern USA. *International Journal of Biometeorology, 49*(5), 303–309.

Xu, Z., Gong, T., & Liu, C. (2007). Decadal trends of climate in the Tibetan plateau – Regional temperature and precipitation. *Hydrological Processes, 22*, 3056–3065.

Yadav, R. R., Park, W. K., Singh, J., & Dubey, B. (2004). Do the western Himalayas defy global warming? *Geophysical Research Letters, 31*, 1720.

Yao, T., Wang, Y., Liu, S., Pu, J., & Shen, Y. (2004). Recent glacial retreat in high Asia in China and its impact on water resource in Northwest China. *Science in China Series D Earth Sciences, 47*, 1065–1075.

Zhou, L., Kaufmann, R. K., Tian, Y., Tucker, C. J., & Myneni, R. B. (2003). Relation between interannual variations in satellite measurements of northern forest greenness and climate between 1982 and 1999. *Journal of Geophysical Research – Atmospheres, 108*(D), 4004.

6 Is Technology the Nexus Between Climate Change-and Disaster-Induced Human Displacement?

Shambhavi Krishna and Shailendra Rai

Introduction

The technological advancement vary from country to country, and therefore, some are developed countries while others are developing countries. Some of the technological advancements which strengthen the economy and infrastructure of a country include hydropower, connectivity in terms of transportation, increase in production for export, and many more which require the investment of time, money, and geographical area (land) in the developmental projects to make use of these technologies to provide electricity; to build up national highways, roads, and railway tracks for trains and metro; as well as to establish industries and factories and carry out mining to increase their export respectively. But, there is boon as well as bane connected with the increase in the developmental projects. These projects are responsible for degrading the environmental and social standards of a particular area where the developmental project has to set up.

Since these projects will alter the ecosystem of that area in many ways such as deforestation, change in course of river, sedimentation in rivers and saltation in the bank of rivers, killing of aquatic lives, increase in air pollution and water pollution, change in weather pattern, temperature, and water cycle of the area, and others, these will naturally affect the animal and human population living in that area. Not only this, these projects might also become the reason behind various natural disaster such as floods, droughts, earthquake, and many more in these areas due to increase in the ambient noise and vibration; also these projects require huge land investment. The environmental degradation being brought about by the developmental projects are responsible for affecting the social, political, and economic aspects of human displacement leading to alternation in the people's ability to get displaced.

Technological achievements in terms of developmental projects help the country to overcome the issues of poverty, unemployment, illiteracy, and inequality but make the country exposed to threats of climate change- and disaster-induced displacements (Terminski, 2015). Though this issue is not new, it is definitely relevant in the present scenario where in the race of development, the countries are neglecting the importance and conservation of their natural resources along with the human resource. Development is necessary for the social and economic growth of the

S. Krishna (✉)
K. Banerjee Centre of Atmospheric and Ocean Studies, University of Allahabad, Prayagraj, India

S. Rai
K. Banerjee Centre of Atmospheric and Ocean Studies, University of Allahabad, Prayagraj, India

M. N. Saha Centre of Space Studies, University of Allahabad, Prayagraj, India

A. R. Siddiqui, A. Sahay (eds.), *Climate Change, Disaster and Adaptations*, Sustainable Development Goals Series, https://doi.org/10.1007/978-3-030-91010-5_6

country, but if it is done sustainably, then it is also equally beneficial for restoring the quality of the environment of the country and its people along with development going side by side.

India: New Perspective on Developmental Projects, Climate Change, and Human Displacement

Soon after the independence, many developmental projects were introduced in the country so as to satisfy the socioeconomic needs of various people living in different parts of the country. Within no time, India became the fastest growing economy of the world as well as fastest growing developing country. The technological advancement gained through developmental projects in India was responsible for bringing various natural calamities, causing degradation of the ecosystem of geographical areas, thus causing human displacements; it is the main reason behind the protest against the developmental projects.

Many projects became controversial because of displacement of local people in the absence of compensation and proper relocation and resettlement process such as dam construction which leads to loss of ecosystem and degradation of natural resources present at the site of construction (Judge, 1997). Similarly, the urban infrastructure development has been responsible for the increase in air and water pollution in many cities, which is the same case in settlement of industries and factories in a particular area which also involve acquiring land from people without providing correct amount of compensation to them. The mining projects cause degradation of ecosystem of the areas of the site followed by displacement of tribal and indigenous people in search of new livelihood because of starting mining in opencast field to exploit low quality ore especially in case of coal. The Metro Rail Project was very important to India, but the construction of this project increased the seismic activity especially in Delhi, as well as many slums were destroyed to acquire land for the construction of tracks for Metro Rails leading to displacement of the poor.

All the three aspects – technology, environment, and humans – are important for a country's overall development, but giving importance to only technological advancement and neglecting the other two aspects will prove disastrous for a country's growth. Though we cannot deny from the fact that the developmental projects will have some impact on the environment and population of a country, efforts should be made to minimize these impacts by encouraging the concept of sustainable development. The time, money, and mind that the government invest in the developmental projects should also be spent in making policies in terms of resettlement, relocation, and mitigation measures for the displaced human population and those who get effected by natural calamities due to developmental projects, as well as proper assessment must be taken to minimize any negative impact on the environment of the area where the developmental projects are set up (Wet, 2006).

Data and Method

The data related to the construction of dams in India especially large dam is taken from the National Register of Large Dam (NRLD) which is a report concerned with the compilation of the dams in the country prepared as per information received from the state government/authority concerned. In order to study the Metro Rail of the country, the data was collected from the Ministry of Urban Development. The Indian Bureau of Mines (IBM) was the source of collection of data showing the number of ores present in the country with the area occupied for the mining process. The data related to the number of national highways and industries or factories constructed in India and which are under construction was collected from Ministry of Statistics and Programme Implementation, Government of India. The data concerned with the number of people displaced by various developmental projects is collected from a research report presented in 2013 to Lok Sabha. On the other hand, many climate change-related facts concerned with the developmental projects and displacement have been taken from many articles and journals which are a part of literature work. The analysis basically includes the presentation of statistics related to many projects in terms of their construction in India state-wise and how much displacement they have caused in India.

Result and Discussion

Load of Developmental Projects in India

Firstly, Fig. 6.1 shows the number of dams which are already constructed in India and also which are under construction state-wise. As we analyze from the figure, each and every state in India has dams with Maharashtra having highest number of dam already working, and also more projects are yet to begin working. Other states like Andhra Pradesh, Chhattisgarh, Gujarat, Karnataka, Madhya Pradesh, Odisha, Rajasthan, Telangana, and Uttar Pradesh are the states which have more numbers of dams. Secondly, Fig. 6.2 shows the details of construction related to the Metro Rail Project in India with number of lines and stations. The analysis shows that the cities having metro running in the present time include Delhi, Mumbai, Bengaluru, Kolkata, Chennai, Lucknow, Kochi, Nagpur, Hyderabad, Jaipur, and Gurgaon as per 2018, whereas the cities which will have metros in the future are Kanpur (likely by 2021), Visakhapatnam, Surat, Guwahati, Patna, Kanpur, Coimbatore, Thiruvananthapuram, Indore, and Varanasi with the number of lines and station in each city. The Metro Rail first started in Kolkata, but New Delhi being the capital of India has more construction of metro followed by Hyderabad, Bengaluru, Chennai, and Jaipur, whereas future projects of metro show more growth rate to be seen in Pune, Navi Mumbai, Nagpur, Gujarat, and Noida. Thirdly, from the point of view of connectivity and transportation, another important developmental project is the buildup of national highways. Figure 6.3 shows the construction of national highways in different states of India in terms of length in kilometers in different years. The states with maximum number of highway construction include Uttar Pradesh, Rajasthan, Andhra Pradesh, Maharashtra, Karnataka, Tamil Nadu, Odisha, Gujarat, and Madhya Pradesh, and the years in which maximum construction worked is done are 2014–2015, and by the end of the year 2015, 101,010 km is the total length of the highways constructed all over India. Fourthly, to boost the economy of the country, mining is an important developmental project, and Fig. 6.4 shows that the states where more mines are on lease are Madhya Pradesh, Gujarat, Andhra Pradesh, Tamil Nadu, Karnataka, and Odisha, and in the year 2017, more numbers of mines were on lease in these states where

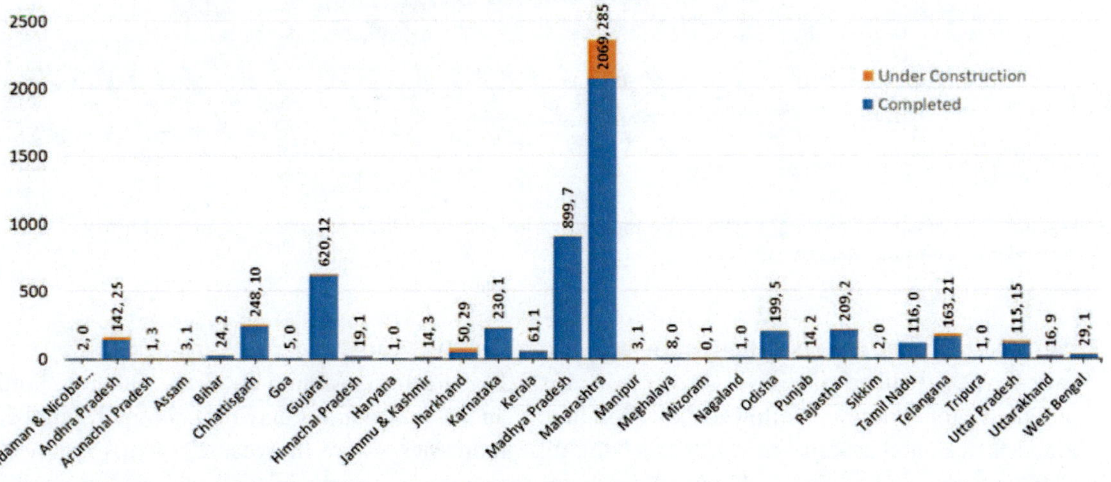

Fig. 6.1 Number of dams in India in different states. (Source: NRLD)

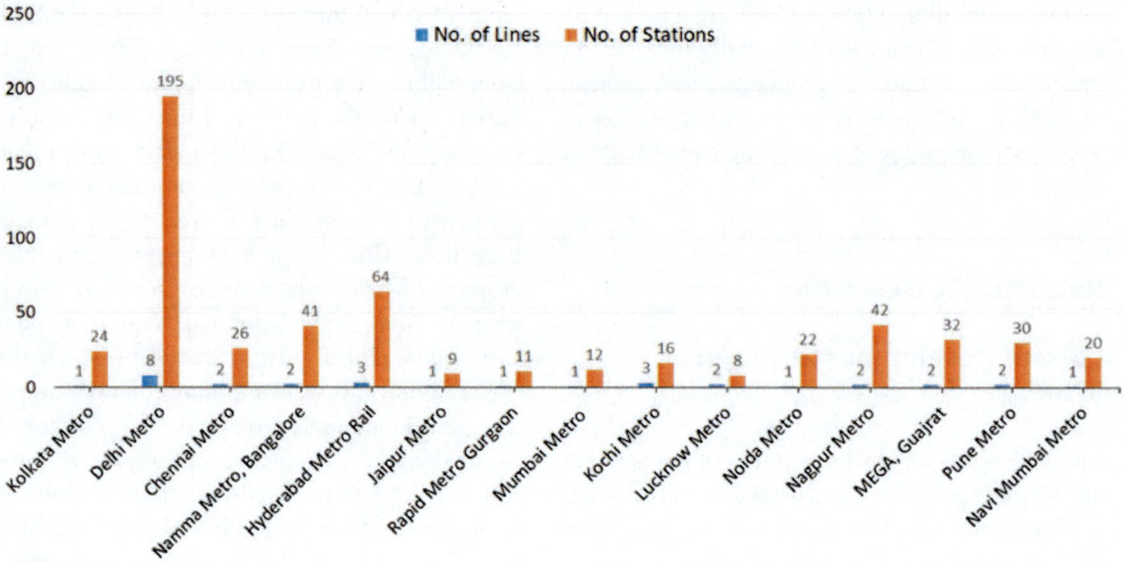

Fig. 6.2 The Metro Rail network in different cities in India with number lines and stations. (Source: Ministry of Urban Development)

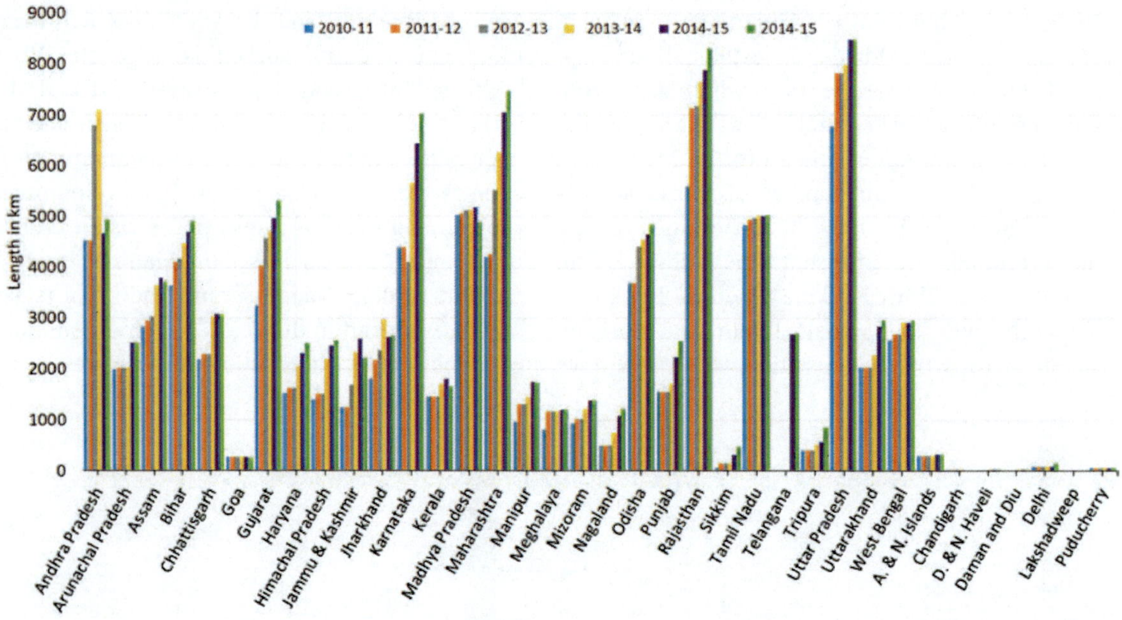

Fig. 6.3 The growth of national highways in India in various years in different states. (Source: Ministry of Statistics and Programme Implementation)

Fig. 6.5 shows the area of mining provided also varies with Odisha having more land area provided for mining followed by Madhya Pradesh, Karnataka, Rajasthan, Andhra Pradesh, Chhattisgarh, and Gujarat. Figure 6.6 shows growth industries in India in different states in various years. The states, namely, Tamil Nadu, Maharashtra, Andhra Pradesh, Gujarat, Punjab, and Karnataka, are having maximum number of industries where the year 2014–2015 shows the maximum growth in the industries in nearly all the Indian states. This shows the load of develop-

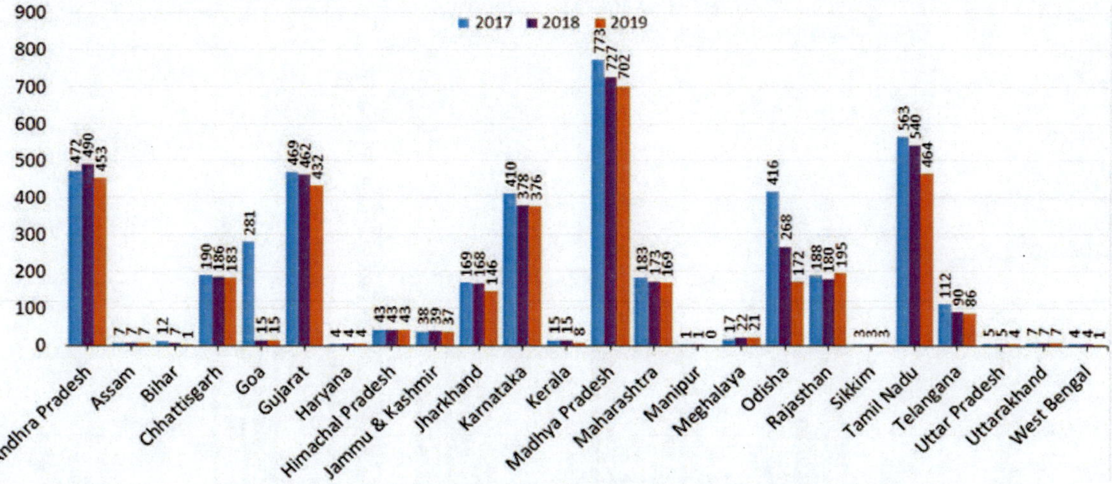

Fig. 6.4 The total number of mines on lease in different Indian states. (Source: IBM)

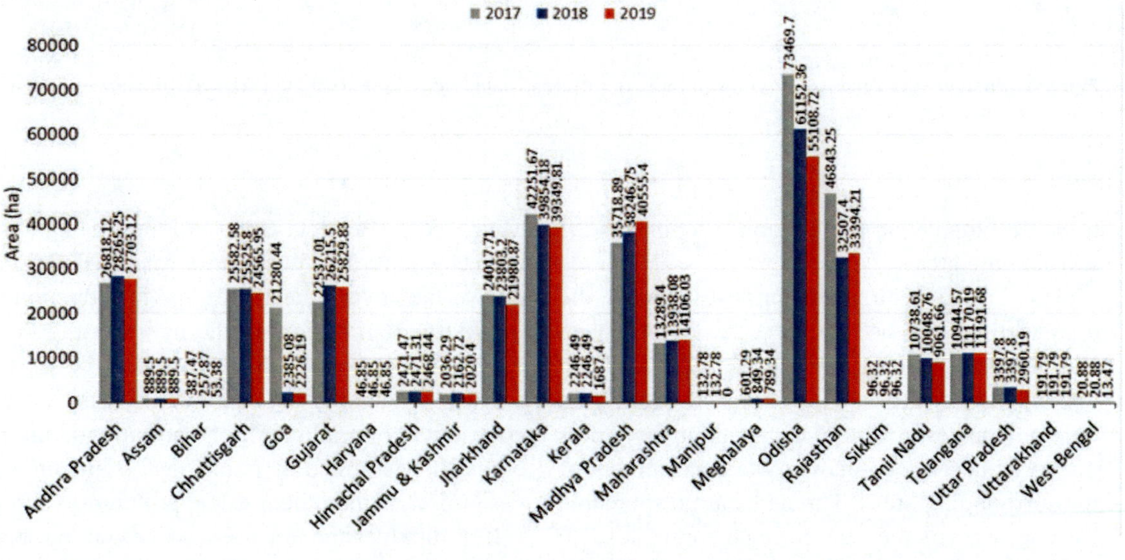

Fig. 6.5 The area of different mines in various states of India on lease in three consecutive years. (Source: IBM)

mental projects in India in different states in various years which will definitely impact the environment, cause climate change, and lead to human displacement.

Effect of Developmental Projects on the Environment and Cause of Climate Change

The results mentioned above provide us the present scenario and status of developmental projects in India. Since India is a developing country, to become a developed country, it is required to increase its growth rate in terms of technology and developments, which means more investment in infrastructure development, business development, and urbanization. This will definitely increase stress on the environment of the geographical areas where these projects are going to be established and will cause change in the ecosystem, thus leading to climate change in the future. Many literature work have presented various kinds of environmental stress as outcome of

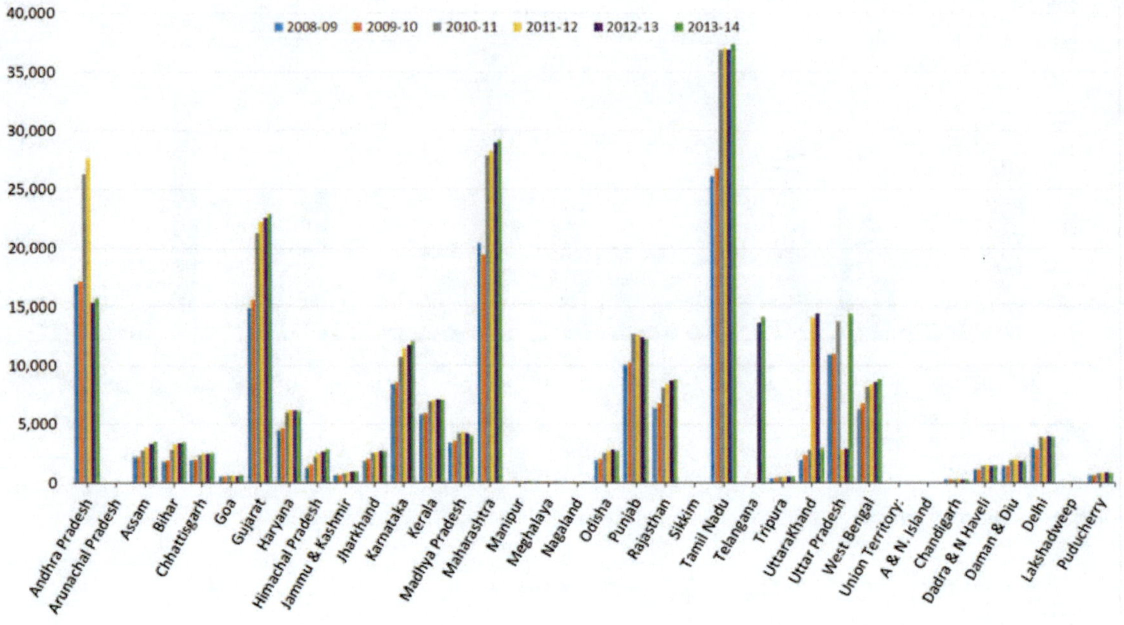

Fig. 6.6 The growth of industrial sector in India in different states from 2008 to 2014. (Source: Ministry of Statistics and Programme Implementation)

these developmental project constructions in the surrounding areas.

There have been many protests against the construction of the dams such as Narmada Bachao Andolan against Sardar Sarovar Dam under Narmada River Valley Project, Tehri Dam on river Ganges in Uttarakhand, and many others due to its negative impact on the environment of the surrounding areas. The dams are responsible for causing seismic effect which might lead to earthquake occurrence in the future as well as cause scouring of the riverbed which affects aquatic life along with loss of river banks (Thukral, 1992). The dam affects temperature of water, saltation, and change in oxygen content which leads to change in ecosystem and destruction of species which will cause climate change in the long run. The other effects include change in course of river and water cycle and the collection of wooden garbage and other cause water pollution with spread of many diseases like as typhus, typhoid, malaria and many more.

The developmental projects which are important from the angle of urbanization, namely, Metro Rail Projects and buildup of national highways, are also the factors for the deterioration of the environment. But, the introduction of the Metro Rail system in the majorly important cities of the country has proved to be a step toward sustainable development because it has contributed in minimizing the pollution from the cities. It has aimed at reducing the greenhouse gas emission and other critical pollutants such as sulfur dioxide and nitrogen oxides along with minimizing traffic which causes a lot of air pollution. According to the reports of Delhi Metro Corporation, the running of metros in Delhi in Phase I has reduced 529,043 tCO_2/annum. But, the negative impacts of construction related to Metro Rail such as construction of tracks, stations, and subways caused deforestation of about 25,507 tress in the Phase I and Phase II construction. It also caused increase in ambient noise and vibration and cracks in the nearby buildings making more and more people prone to earthquake as Delhi lies in zone of high seismic activity. Similarly the construction of national highways causes a lot of cutting of tress, forcefully acquiring the agricultural and farmlands; the construction activities cause increase in suspended particulate matter in the air, thus leading to air pollution and soil erosion; and materials used in construction if washed away by rainwater and absorbed by the ground cause water pollution as well.

Another important developmental project is mining which plays an important role in strengthening the economy of India. But, there are major negative effects of mining on the environment. It is responsible for causing soil erosion and soil pollution in exposed hillside area due to absorption of harmful chemicals used during the process of mining which degrades the nearby soil quality, thus affecting agricultural lands. The other effects include water pollution because of the use of water in the extraction process and other processes in which many chemicals and metals get dissolved such as arsenic, sulfuric acid, and mercury. When this water reaches to nearby water body, it kills the aquatic life like algae, diatoms, zooplanktons, and many more, and on being absorbed by the ground, it also pollutes underground water. Mining also leads to saltation of the drainage areas and these problems with the passage all time totally damage the ecosystem all that area and bring about climate change.

Similar are the effects of setting up of any industry or factory in any area. Though there are many advantages of establishment of industry and factory in any areas such as employment, infrastructure growth, economic growth, and many more, it also comes with a great disadvantage to the surrounding environment. The discharge of the untreated effluents from the industries and the various emissions from the chimneys of these industries as a resultant of combustion and chemical reaction basically include acids, bases, benzene, heavy metal solutions, radioactive waste, toluene, sludge, carbon monoxide, nitrogen dioxide and many other toxic and hazardous waste as responsible for causing air and water pollution and degrade the natural; resources present in the area destroying the natural ecosystem, thus leading to climate change in the near future.

Climate- and Disaster-Induced Displacement by Developmental Projects

As it is observed in this study that in order to make India a developed country, a lot of priority is given to technology through developmental projects which are also causing the degradation of the environment and thus promoting climate change. Not only this, but to establish these developmental projects, a large size of land is also required. Thus, these two are the main reason for the displacement of the local, tribal, and indigenous people from the site of developmental projects (Fernandes, 1991).

Those who are either forced to leave their land or sacrifice their land without getting displaced or loose common property like forest, fisheries affecting their livelihood or disposal of waste of the developmental causing pollution or fear of natural calamities due to developmental make them leave their land and .livelihood are called as Internally Displaced People. The reason behind the concern of these internally displaced people is the consequences being faced by them after getting displaced from their land and property in the search of new place to live and to survive. The problem faced by them after displacement basically include homelessness, unemployment, marginalization which is a downward mobility due to economic crisis, food insecurity, increased morbidity and mortality due to decline in health status because of stress and psychological trauma because of the situation being faced by them, loss of access to common property, and social disarticulation. These are the relevant problems in the case of displacement due to construction of dams, Metro Rails, and national highways, but in the case of mining and industrial setup due to air and water pollution, degradation of natural resources gives rise to various respiratory, cardiovascular, neurological, and skin diseases or even causes cancer which also leads to human displacement.

The statistical data gathered from various literature work show that millions of people have been displaced from the site of developmental projects. Figure 6.7 shows that the dams are responsible for the maximum number of human

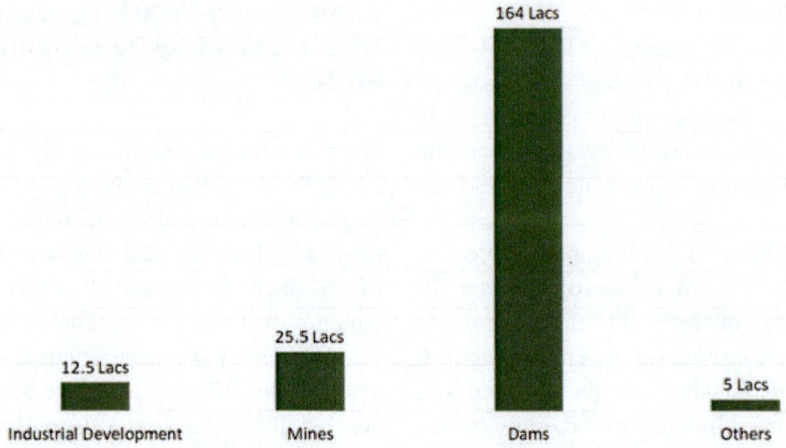

Fig. 6.7 Presentation of "internally displaced people" through different developmental projects in India

displacement of nearly 164 lakhs of people followed by mines, industrial development, and others which basically include metro projects, buildup of national highways, and many more which also highlight the fact that since dams are present in each and every states in India along with other developmental projects, people have got displaced from every corners of the country and huge geographical area of country is also vulnerable toward climatic variability. One of the most controversial developmental projects was Narmada River valley project in which 3300 dams were to be built in the states of Maharashtra, Gujarat, and Madhya Pradesh acquiring about 37,000 hectares of land to provide electricity and irrigation. But, the land required for the project provided by the stakeholder was forcefully acquired, and compensation was not provided to the landowners and most of the lands were agricultural lands which led to displacement of about more than 2 lakhs of people. Similarly the dam over the river Ganga in Uttarakhand called the Tehri Dam built in 1979 lead to 85,500 people being displaced from their homes. On the other hand, the Metro Rail Project in Delhi caused the displacement of 2505 slums for the construction of tracks, stations, and subways (Randhawa, 2012). The other example includes the establishment of the Tata Nano Factory in West Bengal in the district of Singur which required 997 acres of land which mainly included the farmlands; thus, the factory in the future will also affect the ecosystem of the air and cause air, water, and soil pollution. Another example includes the Ring Road Construction in Jaipur City which included 47 km road construction in Phases I and II but 97 km road was supposed to be built in Phase III, thus leading to demolishing of many houses and shops and illegally acquiring farmlands and due to lack of correct compensation, causing the displacement of people.

Hence we observe that fear of natural calamities, degradation of the ecosystem which causes loss of livelihood to locals, and forcefully acquiring land of farmers and the poor by big landowners and stakeholders are the main reasons for the displacement, and the nexus becomes the technological advancement in terms of developmental projects.

Conclusion

The study shows that technology in terms of developmental projects is the nexus between climate change- and disaster-induced human displacement because the developmental projects lead to deterioration of the environment and ecosystem which may cause climate change, as well as they increase the chances of natural calamities at the site of their construction along with acquisition of land, becoming the reason for human displacement. It is observed while doing the analysis of the data collected that the states which have more developmental projects running are Maharashtra, New Delhi, Gujarat, Rajasthan, Tamil Nadu, Karnataka, Andhra Pradesh, and West Bengal and the project which is responsible for displacing more people is the dams; thus, we can say that technology definitely is the main link which causes damage to the local ecosystem, and thus people due to loss of livelihood, land, and basic amenities are forced to leave their homes and get displaced. In the era of the technological development, a country like India being the fastest growing economy of the world cannot put a complete stop to the developmental projects to avoid the risk on the environment as well as to avoid human displacement, but by keeping in mind the goal of sustainable development and proper implementation of planning of the government in terms of compensation, resettlement, relocation, and rehabilitation, the sufferings of the internally displaced people can be minimized.

The very first impact of any developmental project is on the environment, and to avoid climate change or occurrence of any kind of natural calamity in the future, the process of environmental impact assessment should be followed strictly which will ensure that the proposed developmental project will not have any adverse effect on the surrounding environment or cause any alternation in the natural ecosystem of the area and will also not harm socioeconomic, cultural, and health aspects in that area and also to the human population in that area. Each and every developmental project requires a particular suitable site for its construction, and many a time the site will include land of poor people, a forest, farmland, and agricultural land. So, providing correct amount of compensation to the legal landowners as well as getting the displaced population relocated to new location and assuring the total resettlement and rehabilitation should be the primary duty of the government to avoid suffering of the displaced population.

References

Fernandes, W. (1991). Power and powerless: Development projects and displacement of Tribals. *Social Action, 41*, 242–268.

Judge, P. S. (1997). Response to dams and displacement in two Indian states. *Asian Survey, 37*(9), 840–851. *JSTOR*, www.jstor.org/stable/2645701. Accessed 31 July 2020.

Randhawa, P. (2012). Delhi metro rail- beyond mass transit. *Economic & Political Weekly, 47*, 1–5.

Terminski, B. (2015). *Development-induced displacement and resettlement: Socio-legal context*. Ibidem Press.

Thukral, E. G. (1992). *Big dams, displaced people: Rivers of sorrow, rivers of change*. Sage Publications.

Wet, D. (2006). *Development induced displacement-problems, policies & people*. Berghan Books.

Wikipedia. *Singur Tata Nano controversy*. Retrieved from Wikipedia Singur_Tata_Nano_.

7 Climate Change Hazard Perception and Gender Empowerment for Resilience in the Regions of Tropical Asia

Anwesha Haldar and Lakshminarayan Satpati

Introduction

The term 'climate change' usually refers to the significant differences in weather pattern over an extended period of time that links global warming, carbon dioxide emissions and other greenhouse gases; and an increase in the frequencies of extreme weather events and associated climatic hazards. The tropical regions are frequently plagued by climatic hazards such the droughts and water stress in south Central America, northern Mexico, major parts of Africa and Middle-Eastern Asia and the Indian subcontinent; floods in South and South-East Asia, (WRI, 2020); cyclones, hurricanes and tornadoes across the entire tropical regions; and lightning across the entire equatorial regions; while heat waves and extreme temperatures exceeding 150 days per year in sub-Sahara, around the Gulf of Mexico, Amazon, South Asia and South-East Asia. (EMDAT, 2020). The equatorial countries are also often termed as the Global South, which were traditionally termed as underdeveloped but had made immense socio-economic progression in the recent years (Rigg, 2015).

In 1830s the global population was 1 billion, which increased to 2 billion in 1930s and stands at 7.8 billion in August 2020. About 20% of the world's population resides in South Asia that comprises of India, Pakistan, Bangladesh, and Sri Lanka, with India being the second most populous country in the world. Hence it is observed that the world population is growing in leaps and regions of rapid expansion of inhabitants are mostly those having lower developmental rates. This has disseminated to increased vulnerabilities both environmental and social, causing degraded conditions of living. The level of vulnerability is often determined by the intensity or frequency of exposure to natural hazards, coupled with factors like disproportionate age-sex structures; limited adaptive capacity; inequality in resource availability; lack of food, shelter, medical supplies and other timely aids; no access to adaptation technologies; heavy dependence on natural resource base and maladjustment with livelihood opportunities; unplanned urbanization; and poor management facilities during the periods of crises. These issues are grossly common in most hazard-prone developing countries of the Global South as revealed from the graphical representations of the 'Our World in Data' webpagesSource: EMDAT (2020): OFDA/CRED International Disaster Database, Université catholique de Louvain – Brussels – Belgium. http://ourworlddata.org/natural-disaster). June 29 is marked as the International Day of the Tropics

A. Haldar (✉)
East Calcutta Girls' College, Kolkata, India
e-mail: ahgeog_rs@caluniv.ac.in

L. Satpati
Department of Geography, University of Calcutta, Kolkata, India

A. R. Siddiqui, A. Sahay (eds.), *Climate Change, Disaster and Adaptations*, Sustainable Development Goals Series, https://doi.org/10.1007/978-3-030-91010-5_7

by the United Nations in 2016 (UN, 2016) as they have rightly conceived that the 'future belongs to the tropics'.

Over the last decade, these huge impacts of climatic hazards and disasters have compelled the political willingness in the countries of these regions at all levels to include climate impacts as an important agenda. Hence attention has been drawn to increase effective mitigation and adaptive measures for the inflicted communities. While mitigation refers to the efforts that directly address the cause of climate change, adaptation is the adjustments in practices, processes or structures to moderate or change the risks of expected or experienced climatic abnormalities, thus empowering the communities under threat (Lambrou & Grazia, 2006). The UN IPCC has rightly stated that, 'those in the weakest economic position are often the most vulnerable to climate change....they tend to have limited adaptive capacities, and are more climate dependent on climate sensitive resources such as local water and food supplies' (IPCC, 2007). In India too, the National Action Plan on Climate Change (NAPCC) has targeted the protection of the 'poor and vulnerable sections' of the society through 'an inclusive and sustainable development strategy, sensitive to climate change'. It has been projected that by 2050, the tropical region will have to provide quality life to the two-thirds of global children residing therein as well as combating the natural forces of destruction. But the older adults are at a greater risk from extreme weather events including increasing temperatures and rainfall, exacerbated vector-borne diseases, compromised agriculture, reduced availability of fresh water, increased exposure to toxicity and decreased ability to recover physically and mentally (Filiberto et al., 2010). Therefore, every human requires specialized support system to overcome the stressors, and so, gender studies can no longer be alienated from environmental issues, especially climate change hazards.

As 'climate change' is more of a perceived term, it is more relevant to understand the perceptions of the affected population rather than the empirical assessment of the climate change itself. The adaptations to extreme climatic events require changing thresholds of human behaviour; thus, necessitating psycho-social strategies at personal and community level is of utmost significance. This narration tries to explore the gender differences in climate sensitivity and hazard resilience, and then goes on to reviewing the governmental and nongovernmental reports and their outcomes in integrating gender development with climatic hazard issues at both global and local levels to bring out the people's perceptions for accentuating a women-centric approach to climatic hazard management.

Hazard Perceptions

A recent study was conducted by the authors through crowdsourcing online polls using Web 2.0 platforms on the opinions regarding women potency in climatic hazard recovery on a global basis by a tagged cloud diagram. A 'word or tag cloud' is a visual representation of text data, typically used to depict keywords, which are shown in differing sizes in order of their repetitive occurrences or importance.

The target participants, having basic knowledge of environmental challenges, were requested to suggest three most important issues of their concern pertaining to climatic hazard management issues, and on receiving the information, the responses were rephrased to suit the purpose of representation. Although not necessarily the size of words or phrases of this qualitative diagram reveal higher significance, as these represent the frequencies of occurrences, it definitely exposes which terms are being most commonly thought of by the respondents. Out of the total 201 keywords/phrases presented in the diagram (Fig. 7.1), the most prominent ones are: 'Forecasting', 'Preparedness', 'Mitigation', 'Adaptation', 'Relief', 'Resilience', 'Awareness', 'Efficient System', 'Management' and 'Alternative Approach'. The interference that can be drawn out of this representation is that people are aware about various issues concerning climate change, climatic hazards, preparedness and adaptation strategies through alternative planning for the disaster-prone areas. Relief and efficient

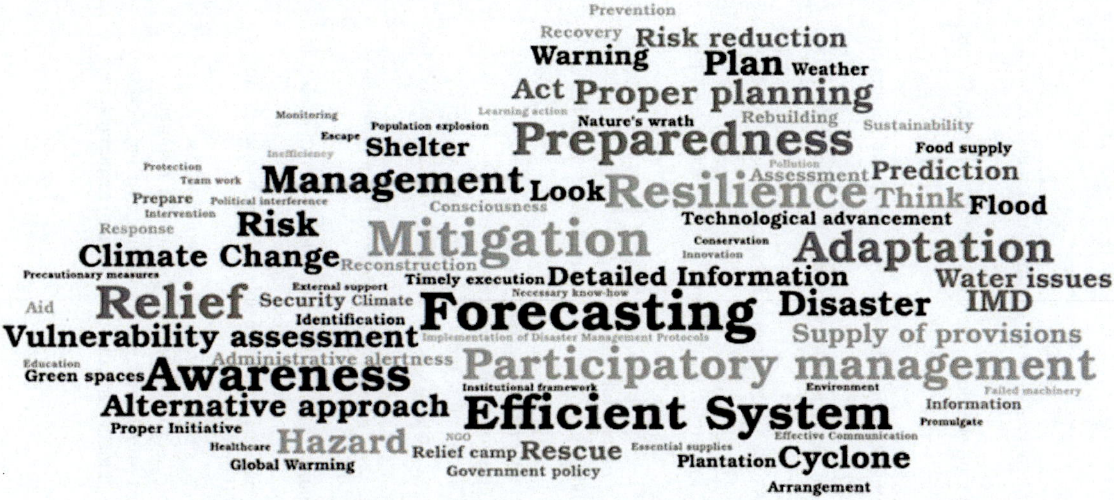

Fig. 7.1 Word cloud of the key texts generated through responses received from the targeted sample respondents representing various cross-sections of population. (Source: prepared by the authors)

system of participatory management are required to overcome lack of adequate livelihood support systems causing financial incapacity and overdependence of the government institutions or elected local bodies at the time of a disaster. It also signifies that the responses are highly representative to the ground reality as well as contemporary knowledge domains of climatic hazard studies.

Gendered Approach to Climatic Hazard Management

The gender dimension in climate change is critical since it has been observed mostly in developing countries that women are considered as lesser mortals and have been discouraged in many regions to build their capacity to combat any form of challenges. It is said that although women worldwide have made great progress in all spheres, the social inequalities, gaining equal rights and socio-economic status are far behind expectations making them disproportionately vulnerable to any form of crisis (WEDO, 2008). While the 2015 Sustainable Development Goals prescribes gender inequality in every sphere, the actual scenario in 2020 is far from the objectives and more importantly in the disaster management arena.

It is most often true that in regions where gender inequalities are deep set, women face quite a lot of hindrances to single-handedly improve their status in the face of any climatic vulnerability due to their physiological issues (Table 7.1). Most women in these local communities do not have equal opportunities to educational excellence, mental and physical fitness trainings, freedom and rights to equal work and payments, shared domestic duties and decision-making power, among other forms of subjugations. Girls from a very young age have been conditioned to think their brothers or male friends as superior and better in every assessment and are ascribed feminine gender roles of a caregiver (IMOW, 2009). While many women of the rural remote communities find empowerment through taking control over the household, while the others have voiced opinions on the neglected plight of women especially to take on societal challenges. The only plausible way out is increasing participation of women in every sphere.

A structured questionnaire survey using 'Google Forms' was conducted in July 2020 to get the responses from a varied type of population that can be categorized by gender, and also

Table 7.1 Climate change impacts and women vulnerability

Climate change impacts		Women's vulnerability issues
Natural impacts	Temperature fluctuations	Health issues, inability to plan for seasonal work, women have no access to information and decision-making power
	Decreased soil fertility and crop failures	Loss in agricultural livelihood, increased hunger and food shortages, increased laborious work at home and field, economic burden on households, loss of security, restricted land rights and poor school enrolments
	Cyclonic storms	Loss of life and properties, loss of shelter, stress for children and elderly women, no special healthcare available for pregnant, lactating and ill
	Extreme rainfall and flooding	Loss of life and properties, loss of shelter, stress for children and elderly women, economic burden on households
	Droughts and heat waves	Loss of agricultural land and livelihood, health issues, lack of safe drinking water, reduced calorie intake, economic burden on households
	Forest fires	Loss of shelter and livelihood, nature and women are symbiotic, recreational facilities and privacy lost, have to go afar for wood collection
	Coastal erosion and embankment breaching	Loss of agricultural land and livelihood, loss of shelter and property. Loss of security
	Glacial melt, avalanches and landslides	Loss of life and property, stress and trauma, hardships of rebuilding houses
	Drying rivers and non-sustenance of ecological flow	Women have to carry water from afar for domestic and irrigational purposes; unclean water affects health, loss of livelihood dependent on the river
	Sea level rise and other long-term impacts	Comparatively less impact, psychological stress from an impending disaster
Social impacts	Scarcity of clean and safe water supply	Waterborne and infectious diseases, lack of proper sanitation leading to health issues
	Inequality in infrastructural and financial facilities	Sense of deprivation and insecurity, crime and social humiliation, reduced productivity and no opportunity to improve; school dropouts, early marriage, marital discord, domestic violence, stress, has to strive harder for financial stability
	Clean fuel and electricity shortages	Food-fuel conflict, pollution, suffocation and lung diseases, accidents, tiredness and poor health, economic burden on households
	Diseases	Increased morbidity and mortality, the children, pregnant, menstruating, and elderly women are left in unhygienic conditions, still births and neonatal deaths, deaths due to taboo on treatments, immunization, family planning methods, ambulance services are unavailable, lack of women doctors/health workers leads to no treatment for women in certain societies
	Displacement and climatic refugees	Poor shelters, no health and hygiene facilities, conflicts, stress, trauma, sexual violence, rise in female-headed families, loss of security, no insurance, discrimination, competition for resources
	Conflict and civil wars	Girl trafficking, sexual violence and trauma, displaced from home and family, pregnant and sick are not cared for, loss of life, domestic violence, rise in inequality and discrimination to resource access

Source: composed and modified from UNFPA and WEDO (2009)

Table 7.2 Reasons for less women participation in climate talks (Figures show percentages of respondents in each spatial category)

Responses	Inter. male	Inter. female	Nat. urban male	Nat. urban female	Nat. rural male	Nat. rural female
Physical incapacity	25.00	8.33	22.58	11.67	20.83	21.05
Lack of decision-making rights	12.50	25.00	32.26	40.00	20.83	15.79
Poor access to education and healthcare facilities	25.00	25.00	35.48	31.67	25.00	47.37
Overburdened with household affairs	37.50	33.33	32.26	48.33	37.50	26.32
Fear of subordination	12.50	16.67	25.81	26.67	8.33	21.05
Lack of confidence in their thinking abilities	12.50	25.00	16.13	35.00	29.17	26.32
Have no exposure to such topics	0.00	25.00	19.35	30.00	16.67	21.05
Is reluctant to show interest in such matters	12.50	25.00	16.13	11.67	12.50	0.00
Never thought about it	25.00	8.33	6.45	5.00	16.67	15.79

Source: Complied and computed by the authors from online response sheets

by residential location type such as rural and urban inhabitants in India, and residents outside India. Among the international respondents, there are NRIs and natives of the United Kingdom, the United States, Canada, Germany, Portugal, Japan and Jakarta, while the national participants were mainly concentrated in West Bengal with a few responses from Delhi, Visakhapatnam, Noida, Mumbai, Bengaluru, Chennai and Agartala. Of about 200 samples, the mean age of males is 36 while that of female respondents is 32 that gave a more present-day practical perspective on the questions. It was interesting to note that when asked which phase of life the respondents were in, the men made formal comment on their employment while most women responded that they were still struggling to achieve their goals (Table 7.2).

With reference to the above, it has been found that many of the responses are highly gender sensitive. For example, physical incapability signifies stereotyping gender roles and that most of the Indian urban and rural males and rural female responses accepted it as a major hindrance to women participation in climatic disaster issues. Right to decision-making gives socio-psychological and political empowerment; however, the lack of it is recognized more by the urban respondents especially women. The rural Indians seem oblivious of the need, and the international respondents face no dearth of it. Poor access to education and healthcare is a matter of concern for all, but rural female population suffers the highest. Contrarily, the rural females do not consider household duties to be overburdening and hindering empowerment, while urban females who perceive this as the most important factor. In conversation with rural female respondents, it was observed that they feel empowered if they have full control of the household and think that this is where they belong. Fear of subordination, restricting their opinions to be publicly placed, lack of equal opportunities and little or no encouragement from the family have also been opined mostly by the Indian urban females. This is also the cause of a major deficient in confidence of adult and elderly women as responded among all categories. Unlike the men, the women in all categories felt that they had no exposure to in-depth knowledge about climate change on a regular basis and hence are incapable to participate in climate-related discussions. While this might be untrue in case of the international and urban Indian females, personally they have an opinion that the rural women have no awareness regarding such matters. But it was observed that the rural female population have much eagerness to know about climatic hazards as they are the worst affected. They possess indigenous know-how and expertise to combat climatic stresses regularly unlike their secured urban counterparts. Therefore from the responses received, it is observed that Indian women, mostly rural, are perceived as the downtrodden and oppressed by both the males and urban females, giving a wrong picture that they are the weaker sections unable to take on hardships and challenges. Thus it can be inferred that with just improved medical and educational empowerment, rural women are ready to overcome all other odds, rebuild their confidence to perform and step forward in climate awareness and join the disaster management processes. However, it is also interesting that it never struck the international males that women were less represented in climate talks worldwide.

The above discussion clearly points some important issues that (a) urban males and females are more sympathetic but not much proactive, (b) urban Indians as well as international males are sympathetic towards women's involvement in dialogues on climate risks and hazards but not active in bringing them forward, (c) rural female respondents are quite capable of being resilient as they are willing to learn, want to be exposed and don't think these as burdens. These are mere general findings based on limited dataset, but the responses speak about a pattern which may be useful for further research on the problem (Table 7.3).

A set of 11 questions were provided to capture the responses of the representative population in Likert scale with the numbers 0, 1, 2, 3, 4 and 5 against the responses denote 'no answer, strongly disagree, disagree, no idea, agree and strongly agree', respectively. All the responses have been clubbed together into two major categories, i.e. male and female, to understand whether there are major differences in the responses in terms of gender. The following points are given to enlist brief description of the findings, question wise.

(a) More women than men strongly believe that women are truly empowered to bring about changes in the society. However, a large proportion of men (mostly urban Indian and international) also agreed with this statement.
(b) Similarly, more women strongly opined that women understand nature better than men, although a few urban men also supported this idea.
(c) Regarding the often used term 'man-made' and its implication on whether it causes hindrance to women participation in risks and hazards issues, it has been found that both men and women did not have clear opinion but slightly agreed with the statement.
(d) The concept of gender neutrality is accepted by the maximum number of the respondents, mostly agreeing strongly in which the share of women is the maximum which means that women do not prefer this segregation.
(e) In respect to participation of women in mitigation of climatic hazards, more women felt it could be better handled by them than the men.
(f) Female respondents are found to be more agreeable than their male counterparts on the issue of whether climatic vulnerability affect women more. In this regard, the maximum percentage of male respondents prefers to be neutral.
(g) On the question of authority of environmental conservation mostly vested with men in the so-called male-dominated society, there is no clear formation of opinion in favour or against the issue for both sexes. This means most of the respondents are either not in a position to accept the concept of male-domination in the society or they don't like to be judged about the authority of environmental conservation.
(h) Both male and female respondents agree, at different levels, in favour of the statement that for holistic solution of long-term persistent problems, women can be the better actors, although majority of male respondents prefers to remain natural.
(i) Regarding tackling the immediate shocks of a disaster, there are significant differences in the opinions of men and women. While a

Table 7.3 Reponses of the sample population in Likert Scale

	Percentages of the respondents[a]											
	Male						Female					
Agenda/issue	*0*	*1*	*2*	*3*	*4*	*5*	*0*	*1*	*2*	*3*	*4*	*5*
Are women truly empowered to take on any challenge and bring a difference to the society today?	1	1	8	9	47	33	0	1	3	11	32	54
Can women understand nature more sensibly compared to men?	5	4	9	26	32	24	0	2	1	16	35	47
Is the term 'man-made' causes a hindrance in women involvement?	6	6	18	28	29	13	5	7	15	36	23	15
Do you believe in the concept of 'Gender Neutrality'?	1	1	7	14	23	53	4	2	0	9	19	66
Do you think climatic hazard mitigation measures can be more effectively executed and operated by women participants?	6	4	11	28	29	22	0	0	3	17	44	36
Is climatic vulnerability affecting women more?	8	2	7	34	33	16	5	5	9	19	34	29
In a society where males are still considered as the stronger gender, do you feel environmental conservation and disaster preparedness predominantly lie in the hands of only the male members?	5	13	18	21	34	8	6	25	16	14	27	13
Do you think women deal with long-term problems in a holistic way?	1	5	0	38	37	19	5	0	7	18	34	37
Do you think women are more capable to tackle immediate shocks during the disaster?	4	7	14	47	16	11	2	0	5	36	27	30
In a post disaster scenario, do you think women are more adaptive and resilient to an altered lifestyle?	7	1	6	20	42	24	4	1	2	21	33	39
Will you be more comfortable with women-led awareness campaigns and disaster management teams?	0	1	6	14	47	32	1	1	1	21	27	49

Source: Computed by the authors from response sheets
Note: [a] Refers to the opinions in percentages of crowd sourcing respondents. The numbers 0, 1, 2, 3, 4 and 5 against the responses denote 'no answer, strongly disagree, disagree, no idea, agree and strongly agree', respectively

good percentage of female respondents agree with the statements, majority of them, along with majority of their male counterpart, prefer to remain neutral. So, this issue should be considered as gender-neutral.

(j) Women are opined to be more resilient in the post disaster situation, as this statement has been affirmed by both sexes of the respondents at various levels of agreements.

Gender equality is said to have a multiplier effect across the spectrum of development (UNDP, 2016). This too is observed in case of disaster management studies. While most literatures are on the age-old attitude of women in danger during disasters, there have been instances in communities across the world where fatalities or hardships suffered are independent of gender. Both eagerly work hand in hand to overcome the crisis after a disaster. Thus this necessitates a community-based approach in hazard management where the expertise of both men and women are appropriately used for a holistic social and environmental development.

Across the world, reports have shown that men and women are differently affected due to differences in cultural taboos, traditions, resource use patterns, gender stereotypes and responsibilities in the face of a climatic hazard. The susceptibility arising from preventing women from voicing their own opinions and decisions, land ownership, political preferences, formal economic opportunities and right to proper healthcare, education, information and living facilities have significantly hindered resilience capabilities. However, it is not always that only women are the sufferers of gender discrimination and imposed gender roles. In the face of natural disasters, men are expected to bear the direct hit and come forward in high-risk rescue activities, unnecessarily causing the community to grow even more vulnerable. Kavita Ramdas of Global Fund for Women had once remarked, 'The gender inequalities that define [women's] lives prior to a disaster are really what puts them at such greatly increased risk after a disaster' (UNFPA and WEDO, 2009). Thus, where the established taboos of masculine or feminine roles in a society are widely prevalent, welfare and development is greatly impeded.

Time and again, it has been proved that women have a close bonding with nature and its resources while caring for plantations, agriculture or collecting firewood, leaf, water and fish and therefore understand the abnormalities much more closely than men. Women's innovative capacities have been already experienced in sectors such as water, energy and reforestation—all of which are relevant to climate change issues. Their efforts must be incorporated into climate change policies and promoted through capacity building (UNFPA and WEDO, 2009). Various women-led environmental movements have been continuously working towards conserving the climate over the years. Environmentalists like Vandana Shiva (India), Mei Ng (Hong Kong), Wangari Muta Maathai (Kenya), Elsie Gabriel (Brazil), Maria Cherkasova (Russia), Rachel Carson and May Boeve (America), Greta Thunberg (Sweden), Patricia Espinosa (Mexico) and Bindu Bhandari (Nepal), to name a few female activists, have created climate awareness on a global scale. Many village women depend on the local forest for essential needs like food, water, fodder and firewood, all of which are threatened by climate change and hazards. Natural resource conservation was a matter of community concern, where both the males and females worked as partners. The women's initiative resulted in reduced damage from floods and landslides, extensive reforestation and less drudgery for the women. These movements triggered changes in government policies and traditional assumptions about gender roles, as women effectively demonstrate their leadership abilities and improve the sustainability of their environment.

Policy Perspectives

Preceding the Sustainable Development Goals in 2015, globally a number of policies were laid down for women in natural disaster mitigation. Notable among these are: (1) the United Nations Conference on Environment and Development (UNCED), the Earth Summit (1992) where the Agenda 21 advocates gender equality in areas of land ownership, resource stewardship, education and employment and recognizes the economic, social and environmental contributions to environmental management. It also targeted that women's biodiversity knowledge is crucial because of the link to household-level food security (UNDESA, 1992; GDRC, n.d.; Lambrou, 2005); (2) Millennium Summit and the Millennium Development Goals (MDGs) (2000) where again gender equality was focussed in the light of climate change. It was stated that the declaration aimed at empowering women in effective ways to combat poverty, hunger and diseases and to stimulate sustainable development. (3) World Summit on Sustainable Development (WSSD), Johannesburg (2002), (4) Hyogo Framework for Action (HFA) (2005), stated that 'a gender perspective should be integrated into all disaster risk management policies, plans and decision-making processes, including those related to risk assessment, early warning, information management, and education and training'. And (5) the United Nations High Level Focus on Climate Change (2007) the November International Women Leaders Global Security Summit acknowledged that 'climate change poses significant security risks, particularly for women, and that women have to be included in decision-making at all levels' (WEDO, 2008). The Human Security Network, under the Hellenic Presidency for

2007–2008, concentrated on the impacts of climate change on vulnerable populations.

Many of the policies that affect population trends, for instance, better academic prospects for girls, greater and equal economic opportunities for women, wide access to reproductive health and family planning, more climatic awareness programmes and introduction to 'green technologies', can reduce vulnerability to climate change impacts to a great extent. Women are better in networking and convincing to management strategies emotionally; thus they can perform better in roles that involve community participation in generating climatic awareness, evolving indigenous hazard combating ideas and planning to achieve holistic resilience against disasters at the community level. Such responsive and responsible practices and initiatives from the women in the affected societies can slow the damages due to greenhouse gas emissions, pollutions and disturbances in natural weather profiles and thus help to ensure adequate and sustainable development for all (PAI, 2009).

They need to believe the fact that they are well positioned to be agents of change through mitigation, management and adaptive activities in their households, workplaces, communities and countries. Various community reports have hinted that women can be effective leaders within their community when it comes to addressing the harmful effects of climate change (Weist et al., 1994; Enarson et al., 2003; GoI-UNDP DRM, 2008; Alam et al., 2015). It is also stated that where women help devise early warning systems and reconstruction efforts, those communities largely perform better when natural disaster hits (IUCN, 2007). Not only are they the agents of change but have repeatedly proved to possess knowledge and capabilities to manoeuvre towards a resilient community. Soliciting and encouraging their leadership to address climate change and inform policy is one way to ensure that a gender perspective is included otherwise climate policies will fail to be effective.

More and more women should be engaged in sectors concerning health and education issues, local livelihood building for financial stability and awareness-generating programmes to draw in more women participant in building a resilient and empowered community and in local resource conservation practices to managing disasters more effectively. Greater attention to women-centric national adaptation programmes of action and climate policies, poverty alleviation programmes, prioritizing funds in women empowerments, engagements and employments starting from the grass-root levels have to be undertaken by the different national agencies to achieve a holistic climatic resilience for the entire community.

Concluding Remarks

From the narrative, it can be inferred that climatic extremes have an increasing trend in the tropical regions and South and South-East Asian countries with their teaming millions which are being susceptible to escalating vulnerabilities. Although advanced climate forecasting technologies have vastly improved, tropical location boosts the hazards unprecedently. Coastal and riverine plains are resource-rich region to facilitate habitation of high-density population, many of whom are living at the margin due to their less access to capacity building support systems making them even more exposed to extreme climatic stresses. In most parts of Tropical Asia, at the community level, even though information is availed from various media, the ground reality forces them to take decisions on the basis of their daily experiences and the perceptions of climatic change at the local level. Since the mobility of women in the susceptible locations is still comparatively less than the males, the in situ risks of the area affects them more, at least psychologically; therefore, they are more serious in discussing climatic issues and getting informed about subsequent impacts. They are eager to be empowered on these lines which are perceived to make them habituated to take on the realistic decisions against the adversities. Thus, it has been observed that women are not to be secluded from the disaster management advisory and participatory bodies any longer and they are willing to come forward to lead climate change mitigation challenges from the front.

References

Alam, M., Bhatia, R., & Mawby, B. (2015). *Women and climate change: Impact and agency in human rights, security, and economic development*. Georgetown Institute for Women, Peace and Security.

EMDAT. (2020). *OFDA/CRED international disaster database*. Universitécatholique de Louvain. http://www.emdat.be/. Published online at https://ourworldindata.org/natural-disasters. Accessed 18 July 2020.

Enarson, E., Meyreles, L., González, M., Morrow, B. H., Mullings, A., & Soares, J. (2003). *Practical guidelines for assessing local disaster risk*. International Hurricane Center, Florida International University.

Filiberto, D., Wethington, E., Pillemer, K., Wells, N. M., Wysocki, M., & Parise J. T. (2010). Older people and climate change: Vulnerability and health effects. Winter, 2009–2010 issue of *Generations,* ASA's quarterly journal; the entire issue was devoted to the topic. In *Gray and green: The intersection of aging and the environment*. https://www.asaging.org/blog/older-people-and-climate-change-vulnerability-and-health-effects.

Global Development and Research Center (GDRC). (n.d.). *Agenda 21: An easy reference to specific recommendations on women*. http://www.gdrc.org/gender/a21/agenda1.html.

GoI-UNDPDRM. (2008). Women as equal partners: Gender dimensions of disaster risk management programme: Compilation of good practices.

Intergovernmental Panel on Climate Change (IPCC). (2007). *Climate change 2007: Synthesis report. . Summary for policy makers*. IPCC and Cambridge University Press.

International Museum of Women: Exhibiting Change (IMOW). (2009). Swept away: Women and natural disasters. In *Women, power and politics*. Online Exhibition. http://www.imow.org/wpp/stories/viewStory?storyid=1237.

International Union for Conservation of Nature (IUCN). (2007). *Gender aspects of climate change*. IUCN. Cited in WEDO (2007).

Lambrou, Y. (2005). *Gender perspectives on the conventions: Biodiversity, climate change and desertification*. Gender and Population Division. United Nations Food and Agriculture Organization (FAO). http://www.fao.org/sd/dim_pe1/pe1_050301a1_en.htm.

Lambrou, Y., & Grazia, P. (2006). *Gender; the missing component of the response to climate change*. FAO.

Population Action International (PAI). (2009). *The importance of population for climate change challenges and solutions*. Factsheet 37. 8 June. http://www.populationaction.org/Issues/Population_and_Climate_Change/Summary.shtml.

Rigg, J. (2015). *The global south. Concepts of the global south –voices from around the world global south studies center*. University of Cologne. http://gssc.uni-koeln.de/node/452

UN Department of Economic and Social Affairs (UN DESA). (1992). Global action for women towards sustainable and equitable development. In *Agenda 21*, Chapter 24. http://www.un.org/esa/dsd/agenda21/.

United Nations (UN). (2016). *Resolution adopted by the general assembly on 14 June 2016*. 7th Session, Agenda Item 20. A/RES/70/267. https://undocs.org/A/RES/70/267. Accessed on 17 July 2020.

United Nations Development Fund for Women. (2009). *Adivasi women: Engaging with climate change*. UNIFEM, IFAD and the Christensen Fund.

United Nations Development Programme (UNDP). (2016). *UNDP support to the integration of gender equality*.

United Nations Population Fund (UNFPA) and (WEDO). (2009). *Overview: Women at the forefront: A resource kit on climate, population and gender*. https://www.unfpa.org/publications/overview-women-forefront.

Weist, R. E., Mocellin, J. S. P., & Motsisi, D. T. (1994). *The needs of women in disasters and emergencies*. University of Manitoba and United Nations Development Programme.

Women's Environment and Development Organization (WEDO). (2008). *Case study: Gender, climate change and human security: Lessons from Bangladesh, Ghana and Senegal*. WEDO. http://www.wedo.org/wp-content/uploads/hsn-study-final-may-20-2008.pdf.

World Resources Institute (WRI). (2020). https://www.wri.org/aqueduct/data. Accessed on 17 July 2020.

Climate Change and the Rising Hazard Potential in Lahaul and Spiti: Himachal Pradesh

8

Varuni Pathak and Milap C. Sharma

Introduction

The Himalayan glaciers cover approximately an area of three million hectares, or 17% of the mountain area, known now as the Third Pole. There are over 16,000 glacier over the Trans Himalaya, but 9575 glaciers of varied dimensions that still occupy the India Himalaya, forming a lifeline for millions of people in and outside the mountain system and supporting perennial river (IPCC, 2007; Raina, 2008). Many of these rivers are associated with the great civilizations of the world, inferring the continuity over the millennial time scale. However, the Himalayan mountain system is one of the most fragile and sensitive ecosystems, being the most recent in tectonic upheavals, and thus vulnerable to several environmental issues, making living in the mountainous areas a challenge. This situation is seriously compounded by the changing climate in the recent decades. A holistic analysis of the natural resources upstream is the need of the hour for a robust scientific planning for the teaming millions downstream.

On the one hand, the Himalayan volume of the glacier is projected to decline by the end of the twenty-first century in response to decreased snowfall, increased snowline elevations, and longer melt seasons due to human-induced climate change due to steady pace of increasing population and as a consequence, the vulnerable areas and the no-go areas until very recently being brought under human occupation. Even if the warming is limited to the ambitious target of +1.5 °C (Paris Climate Deal 2016), this might lead to a 2.1° rise in the HKH region due to elevation-dependent warming (2016). In such a scenario, the region will lose 36% of its glaciers, with more than half of glacier ice lost in the eastern Himalayas. This situation is further aggravated by the reckless anthropogenic activities leading to widespread disturbance of homeostatic setup of alpine glacier regions. Natural disasters like frequent earthquakes, landslides caused by seismic activity, snow avalanches, cloud burst, floods, and heavy variable precipitation in the western, central, and eastern Himalaya are considered to be triggered by such climatic variability in one way or the other. Most of the Himalayan glaciers are receding and losing mass balance except a notable mass gain in the Northwestern and Karakoram Himalayas (Bolch et al., 2008), called the Karakoram Anomaly. This disturbing mass balance of the glaciers will not only affect the availability of water for sustenance of life but also create acute socio-cultural, political, and economic implications on a trans-boundary scale. These natural disasters can prove to be

V. Pathak (✉)
C.S.R.D. Jawaharlal Nehru University, New Delhi, India

M. C. Sharma
Jawaharlal Nehru University, New Delhi, India

A. R. Siddiqui, A. Sahay (eds.), *Climate Change, Disaster and Adaptations*, Sustainable

catastrophic in nature crossing the national boundary limits, thus instigating nations to an armed conflict on suspicion. The glacial lake outburst floods (GLoFs) are one such natural event with multiplying effect of causing humongous disasters downstream valley area, initiated from the thus serene and beautiful Himalayan environment. The Chorabari flash floods of 2013 may be treated as the tip of an iceberg in the fast changing environmental conditions in and outside the Himalayan realm. One must not forget that the replenishment of the fertile soils now available via these perennial rivers in the Great Indo-Gangetic plains is the handiwork of the otherwise variable glacier meltwaters from the Himalayas over many millions of years.

Climate Change Hazard

The HKH provides two billion people a vital regional lifeline, i.e., water for food (especially irrigation), water for energy (hydropower), and water for ecosystem services via riparian habitats and environmental flows, and also beholds a rich and diverse cultural aesthetic value. The HKH shows a rising trend in extreme warm events, failing trend of extreme cold events, and rising trend of extreme values and frequencies of temperature-based indices. It is further enhanced by short-lived climate pollutants such as black carbon which is emitted in large quantities in the industrial world. Biodiversity loss can only get accelerated by 70–80%, particularly threatening the endemic species. In the energy sector, India is already facing the crunch, with large dependence on the conventional thermal sector. But to achieve the target of Sustainable Development Goals and become self-reliance in clean energy, the Himalayas rivers are the lifelines. The natural gradient and perennial flow in the glacier-fed rivers is certainly a blessing, not being harnessed fully as yet. There is still a 500 GW energy potential untapped.

Glacial lakes are an indirect key indicator of glacier dynamics and climate variability (Gardelle et al., 2011), and potentially unstable lakes can produce glacier-related hazards called the glacial lake outburst floods. Glacial lake outburst flood (GLOF) is the sudden and rapid discharge of glacial lake water downstream with great velocity and ferocity that causes huge destruction in downstream reaches (WWF, 2005). GLOF is also referred to as *jökulhlaup* (Iceland), a *debacle* (French), and an *aluvion* (South America; Mool et al., 2001). Irrespective of the nature and types of causes of such floods, these are extensively damaging more so because of the large boulders and other stored Quaternary sediments carried along downstream with the high-velocity water discharge. These GLOFs also have wide-ranging impact cutting across borders, as most of the Himalayan rivers have trans-boundary extent. Glacial lakes have also been considered as the key litmus test for glacier stability and climate variability (Gardelle et al., 2011). We present the holistic canvass of the related floods and intertwined factors in the recent past in the study area (Fig. 8.1).

GLOF in Himalayas

There are scores of historical evidences of infrequent glacial lake outburst floods that occurred in the Himalayas. However, the information related to its causes and mechanism is limited and perfunctory. In 1841, a flood from a lake occurred on the Indus River, to the north-east of Nanga Parbat, in Kashmir, annihilating a Sikh Army camp at Attock. Again in 1858, the breach of a landslide blockade across the Hunza valley had caused a 9 m rise of river level at Attock in less than 10 h (Owen et al., 1995; Reynolds & Richardson, 2000). The first recorded GLOF event in the Indian Himalayas is documented in 1926, from Shyok glacier that had destroyed Abudan village 400 km downstream from the source (Mason, 1929). Coxon et al. (1996) have reported an evidence of Batal lake outburst flood which they considered to be a post-last glacial event. The oral history of the regions considers this event related to the surge of the Bara Shigri glacier somewhere around 1860s, causing devastating floods in the Chandra river upon breach of ice blockade. This flood caused large-scale

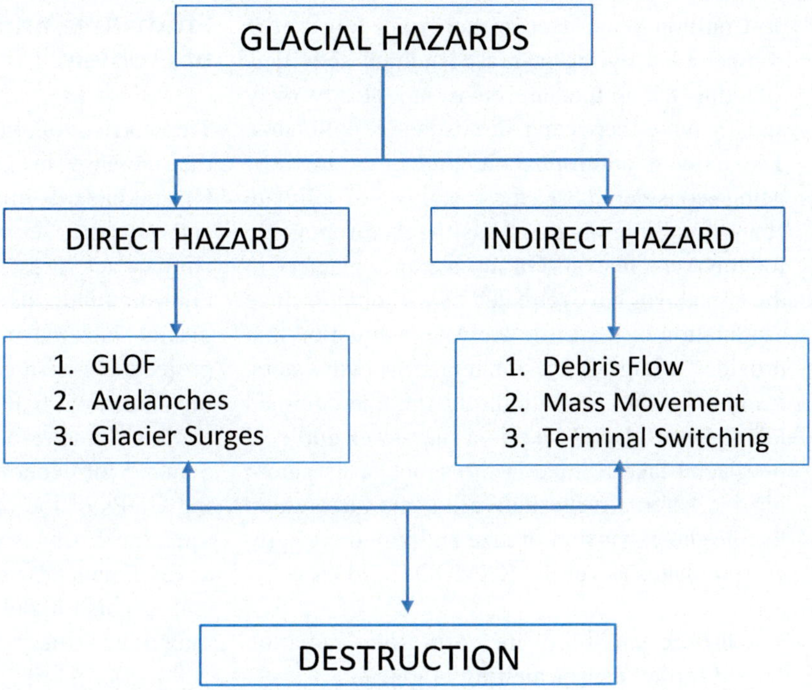

Fig. 8.1 Glacial hazard classification

re-sedimentation and landscape modification within the Chandra valley, possibly many 10s of miles downstream. The Dig Tsho lake outburst of 1985 turned out to be a catalyst in ushering glacial hazards research in the Himalayas, with focus into catastrophic floods from glacial lakes. Subsequently, this event of GLOF made the government and the community at large, aware of the risk, glacier retreat, and climate change (Reynolds & Richardson, 2000). Prior to this event, only a scanty work published appeared on the identification and assessment of the potential hazards and to predict flood behavior (Reynolds & Richardson, 2000). The studies on the Himalayan glaciers (since the 1970s) and GLOFs (since the late 1980s) have been region specific, yet with a limited temporal data on the hydrometeorology.

Glacial Lakes in HP

Glacial lakes can form in either episode of advance or retreat. But it is the retreating glaciers where catastrophic events become beyond controllable as large number of lakes may form in a short duration, given the temperature-driven obliteration of ice bodies. This is the present day norm in the mountain landscapes all over, having the ice bodies. Formation of glacial lakes is multidimensional and multi-faceted. There are some prominent geophysical and other natural factors responsible for the formation of glacial lakes (ICIMOD, 2011), besides local factors depending upon the geology, topography, climate, and active local human interference. In Himachal Pradesh, tens of scientific studies have revealed the evidences of glacial advance and retreat in particular. Four glacial stages have been identified in the study area (Owen et al., 1996, 1997), each one many degrees smaller than the previous. Large moraines formed at the margins of many of these glaciers as a result of a variety of advances and recessions that evolved lakes from the glacier meltwaters either as proglacial, supraglacial, or moraine-dammed ablation valley lakes. But in the recent past, Saini et al. (2019) have suggested that the local "Little Ice Age" had a rather late beginning in the Lahaul Himalaya that spanned into the middle of the nineteenth century.

Until the recent geological time, glacial and inter-glacial events were rather a cyclic natural process that has occurred multiple times during past million years. But in the recent times, it is compounded by the increased human activities, affecting the immediate environment adversely and in turn, increasing atmospheric pollutants. The modern anthropogenic processes are now being considered as the catalyst of climate change, calling this period as the Anthropogenic for this very reason. On an average, glaciers in the Himalayas have retreated by a kilometer since termination of the Little Ice Age, a situation that provides a large space for retaining meltwaters, leading to the formation of moraine-dammed lakes. Sissu glacial lake is a classic example of proglacial lake formed at the snout of its parent glacier. Alternatively there are some factors that lead to the expansion of size and volume of pre-existing lakes as per the ICIMOD (2011), viz.:

- Climatic variability increases solar radiation and further causes melting of glaciers.
- Persistent precipitation events or increase in snowmelt.
- Constriction of regular outlet because of re-adjustment of moraine boulders and imbalance inflow and outflow. This sedimentation of silt from the glacier runoff, enhanced by the sand/silt deposit, obstructs the water channel of the lake.
- Blocking of ice conduits by sedimentation or by enhanced plastic ice flow on the supraglacial surface.
- Thick layer of glacial ice (dead ice) weighed down by sediment below the lake bottom on resettling of iceberg which stops subsurface infiltration or seepage from the lake bottom.
- Shrinking of the glacier tongue higher up, causing meltwater to accumulate within end moraine complex for increase in proglacial lake.
- Blocking of an outlet by an advancing tributary glacier.
- Landslide at the inner part of the moraine wall, or from slopes above the lake.
- Melting of ice from an ice-cored moraine wall (dead ice).
- Melting of ice due to subterranean thermal activities (tectonics).
- Inter-basin subsurface flow of water from one lake to another due to height difference and constriction in flow path.

Study Area and Statement of Problem

The worrisome issue in the Himalayan regions is the increase in glacial hazard's vulnerability. Glacial hazards impose (a) the risk of loss of life and (b) the serious threat to infrastructures. Human activities have reached further into the high mountainous regions of the world, making glacial hazards more apparent. Development projects in Asian countries are invariably intruding into areas vulnerable for glacial hazards to exploit alpine resources of land and water. Water demand for economic development of the growing GDPs of the nation is growing at geometric rate. Increasing settlement and proliferation of development of roads, highways, bridges, artificial canals, hydel projects, and tourism have magnified the threat of GLOF (Ghimre, 2004–2005).

All four Himalayan states, viz., Sikkim, Himachal Pradesh, Jammu and Kashmir, and Uttarakhand, are potential suspects to direct impact of GLOFs. Out of which Himachal Pradesh is the most vulnerable not just due to intrusion of human activities but at the same time because these lakes are the results of the Little Ice Age formed between the moraine dams of the boulders brought in there by the previous glaciations, making these less stable for being new in origin. Lahaul and Spiti which is barren, deserted, and secluded with the least population experienced tremendous infrastructural developments in the past few decades in the higher altitudes. There are a number of small hydel power projects, resorts, national highways, helipads, small dams, canals, and other allied economic enterprises. Also tourism has a major role in making it a popular destination for hiking, skiing, and trekking. Its picturesque beauty, distinct culture of peace, monks and monasteries, and an en route location to Leh have made it a population tourist destination. There are a flock of tourist resorts, home stays, and tent camps all across the region. Outburst of the lakes can have catastrophic floods with trans-boundary impacts dismantling the entire infrastructural and natural setup. Therefore, it is advisable not only to map and assess but also to monitor these potential glacial lakes as a tool of advance contingency plan to minimize future devastation. These inventory maps and continuous assessment can aid in making disaster-resilient lifestyle and infrastructure. This will further help in coming up with comprehensive disaster adaptation and mitigation executive plan policies and programs by the different stakeholders involved.

Location and Description of the Study Area

It has two physiographic and cultural divisions which are Lahaul and Spiti. The Lahaul division of Lahaul and Spiti district of Himachal Pradesh has an aerial extent from 32°8′N to 33°19′57″ and 76°46′29″E to 77°55′E meridians. It has an estimated area of about 6849 km^2. The maximum length is about 115.873 km and maximum breadth is about 86.904 km (Fig. 8.2). Lahaul has bare and rough terrain and steep slopes except in the western part where one encounters pine vegetation. The region has two prominent physiographic units that are *the mountain ranges* and *the river valleys*. The *Lesser Himalayas* is in the north-west to south-east direction and the *Greater Himalayas* in the north-east. Pir Panjal range of the Lesser Himalayas and Zaskar of the Greater Himalayas give striking intra-physiographic variations. The Pir Panjal range has an average height between 3500 and 5000 m. However, many peaks exceeding 6000 m are Dara Goh, Khodo Goh, and Makar Beh. The wall created by this range obstructs monsoon and reduces the amount of rainfall and vegetation in this region. The continuous wall of the Pir Panjal range restricts the accessibility only through the Rohtang Pass (3955 m). Geologically, the Middle Himalaya of this part is composed of a zone of highly compressed and altered rocks of various ages. The Greater Himalayan range, in the north-east of this region, consisting of intricately ramifying glaciated ranges of crystalline rocks is an important range in the region. The average height of this range is 5000 m which finally merges into the Pir Panjal range in the eastern part. Some fringes of this range run from north to south. The Greater Himalayan part constituted of more glaciers as compared to the Pir Panjal. However the Pir

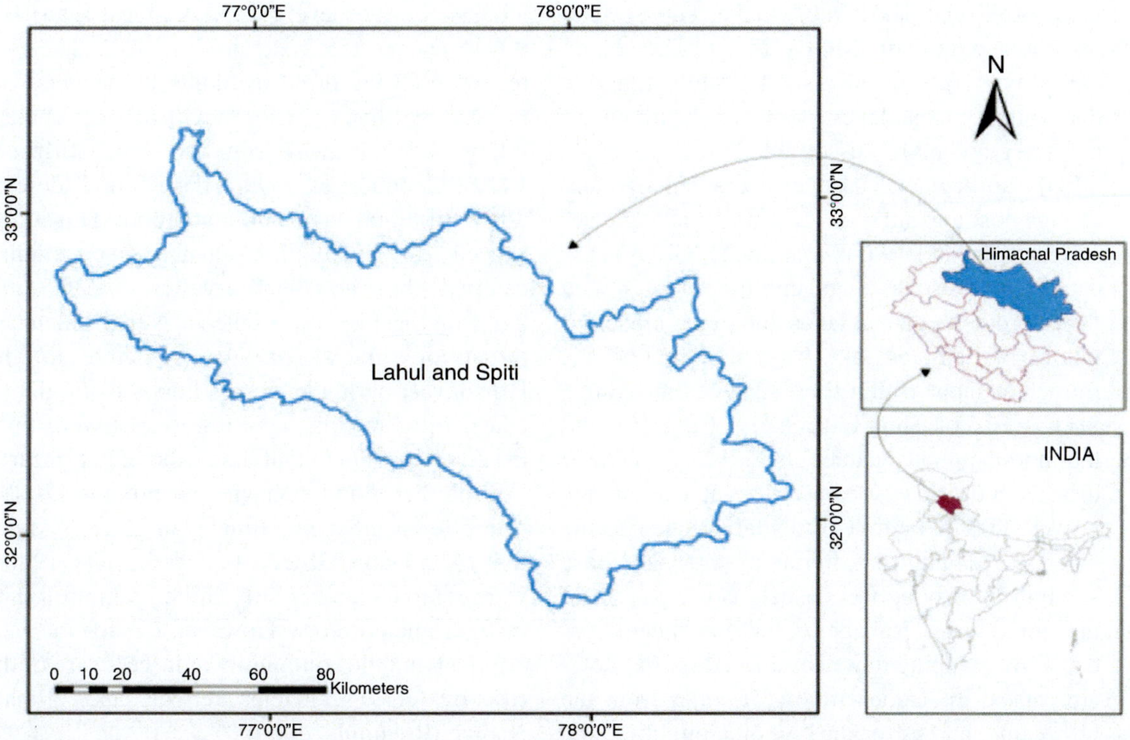

Fig. 8.2 Location map of Lahaul and Spiti, Himachal Pradesh

Panjal contains one of the largest glaciers, i.e., Bara Shigri right close to its merger with the Greater Himalayan range at Kunlun Pass. Lahaul region is further divided into five valleys, namely:

- The Chandra Valley
- The Bhaga Valley
- The Patan Valley
- The Chenab Valley
- The Central Triangular Mass

Literature Review

Climatic variability and fluctuations in the past few decades have led to accelerated melting of glaciers. As a result, numerous meltwater fresh lakes can be seen in the vicinity of the source glacier lakes. New lakes tend to be more active, vulnerable, and variable in the geometry and morphology (Mool et al., 2001). These glacial lakes are further categorized in various types depending on their location, mode of formation, and other morphometric characteristics with respect to the parent glacier by the ICIMOD (2011) which are blocked lake, moraine-dammed lake, proglacial lake, valley lake, supraglacial lake, and cirque (ICIMOD, 2011). Cirque and valley types are usually stable, while moraine dam and proglacial lakes formed on the tongue of a glacier are more susceptible to the event of GLOF. Supraglacial lakes are small in size and change position and size frequently on the glacier surface. Glacial lake outburst floods (GLOFs) are cataclysmic discharges of enormous water with debris from the glacial lakes due to the breach or failure of dammed lake by a moraine or ice, formed at snout within short span of time frame. Such types of abrupt discharge from ice and moraine-dammed glacier may cause a huge disaster at the downstream valleys at a larger area (WWF, 2005). Failure of a glacial lake occurs when the damming material of the lake loose strength by driving factor such as the weight of the stored water, leakage forces, and shear stress due to overtopping flow. These overtopping flows are caused by sudden inflow of water from the upstream sources or heavy precipitation, and moreover abrupt mass movements, e.g., snow and ice avalanches, debris flows, rock fall, an earthquake, landslides, etc., entering the lake trigger displacement waves (Fig. 8.3). As soon as this overflow starts, dam erosion also starts, forming an initial failure of the lake ultimately leading to a higher outflow that causes progressive breach extension.

GLOF event gets further accelerated by the alternate freeze and thaw, i.e., inter-play, of water and ice itself. Conversion of ice into water releases the latent heat of fusion that reduces pressure on the walls of the lake, while conversion of water into ice expands the volume, thereby exerting more pressure and thrust on the walls. Therefore, it leads to positive feedback mechanism, hence, more chances of outpouring or overflowing of water (Fig. 8.4).

There have been many infrequent evidence of glacial lake outburst floods across the Himalayas. But the information of the causes and mechanism is very limited and not precise. In 1841 a flood from a lake near Indus River to the northeast side of Nanga Parbat in Kashmir, destroyed a Sikh Army camp at Attock. In 1858 the breach of a landslide across the Hunza valley caused a 9 m rise of river level at Attock in less than 10 h (Owen et al., 1995; Reynolds, 2000). The first recorded GLOF event in Indian Himalayas was in 1926 by Shyok glacier that destroyed Abudan village 400 km away from the source (Mason, 1929). Coxon et al. (1996) have found the evidence of Batal lake outburst flood. This flood caused re-sedimentation and landscape modification within the Chandra valley. In 1985 outburst of the Dig Tsho lake in Nepal enhanced thrust on glacial hazards research in the Himalayas, particularly into floods from glacial lakes. Prior to this, very few works were published to recognize and assess potential hazards and predict flood behavior (Reynolds, 2000). These few studies on Himalayan glaciers (since the 1970s) and GLOFs (since the late 1980s) were region specific with limited temporal data on hydrometeorology. However, it made the government and the community at large aware of the risk of GLOFs, glacier retreat, and climate change (Reynolds, 2000).

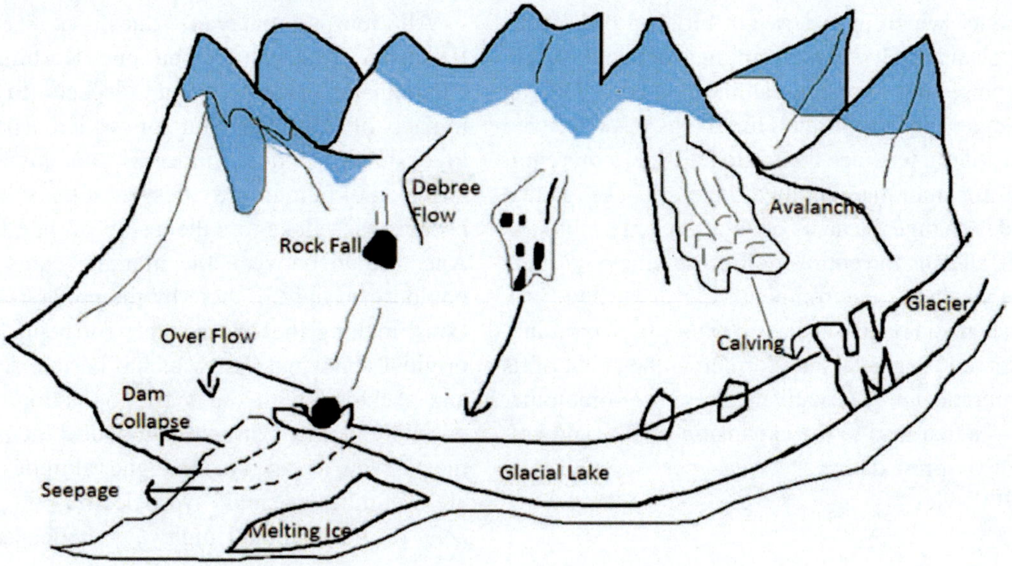

Fig. 8.3 Possible causes of GLOFs (Fujita et al., 2012)

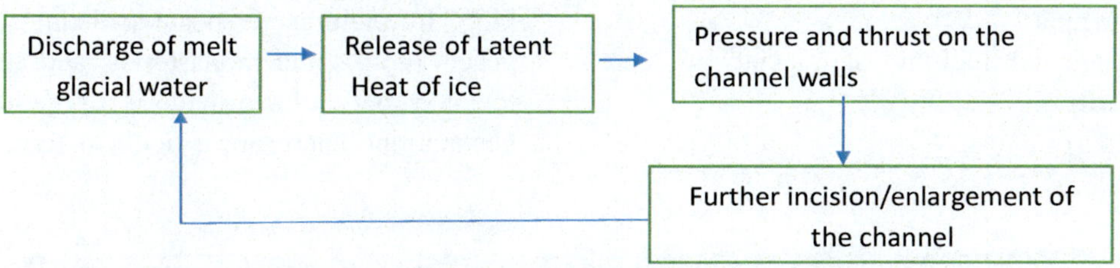

Fig. 8.4 Positive feedback mechanism of catastrophic GLOFs

The worrisome issue in the Himalayan regions is the increase in glacial hazard's vulnerability. Glacial hazards impose (a) the risk of loss of life and (b) the serious threat to infrastructures. Human activities have reached further into the high mountainous regions of the world, making glacial hazards more apparent. Development projects in Asian countries are invariably intruding into areas vulnerable for glacial hazards to exploit alpine resources of land and water. Water demand for economic development of the growing GDPs of the nation is growing at a geometric rate. Increasing settlement and proliferation of development of roads, highways, bridges, artificial canals, hydel projects, and tourism have magnified the threat of GLOF (Ghimre, 2004–2005).

Formation of Glacial Lakes in Himachal Pradesh and Need of the Study

Formation of glacial lakes is multidimensional and multi-faceted. There are some prominent geophysical and other natural factors responsible for the formation of glacial lakes all over the globe as listed by the ICIMOD (2011). Besides some local factor depending upon the geology topography climate and human interference active locally in Himachal Pradesh. In Himachal Pradesh, several scientific experiments show evidences of glacial advance and retreat. During the "Little Ice Age" (AD 1550–1850s), some glaciers grew much larger than what these are today. Large moraines formed in front of the glaciers as a result of this advance which have now got blocked by retreating glacier meltwaters, forming proglacial, supraglacial, and moraine-dammed lakes. On an average, glaciers in the Himalayas have receded by a kilometer since the Little Ice Age, a prerequisite for moraine-dammed lakes. These glacial cycles change the areas of ablation and accumulation, altering the entire equilibrium line of altitude of the glacial environment (Figs. 8.5 and 8.6). Sissu glacial lake is a classic example of moraine-dammed proglacial lake formed at the snout of its parent glacier. Alternatively, there are some other factors that lead to the expansion of size and volume of pre-existing lakes as per the ICIMOD (2011), viz.:

- Climatic variability increases solar radiation and further causes melting of glaciers.
- Persistent precipitation events or increase in snowmelt.
- Sub-glacial melting of glacial by tectonic activities.

All four Himalayan states, viz., Sikkim, Himachal Pradesh, Jammu and Kashmir, and Uttarakhand, are potential suspects to direct impact of GLOFs. Out of which Himachal Pradesh is the most vulnerable not just due to intrusion of human activities but at the same time because these lakes are the results of the Little Ice Age formed between the moraine dams of the boulders brought in there by the previous glaciations, making these less stable for being new in origin. Lahaul and Spiti which is barren, deserted, and secluded with the least population experienced tremendous infrastructural developments in the past few decades in the higher altitudes. There are a number of small hydroelectric power projects, resorts, national highways, helipads, small dams, canals, and other allied economic enterprises. Also tourism has a major role in making it a popular destination for hiking, skiing, and trekking. Its picturesque beauty, distinct culture of peace, monks and monasteries, and an en route location to Leh have made it a populous tourist destination. There are a flock of tourist resorts,

Fig. 8.5 Style of Sissu glacial lake's evolution

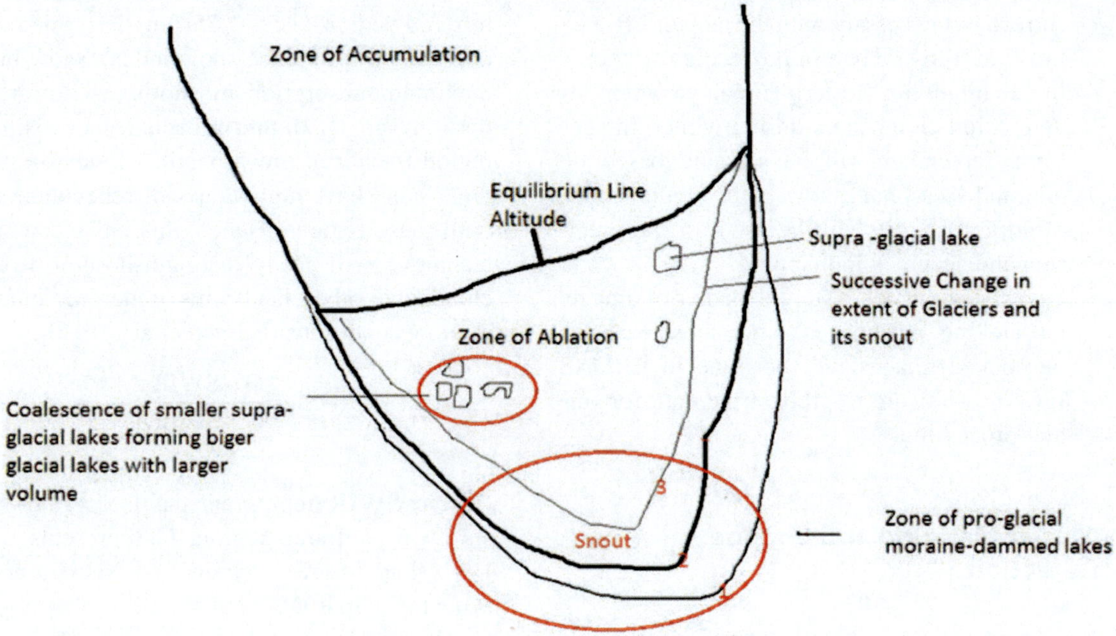

Fig. 8.6 Formation of glacial lakes with respect to retreat movement of glaciers

home stays, and tent camps all across the region. Invariably, the season for such flocking is in the peak summer months. Outburst of any lake upstream can have catastrophic floods with transboundary impacts, dismantling the entire infrastructural and natural setup. Therefore, it is advisable not only to map and assess but also to monitor these potential glacial lakes as a tool of advance contingency plan to minimize future devastation. These inventory maps and continuous assessment can aid in making disaster-resilient lifestyle and infrastructure. This will further help in coming up with comprehensive disaster adaptation and mitigation executive plan policies and programs by the different stakeholders involved.

Materials and Methods

Alpine region puts a constraint on detailed and minute field surveys. Therefore, satellite images provide the crucial information of glacial environment on real-time basis. There are various digital tools used for the identification and delineation of glaciers and glacial lakes done along with visual identification manually.

The following methods and tools will be applied for collecting data, correlating them, and finding out the final conclusion.

Mapping and Change Detection of Glaciers and Glacial Lakes

Remote sensing satellite data has been used for the mapping and change detection of glaciers and glacial lakes as it is not possible to map all the glaciers and lake in high-altitude rugged mountainous topography through fieldwork. Therefore, using these data sets in ArcGIS 10 and Erdas Imagine 14 supervised, unsupervised, as well as spectral ratio methods will be used for this purpose.

Image Preprocessing and Data Preparation

- Radiometric, atmospheric, and geometric errors are the three basic errors which will be corrected if necessary with the help of ERDAS Imagine software before proceeding further.

- In Lahaul, in the Western Himalaya where little precipitation occurs during winter, images from September will be selected based on minimal seasonal snow and cloud cover. Multiple data sets will be used for the precision and accuracy in mapping.
- Layer stacking of different bands of image, mosaicking of images, and subset of the mosaicked images will be done in ERDAS Imagine software to prepare the data for further processing.

Glacier Mapping and Change Detection

In order to achieve to the first objective, the glaciers will be manually delineated using standard false-color composites (SFCC) with band combinations of red, near-infrared (NIR), shortwave infrared (SWIR), and a true-color composite of red, green, and NIR of Landsat MSS/TM/ETM+ and LISS IV images. Some contrast techniques such as brightness and linear stretching will be used, and DEM (digital elevation model), i.e., ASTER (Advanced Spaceborne Thermal Emission and Reflection Radiometer) and SRTM (Shuttle Radar Topography Mission), of the study area will be used for altitude information, and thus slope map will be also generated from it. A shaded relief map will be prepared from the SRTM DEM (90 m) to aid the manual interpretation of glacial features. In addition, the Survey of India maps will be referred for the identification, location, and interpretation of glaciers on satellite images. Glacier boundaries in the accumulation area will be demarcated by "ice divides," which were visually detected and manually digitized based on the shaded relief map (Basnett et al., 2013). Various image enhancement techniques will be used to delineate the glaciers from other features such as the following:

- **NDSI:** Normalized difference of two bands one in green portion of electromagnetic radiation (0.52–0.60 μm) and another in shortwave infrared region (1.55–1.75 μm) is termed as normalized difference snow index. Snow has maximum absorption in shortwave infrared portion and maximum reflectance in visible region (center at green band). Clouds on the other hand have high values of reflectance in both these regions. Hence, this enhancement technique will easily distinguish snow from cloud and other landforms under mountain shadow conditions (Basnett et al., 2013)

$$\text{NDSI} = \frac{\text{G} - \text{SWIR}}{\text{G} - \text{SWIR}} \tag{8.1}$$

(where **SWIR** represents the DN values in shortwave infrared region, **G** represents the DN values in green region, and **NDSI** represents the output DN values).

Glacial Lake Mapping, Classification, and Change Detection

Generally, glacial lakes exist on or along the glacier boundary or at the snout. During the winter season, the upper surfaces of the glacial lakes freeze but again melt in the summer season. Therefore, summer data sets will be used for lake identification and mapping. Thus, glacial lakes will be discriminated from the other features by applying this enhancement technique:

- **NDWI:** The automated classification will be used to detect the glacial lakes over the entire study area in Lahaul Himalayas using the normalized difference water index (NDWI; Eq. 4), applied on the spectral bands TM1 and TM4 of Landsat ETM+ satellite images.

$$\text{NDWI} = \frac{\text{TM4} - \text{TM1}}{\text{TM4} + \text{TM1}} \tag{8.2}$$

- According to several studies, it is believed that shadowed areas get misclassified as lakes which can be seen with the help of DEM. Therefore, slope map (semi-automated extraction from DEM) will also be used to find out water bodies in this region.

Glacial Lakes Change Detection

Change in the areal extent of the lakes will be mapped temporally to assess the vulnerability of the growing lakes with respect to the source glaciers. Glacial meltwater leads to the lakes; therefore, there could be direct relationship established between the glacier dynamism and spatial extent of glacial lakes.

To Assess Potential of Outburst of Glacial Lakes

Locations of lakes with relation to the parent glacier, area, depth, volume, morphological features, blockade type, etc. will be taken into consideration to classify the lakes in order to assess vulnerability of outburst. The following chart shows the criterion and directions for classifying the glacial lakes that have higher vulnerability of outbursts. To complete these objectives, extensive satellite data sets were used for different time scales (Table 8.1).

Results and Discussion

Glacial Lake Inventory

Glacial lakes in Himachal Pradesh are classified into blocked lake, moraine-dammed lake, proglacial lake, valley lake, supraglacial lake, and cirque (ICIMOD, 2011). Cirque and valley types are usually stable, while moraine-dammed and proglacial lakes formed on the tongue of the glacier are more susceptible to the event of GLOF. Supraglacial lakes are small in size and change position and size frequently on the glacier surface. Still, the review of the events of GLOF reveals that some of the moraine-dammed lakes that failed and caused a GLOF were derived from supraglacial lakes. There are 212 glacial lakes of varied size, volume, and depth in Himachal Pradesh (Fig. 8.7). Ten glacial lakes are identified as high potential lakes of the breach on the following criteria put forwarded by the ICIMOD (2011):

1. *Moraine dammed (high)*
2. *Frequent changes in size (high)*
3. *Associated with steep adjacent topography (high)*

Table 8.1 Different data sets used in the study

Primary data				
Field surveying using GPS, clinometer; questionnaire survey				
Secondary data				
	Data	Type of image	Scale and spatial resolution	Sources
Toposheet	Toposheet	Thematic	1:50000	SOI
Satellite images	LANDSAT TM (1989)	Visible, NIR, SWIR, thermal	30 m, 30 m, 30 m, 120 m	
	LANDSAT TM(2011)	Visible, NIR, SWIR, thermal	30 m, 30 m, 30 m, 60 m,	
	SENTINEL-2 (2016)	Visible, NIR, SWIR	10 m, 10 m, 20 m	
	ASTER DEM	Photogrammetric	Horizontal 30 m Vertical 20 m	USGS
	SRTM DEM	Interferometric SAR	Horizontal 90 m Vertical 15 m	CGIAR-CSI
	CORONA (1976)	Panchromatic image	7.5 m.	USGS
Land record data	Year: 2011		Village level	Panchayat records

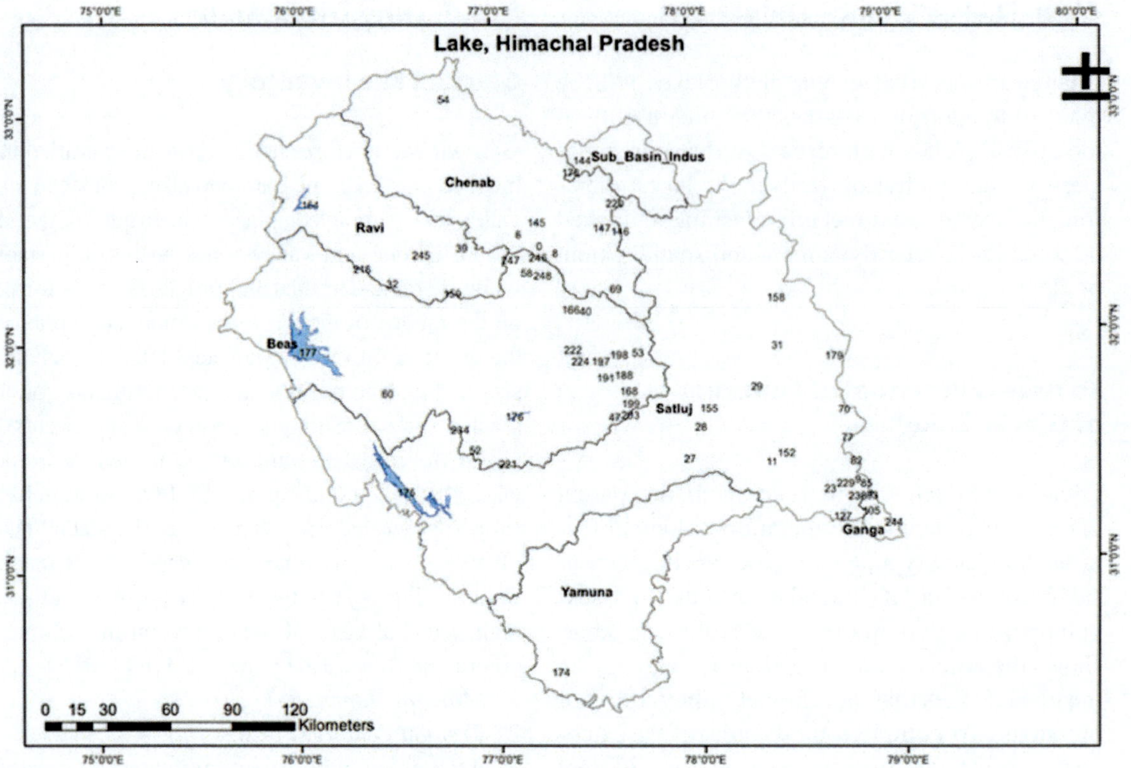

Fig. 8.7 Glacial lake distribution of Himachal Pradesh (2015)

4. *Narrow crest width (high)*

Out of these ten potential GLOF lakes, three are in Lahaul and Spiti districts. These lakes have history of breach, and also studying these on the basis of GLOF parameters demonstrates the future threat as well. Out of the ten lakes having potential to GLOFs, three lakes are in Sutlej and Chenab basins, while four glacial lakes are in Ravi basin. If we look at the type-wise potential threatening glacial lakes baring two that are supraglacial lakes, the rest eight lakes are proglacial lakes. This implies it is proglacial lakes usually at the end point of the parent glaciers that is the most dangerous category. This is because the lake is in direct contact with the parent glacier. Therefore, the meltwater is directly feeding into these lakes round the year. At the same time since these lakes are at the terminus of the parent glaciers, due to the release of the latent heat of water, melting of the parent glaciers further gets accelerated. Proglacial lakes are further more dangerous as these are formed between the present glacial terminuses and the huge boulders, pebbles, and other sediments together called as moraines deposited and consolidated of the past glacial extent in history. And as abovementioned, alongside the downpour of water, it is the release of these damming and holding materials that increase the loss and destruction to enormous scale. Therefore, the proglacial lakes that are dammed by moraine are the most dangerous of all the types of lakes.

Glacial lake inventory of Himachal Pradesh depicts the presence of almost all types of glacial lakes (Table 8.2). There are more than 18 lakes which are blocked types. This implies these lakes are formed in shallow or deep impoundments and blocked by glacial movements or boulders. Blocked type lakes are usually small in size and scattered in distribution while supraglacial lakes are formed at the gla-

Table 8.2 Type-wise number of lakes in Himachal Pradesh

Type	Frequency	Percent	Valid percent	Cumulative percent
Blocked	114	53.8	53.8	53.8
Cirque	23	10.8	10.8	64.6
Moraine dammed	21	9.9	9.9	74.5
Proglacial	22	10.4	10.4	84.9
Supraglacial	14	6.6	6.6	91.5
Valley	18	8.5	8.5	100.0
Total	212	100.0	100.0	

Table 8.3 Size class distribution of glacial lakes

Area	No. of lakes	%
Area less than 1 km^2	70	33.01
Area between 1 and 3 km^2^	30	14.15
Area between 3 and 5 km^2	22	10.377
Area between 5 and 10 km^2	32	15.094
Area above 10 km^2	58	27.35
Total	212	100

cial surface. These are relatively not dangerous unless these small scattered lakes at the glacial surface do not join and form a bigger lake due to coalescence effect. Moraine-dammed lakes though in number are predominantly the most catastrophic one. While in terms of size class distribution, majority share of the lakes have less than 1 km^2 that is 70, while lakes with area more than 10 km are 58. The rest of lakes vary in size between 1 and 10 km^2 (Table 8.3). Out of these 212 lakes, there are 3 lakes with potential threat that is in Lahaul and Spiti. Some of the lakes were investigated in the field directly (Fig. 8.8 and Plates 8.1, 8.2, 8.3, 8.4, 8.5, and 8.6).

In Lahaul and Spiti region, three lakes have been identified with potential threat of GLOF. Together with the lake dynamism, there is a need to understand and study the morphology and setup of the entire watershed of the Sissu Lake watershed. Topography, lithology, and landform developments of the region have a direct relationship on the lake dynamism and a direct impact on deciding the intensity and magnitude of the hazard. It can accelerate or retard the rate and level of threat in situation depending upon the geomorphology of the region concerned.

Geomorphology and Sissu Lake Dynamism

Geomorphic maps for any region act as preliminary tools for deciding the course of action for the management of land. The most important advantage of such mapping is geological and geomorphological hazard risk management. It is because geomorphic map provides us the base to understand the current process operating on the surface or underneath the earth surfaces, besides giving us a proxy of the past processes that have acted upon to give shape to the current landforms. This way one single map can help us understand the processes that are responsible for giving the earth its present shape and tell as to how did these landforms gradually evolved out in its present shape. Once the landforms in any region are mapped properly through geomorphic mapping, it can give us insight on geologic history, structure, and lithology of any region. Geomorphic mapping is a tool through which the analysis and solution of problems concerning land and environment planning and management can be done. Any region will be prone to more floods if it has a topography or slope that can increase the movement of the water downstream. In addition, it has unconsolidated lithology things can further go worse. There

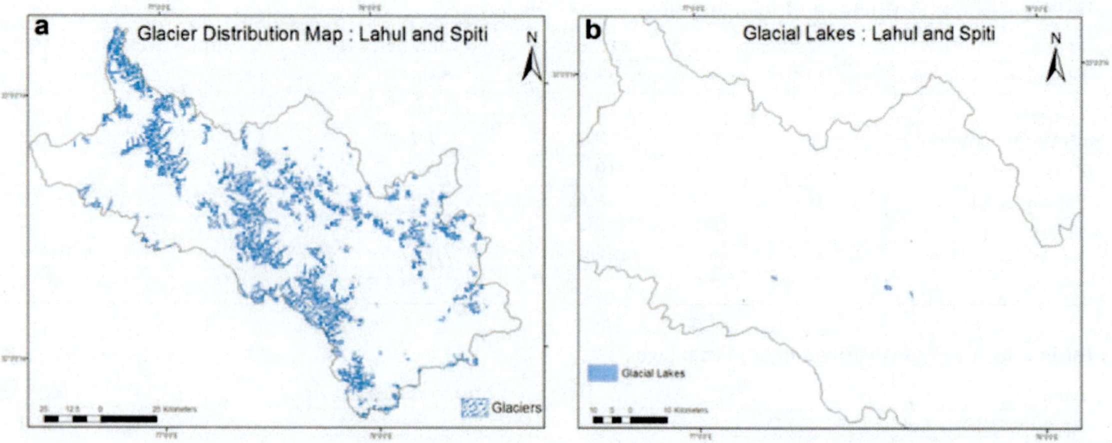

Fig. 8.8 (**a**) Glacier distribution, (**b**) glacial lakes: Lahaul and Spiti, Himachal Pradesh

Plate 8.1 Tarn/Cirque Lake of Dashur on the Pir Panjal. Such lakes are almost permanent features on deglaciated trains and have stability unless a catastrophic rockfall or slide debounches into it

Plate 8.2 Samudri Tapu proglacial lake. These types of lakes are extremely prone to flooding on account of increase in volume of water and resultant pressure

Plate 8.3 Proglacial lake of Uldhampu glacier in the Miyar Basin. Such lakes have history of breaching and destroying farmland and bridges in the past

Plate 8.4 The most catastrophic lakes are formed by glacier surge episodes such as this paleo-lake deposits of Bara Shigri glacier. The glacier is known to surge and block the river valley and flood almost the entire length of the Chenab

Plate 8.5 Kettle and ribbon types of lakes in glacial frontal area are common but purely transitory. This one in front of Miyar glacier (27 km) is one such example. This disappears on the seasonal and yearly time scale

Plate 8.6 End moraine and frost-shattered rocks create conditions for the meltwater to accumulate and form lakes. Summer melt waters in the Neelkanth glacier environment

are several techniques involved for the identification of the accurate landforms and processes involved and then mapping it. These techniques are relief shading, gradient, and profile curvature. Using these techniques, a detailed geomorphic map has been prepared for the Sissu watershed, along with coding and categorization of each unit.

Sissu watershed, in general, is a barren eroded region with rocky outcrop surface (Fig. 8.9). Sissu watershed has plenty of small and large glaciers which are valley as well as hanging type glaciers. There are quite a large number of cirque glaciers as well. The entire region has dense network of small meltwater streams coming out from these glaciers. All these meltwater streams join the main stream. Sissu is also a glacier-fed river, the area that receives scanty rainfall. Although this is a dry area, differential occasions of melting of snow and glaciers have led to the formation of numerous rills and gullies. This watershed is prone to debris flows and mass movements. There are few alluvial cones at the top of the slope. The entire area is an eroded surface with eroded loose and unconsolidated material based on the force and mechanism of gravity. There are large size boulder deposits at the bottom of the hills.

The region shows the extent for glacial advancement in the past in the form of lateral and terminal moraines. The cyclic advance and retreat of glaciers is seen by the alignment of series of lateral moraines and terminal moraines. It is because of this alternate advance and retreat of glacier that led to the formation of a big proglacial lake at the mouth of the source glacier and dammed by its end moraine complexes. These end moraines are of recent origin, i.e., from the Little Ice Age period indicating that morainic materials must not be strong, consolidated, and cohesive.

Mountains of the region are aligned in such a way that they get sufficient amount of sunlight all throughout the year. This will increase the melting of the parent glacier. If this continues, this meltwater will add on to the present lake. This has been responsible for the dynamism of this lake over the years. And this is a serious issue because

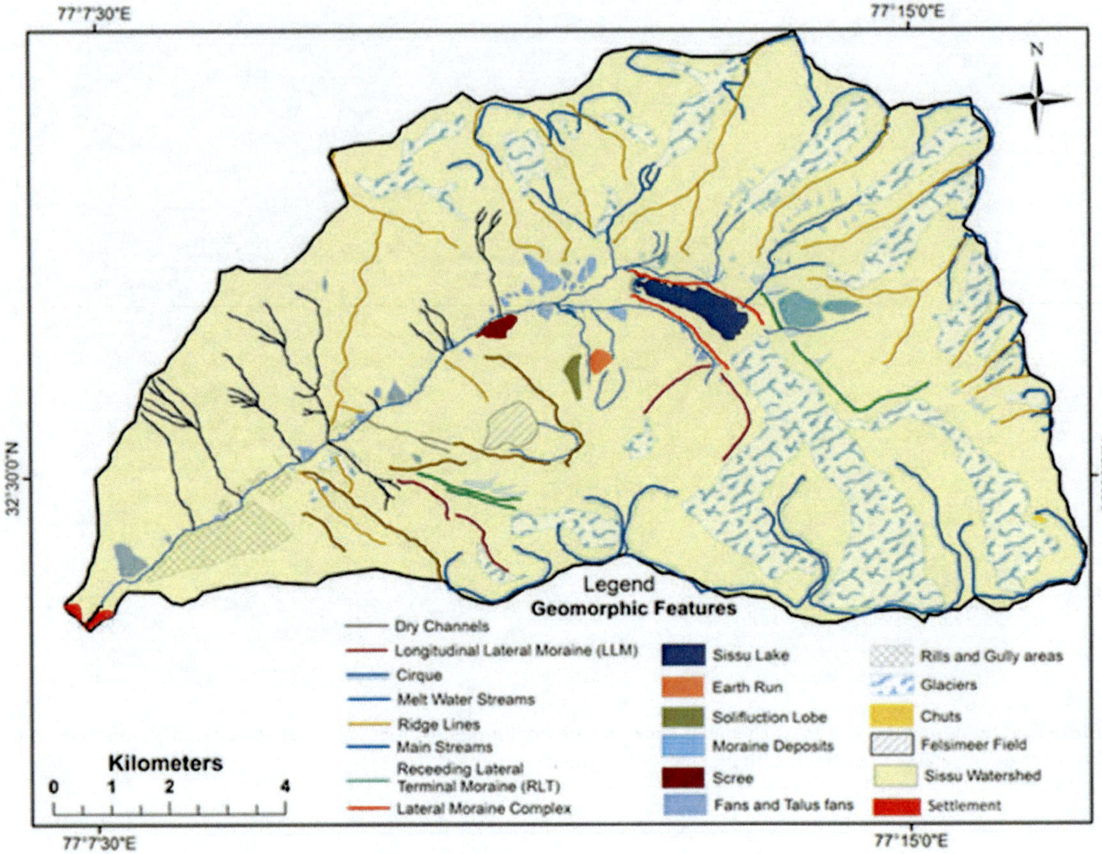

Fig. 8.9 Geomorphology of Sissu Nallah watershed

this region has a large size of settlement within the valley along with other economically important assets. Therefore, it is important to look into the dynamism of this lake. The most interesting fact of the Sissu Lake is that the meltwater of hanging glaciers has eroded the lateral ridges of the previous glacier extent and drains straight into the lake. This process may have accelerated due to the pressured warming.

There is a stark change in *land use and land cover* within the Sissu watershed from the period of 1975 to 2015 (Fig. 8.10a, b). There is a great amount of decrease in glacial area. At the same time, there is a huge increase in the glacial lake as well as the area under human settlement. From here also change in lake area is very much evident.

The higher reaches of this watershed is barren and unstable with prominent signatures of frost-shattered material, debris flow, slump, slide, and rock fall. Rapid melting of the parent glacier has increased the size and volume of the lake in the recent years. This would invariably exert pressure on the damming material holding the water because the increasing volume can break the threshold of this dam freeboard. Associated factors such as these increase the threat of the lake flood. In terms of the general gradient, there is a sharp break at the lake outlet that can increase the velocity of the outburst floods by a great scale.

The entire watershed region has undergone tremendous change in the natural settings since there are intense human interventions within the watershed region. Rapidly increasing temperature has definitely led to the melting of glaciers. Lithologic instability leads to the movement of boulders downslope. This release of the covering material in the form of debris changes the geo-

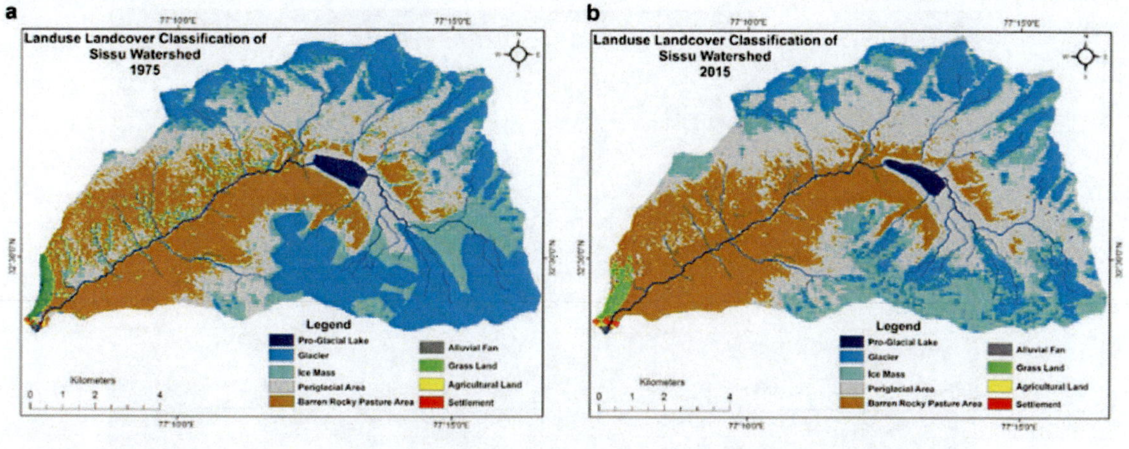

Fig. 8.10 (**a**) Land use and land cover 1975. (**b**) Land use and land cover 2015

thermal settings. Thus penetration of insolation increases to a greater depth, and melting of glaciers gets accelerated at a much faster pace. This region has witnessed rapid human influence in the form of intrusion of economic activities in the natural settings. There are several chains of micro hydel power projects and one big hydropower project that have come up within the watershed area.

The Himalayan glaciers have wide range in altitudinal distribution from few hundred meters to thousands of meters which provide different climatic zones. This influences their melting and ablation rates. This altitudinal variation leads to higher inversion of temperature in the region (with increasing altitude of about 165 m, temperature starts decreasing at the rate of 1°). Therefore, due to greater height in the Himalayan region, there is more inversion of temperature, and there is persistent snow cover. But in the same region there is huge debris cover in the ablation zone as well. Snow avalanching results into the spread of thin sediments on the glacier surface at the glacier surface at the ablation zone. Besides summer, temperatures over the Himalayan glacier are higher on the southern slopes. All of these factors tend to enhance the melting rate to form glacial meltwater. In the nutshell along major climatic factors, micro factors affecting, influencing, or rather accelerating the melting rate of the parent glaciers of Sissu (Bahadur, 2004) are as follows:

- Construction of hydro-electric dams in the lake region
- Siltation effect
- Encroachment on lake catchment area

These factors cumulatively must have led to the accelerated melting of the parent glaciers of the Sissu. It is a highly dynamic proglacial lake above 4000 m altitude experiencing changes in its areal and volumetric extent over the past couple of decades (Fig. 8.11). In the year 1989, it was small in size with 0.172 km^2 area. In 2011 size doubled reaching up to 0.396 km^2 and further got bigger in 2019. Therefore increasing the lake area (Fig. 8.10). The satellite image of the lake depicts the gradual increase in temporal increase in size and area of the glacial lakes over the past 2 decades (Fig. 8.12). Plate 8.7 is the real-time field photograph of the lake.

Overview of the Villages Fed by the Glacial Lake

Sissu and Sashin are two large village settlement to be affected by the most in case of any outburst from the lake. Land record data depicts the predominance of agriculture as a major economic activity

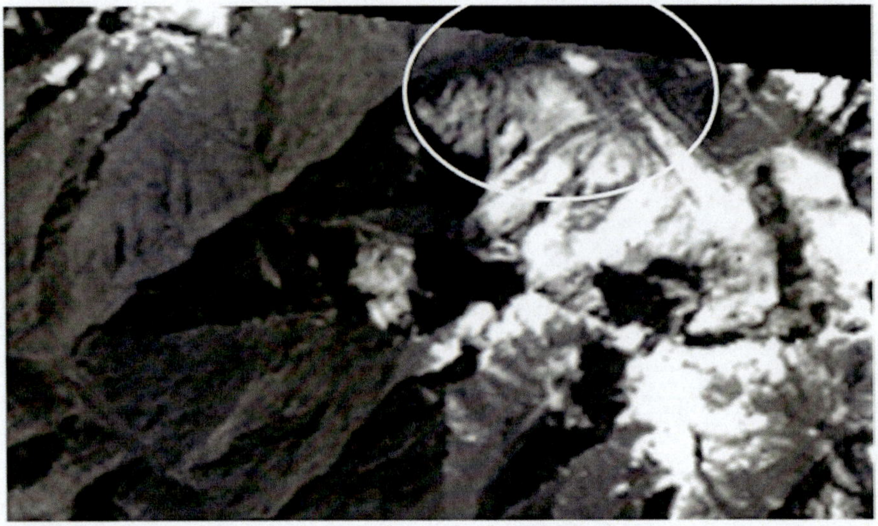

Fig. 8.11 CORONA image of Sissu glacial lake 1976

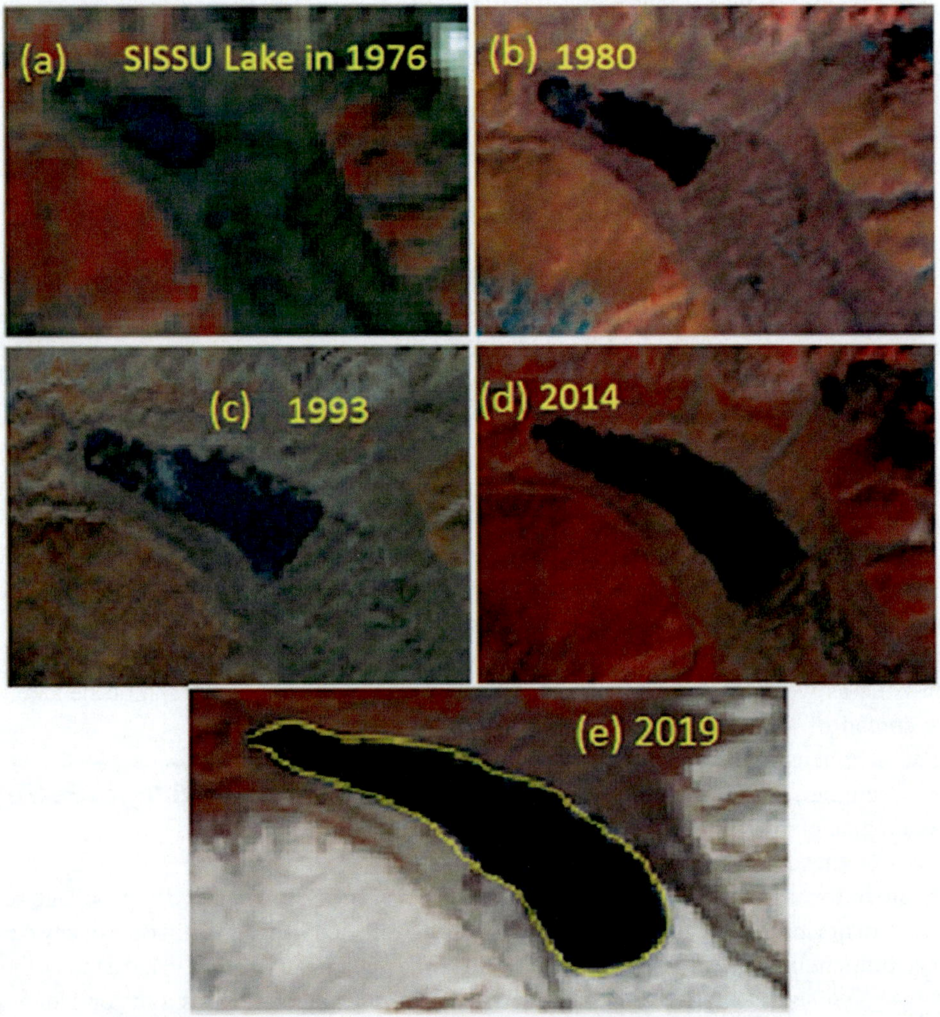

Fig. 8.12 Satellite images of Sissu glacial lake, a spatiotemporal comparison. Note that the lake was partially frozen

Plate 8.7 Field photos of the Sissu glacial lake. (Credit: Jagdish Katoch)

Table 8.4 Land record data of (a) Sissu village and (b) Sashin village

Area under different land use practices	Unit (Bigha)	Area under different land use practices	Unit (Bigha)
Arable land	32,830	Irrigated land	31,860
Nursery/trees	05	Unirrigated land	44,241
Pastures	11,420	Pastures	2936
Orchards	60	Trees	5426
Banjar+tress	80	Banjar	1302
Hospital	08	Banjar + Kheti	1302
Waste/open land	4632	Gair Mumkin	7583
No of plots	156	No of plots	183

while pastures for the livestock provide alternative sustenance in harsh conditions (Table. 8.4).

Tourism provides economic benefits in peak summer and monsoon seasons. In case of any unforeseen circumstances, agricultural fields, pastures, and open forests will get washed away. It has a number of government-, private-, and semi-government-built area infested with a lot of money. Government sites include PWD rest houses, Mahila Mandal and Yuva Mandal buildings, primary and senior secondary school building, PHC, nursing homes, post office, patwari bhawan, 16 homestays, and 8 hotels. All of these will be under the impact of the disaster. Sissu village is also on the only connecting highway that leads to Leh in Ladakh. This highway gives way to 1500–200 vehicles a day in peak season and in off season of about 400–500/day, which provides daily means of communication, transport of food, and other essential supplies to army and the public in general. It is the only route that connects Ladakh and Lahaul Spiti district to mainland India. Destruction of this highway implies cutting of the ties and isolating the alpine population in case of any un-natural situation. Therefore, the disaster will not only have huge ecological, monetary, and infrastructural losses but at the same time deep socio-physiological impacts that might take decades to heal (Fig. 8.13).

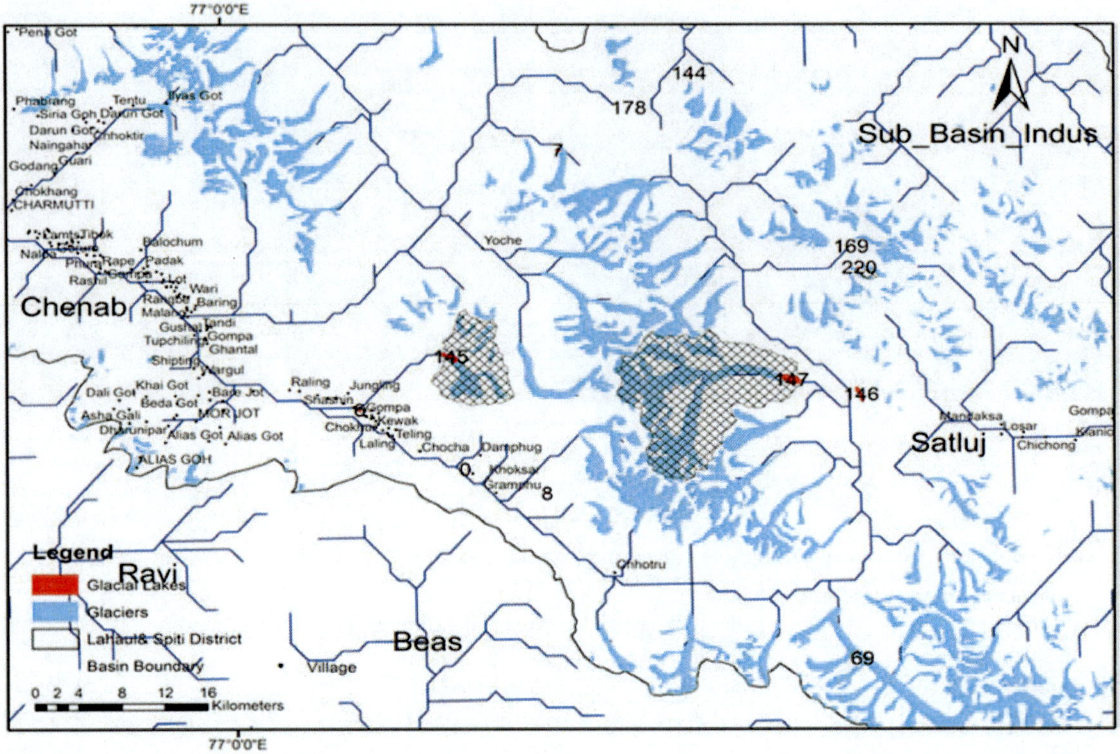

Fig. 8.13 Villages within the buffer zone of the Sissu watershed

Conclusion

In order to develop mitigation plans and policy measures, identification of the hazard potential and it's continued monitoring and assessment is a must. Improved understanding of cryosphere changes and its drivers will help reduce the high risk of mountain hazards. Once the causes, impact, duration, and tentative damaging potential of a hazard is known, quick and accurate measures to contain and reduce the negative externalities of the disaster can be put forward. It will be a crucial step to reduce the damage to minimum. This requires strenuous and collaborative efforts from different governments together because glacial disaster produces trans-boundary impacts globally. The climatic change/variability in recent decades has aggravated impacts on the glacier environment in Himalayan region because Himalayas are geologically young and fragile, which have further been disturbed by human influence and climatic variability. Though the glacial born disasters cannot be predicted and contained, successful monitoring can surely give preparedness time reducing the damage. This required sustained efforts from policy makers and will to tackle it by sharing data sets with other nations. There is an imperative need for accurate precise and continuous monitoring of the glaciers and glacial lakes to formulate risk reducing measures. A comprehensive, concrete and holistic risk reduction strategies are very needed to be formulated. There are several nations who are likely to face this threat in the near future and have focused majorly on structural mitigation approach. This approach focuses heavily on lowering the level of water in the potentially dangerous lakes by draining it out from the lake by keeping a regular watch on the periodic rise of formation and expansion of glacial lakes using the modern state-of-the-art techniques. This has in a way helps out by preparing the inventory of glacial lakes and identifying the potentially hazardous ones. However, these "structural" approaches have only been partially successful in addressing the risks and that too only in some of the identified potentially dangerous glacial lakes.

Therefore, looking at such complex overall phenomenon of GLOFs, it is imperative that local community groups; local, regional, and central government bodies; and various government and NGO bodies should join hands with the scientific community to decide a full-fledged systematic action to suggest ways out to minimize and if not curb the devastation level of GLOF events.

Legislative policy framework needs to focus in depth on sensitive ecosystem of Himalayan region regarding its needs and arrangement. It's a great challenge to designing and more importantly implementing measures to ensure that the risk of the outburst is minimized, vulnerability of nearby communities is reduced, and at the same potential benefits and possible advantages of glacial lakes are restored. The National Environment Policy 2006 though is a much welcome step that focuses on adopting "best practice" norms for constructing infrastructures on mountains and encouraging adoption of appropriate land use planning efforts.

But still there is a long road ahead to ensure the minimal devastation effect borne out of GLOF events requiring strenuous honest efforts so that our mountain system can remain acting our backbone for a peaceful, safe, and healthy life with abundance of resources and no negative impacts.

Acknowledgement Dr. Rakesh Arya for creating the initial database, University Grants Commission for JRF/SRF Fellowship, and the Department of Science & Technology, Govt of India for supporting the field campaigns.

References

Bahadur, J. (2004). *Himalayan snow and glaciers: Associated environmental problems, progress and prospects*. Concept Publishing Company.

Bajracharya, S. R., & Mool, P. K. (2009). Glaciers, glacial lakes and glacial lake outburst floods in the Mount Everest region, Nepal. *Annals of Glaciology, 50*(53), 81–86.

Bajracharya, B., Shrestha, A. B., & Rajbhandari, L. (2007). Glacial Lake outburst floods in the Sagarmatha region: Hazard assessment using GIS and hydrodynamic modeling. *Mountain Research and Development, 27*(4), 336–344.

Basnett, S., & Kulkarni, A. (2012). Monitoring the seasonal snow cover in Sikkim Himalaya using remote sensing techniques. In M. Arrawatia & S. Tambe (Eds.), *Climate change in Sikkim: Patterns, impacts and initiatives* (pp. 69–80). Information and Public Relations Department, Government of Sikkim.

Basnett, S., Kulkarni, A. V., & Bolch, T. (2013). The influence of debris cover and glacial lakes on the recession of glaciers in Sikkim Himalaya, India. *Journal of Glaciology, 59*(218), 1035–1046.

Bolch, T., Buchroithner, M. F., Pieczonka, T., & Kunert, A. (2008). Planimetric and volumetric glacier changes in the Khumbu Himalaya since 1962 using Corona, Landsat TM and ASTER data. *Journal of Glaciology, 54*(187), 592–600.

Fujita, K., Nishimura, K., Komori, J., Iwata, S., Ukita, J., Tadono, T., & Koike, T. (2012). Outline of research project on glacial lake outburst floods in the Bhutan Himalayas. *Global Environmental Research, 16*, 3–12.

ICIMOD. (2011). *Glacial lakes and glacial lake outburst floods in Nepal*. International Centre for Integrated Mountain Development.

IPCC. (2007). *Climate change 2007: Impacts, adaptation and vulnerability*. Contribution of Working Group II to the Fourth Assessment Report of the Intergovernmental Panel on Climate Change.

Mason, K. (1929). Indus floods and Shyok glaciers. *Himalayan Journal, 1*, 10–29.

Mool, P. K., Bajracharya, S. R., & Joshi, S. P. (2001). *Inventory of glaciers, glacial lakes and glacial lake outburst floods: Monitoring and early warning systems in the Hindu Kush-Himalayan Region, Nepal*. International Centre for Integrated Mountain Development.

Raina, V. (2008). Glacier processes and landforms. In V. Raina & D. Srivastava (Eds.), *Glacier atlas of India* (pp. 44–47). Geological Society of India.

Raina, V. (2009). *Himalayan glaciers: A state-of-art review of glacial studies, glacial retreat and climate change* (Discussion Paper). MoEF, Government of India, New Delhi.

Raj, K. B., & Kumar, K. V. (2016). Inventory of glacial lakes and its evolution in Uttarakhand Himalaya using time series satellite data. *Journal of Indian Society of Remote Sensing, 44*, 959–976.

WWF. (2005). An overview of glaciers, glacier retreat, and subsequent impacts in Nepal, India and China, Retrieved 2 September 2009.

Climate Change and Disaster-Induced Displacement in the Global South: A Review

Sk. Mustak

Introduction

Migration is often defined as the movement of people from one place to another. It can be permanent or temporary, internal or international, voluntary or involuntary (Hussain, 2011). Different types of human movements have been defined distinctively in social sciences based on the reasons of migration, the duration of movement, and the distance covered. One of the most significant and dominant migration types is forced displacement (involuntary migration), caused due to climate change, natural or man-made disasters, war, politico-civil disturbance, etc. (World Migration Report, 2020). The UNHCR defines it as "displacement as a result of prosecution, conflict, generalized violence, or human rights violations." In this regard, the persons displaced within the country are known as internally displaced, and those displaced across international borders are known as refugees, asylum seekers, or illegal migrants. As per the World Migration Report, 2020, there are approximately 272 million international migrants in the world (3.5% of the global population), more than 40% of which are from Asia. Although refugees and internally displaced migrants form a relatively small proportion of this number, their numbers have been steadily increasing for the last few decades, and they are often the worst affected. Between 2008 and 2018, 265 million people were displaced from their place of origin due to disasters, out of which more than 85% of the displacement resulted from weather- and climate-related disasters like floods, severe rains associated with El Nino, droughts, bushfires, cyclones, etc.(Ponserre & Ginnetti, 2019).

Disaster-induced displacements affect both developed and developing countries with an average of 24 million displacements every year (between 2008 and 2018). In 2019, 33.4 million new people were displaced globally. 24.9 million of this were displaced due to disasters – while those affected by weather- and climate-induced disasters stood at 23.9 million – taking the total number of displaced people around the globe to 50.8 million. Out of this, internally displaced population due to disasters stood at 9.1 million. Approximately 72% of these were in South Asia, East Asia and Pacific, and Africa. The study of climate change and disaster-induced displacement (CCDID) becomes more significant in terms of the Global South as more than 89% of these displacements (2018–2019) were concentrated in the Global South, with its Asia-Pacific and sub-Saharan region experiencing more than 85% of it. The maximum number of disaster-

Sk. Mustak
Department of Geography, Central University of Punjab, Bathinda, Punjab, India
e-mail: sk.mustak@cup.edu.in

A. R. Siddiqui, A. Sahay (eds.), *Climate Change, Disaster and Adaptations*, Sustainable

induced displacements in the Global South region were observed in Afghanistan (1.2 million), India (590,000), Ethiopia (390,000), the Philippines (364,000), and Sudan (272,000) (IDMC-GRID, 2020).

Keeping in mind the above discussion, in this paper, we have attempted to review the literature available on CCDID in the Global South in order to provide an analysis of human displacement trends, temporal and spatial. The main objective, here, is to not only build a holistic understanding of the concept of displacement but also, to identify the dynamics of climate change, disasters, and displacement. This paper also touches upon the mitigation and adaptation measures being used in the region, attempting to suggest ways to minimize post-disaster and post-displacement losses by integrating decision-making at the institutional and community level.

Methodology

Datasets

This study has been carried out by using secondary data in the form of published and unpublished open-source research papers and reports of governmental and non-governmental organizations like the International Organization for Migration, Internal Displacement Monitoring Centre, United Nations, etc. A qualitative and descriptive approach has been followed to assess the available information on the issue. A brief description of the datasets studied is in the table below (Table 9.1).

Additionally, a series of global database (from IOM, IMDC, UNHCR etc.) and reports and publications like the World Migration Report (2020), Global Report on Internal Migration (2020), Climate change, Migration, and Displacement Report (2017), 2030 Agenda for Sustainable Development, United Nations Policy Brief on Human Mobility, Climate change, Disaster, and Displacement Report (2017), and several additional research papers on vulnerability, displacement, climate change, and disasters have also been studied.

Study Area

This paper focuses on studying displacement in the Global South. The Global South has been defined in a number of ways since the 1970s. Commonly, it is used to refer to a group of countries in Latin America, Africa, Oceania, and developing part of Asia, and not the geographical southern hemisphere of the earth. First used by Carl Oglesby in 1969, the term gained popularity in the twenty-first century and since, has been used as an alternative to "third world," developing world, newly industrializing, countries having been victims of colonialism, and so on. As this is a dynamic concept and no definite boundary has been identified for the same, it is difficult to correctly delineate the study area considering a wide variety of parameters. So, for the purpose of this study, UNDP identified areas in South-South cooperation have been studied under the umbrella term of Global South. These areas aren't as economically and technologically developed as their northern counterparts and are often grappling with issues like political turmoil, poverty, wars, corruption, conflict, etc. (Kaul, 2013).

Methods

This paper has been written by reviewing the available secondary data and analyzing it to determine the most acceptable concept of displacement, its patterns, and catalyst forces in the Global South. The paper identifies different climate-induced disasters and assesses the literature available on the same to provide comprehensive information related to movement in their respect, including outcomes and mitigation practices. Thus, after introduction and methodology, this paper has been distinctly divided into four major sections, namely, defining migration, mobility, and displacement; trends in forced migration due to climate change and disasters; understanding dynamics of climate change, disasters, and displacement in the Global South, disaster mitigation, and emergency response; and conclusion and recommendations.

Table 9.1 Database for the purpose of this study

Authors	Approach	Dataset	Indicator	Result
Huggel et al. (2015)	Quantitative	Study of Emergency Events Database(EM-DAT), DesInventar, and Peruvian National Information System for the Prevention of Disasters: SINPAD)	Reliability of disaster database for climate change study in Peru	Finds enormous limitations in integration of different datasets and points to different outcomes and interpretation having strong implications in policy making and disaster response
Gleick (2017)	Descriptive	–	Droughts in the East Mediterranean region	Provides a commentary on causative effects of drought related to climate change and displacement in the region
Bank and Fröhlich (2018)	Descriptive	Uses UNHCR database on refugees	Forced migration in the Global South (Syria, Oceania, and India)	Establishes the global south to south forced migration trend, and reorients debate toward displacements caused by reasons other than conflicts
Khanam (2018)	Descriptive	–	Climate change-related disasters and climate refugees in Bangladesh	Analyzes linkages between climate change and forced migration, and suggests policy changes and action plan to combat this
McLeman (2018)	Descriptive	–	Sea level rise and migration in coastal regions	Discusses the differences of adaptation and relocation strategies, and stresses on "Migration with dignity" concept to combat the threat without enhancing vulnerabilities of communities
Abel et al. (2019)	Quantitative	Asylum seeking data from 157 nations from 2006 to 2015	Drought and conflict in parts of Western Asia	Climate impacts conflicts and asylum seeking flows with respect to time and place
Adams et al. (2019)	Quantitative	Survey of 1500 households	Sea level rise and floods in south-west coast of Bangladesh	High exposure and vulnerability to risks, and negative climate incidents have a positive feedback relationship with mobility potential
Heslin et al. (2019)	Qualitative	International migration data of Internal Displacement Monitoring Centre	Environmental migration in Pacific Island states, the USA, Japan, India, and West Africa	Provides an understanding of resettlement and displacement years from disaster event, and significance of including challenges faced in return and settlement of climate refugees
Tabe (2019)	Qualitative	Ethnographic study of Pacific islands	Migration and disasters in Solomon, Gilbertese, Phoenix, and Wagina Island	Differentiates between "migration as adaptation" and "migration as displacement," and provides evidence to stress the benefits of involving the local community in decision-making regarding movement and relocation
Zander et al. (2019)	Quantitative	Online survey of 2219 respondents	Heat stress in Indonesia, Malaysia, and the Philippines	Using protection motivation theory, it arrives at the conclusion that different stress events (long term and short term) warrant different behavior adaptation but strong intensions to move almost always end up in action

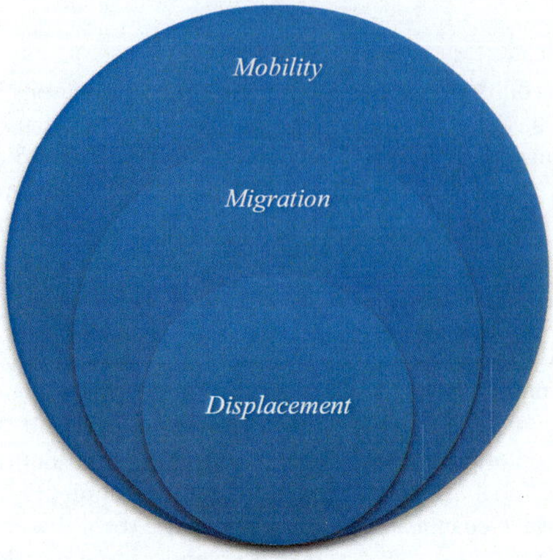

Fig. 9.1 Overlapping relationship of mobility, migration, and displacement

Results and Discussions

Defining Migration, Mobility, and Displacement

In order to understand climate change-induced displacement trends and impacts, one first needs to differentiate between migration, displacement, mobility, and associated terminology and then, identify the relationship between climate change and displacement (see Fig. 9.1). There is no universally accepted definition of either of these terms, and these are sometimes even used interchangeably. However, these are different in their origin and intent, and using one in place of another can lead to misrepresentation of available information.

Migration in general is indicative of voluntary movement caused by push or pull factors, whereas *displacement* or forced migration is an involuntary process caused mainly by disasters (environmental or man-made), famine, conflict, or development projects. Both are social constructs and have no international legal backing (UNHCR, 2016). *Mobility* in general is a broader concept that deals with movement of goods and people over time and space. It can range from local transportation access and commute to migration and thus, is a dynamic concept having different connotations and interpretations based on location, causes, and database considered (Suliman, 2016).

While studying displacement, we are often faced with terms like conflict-induced or disaster-induced, based on whether the source of movement is anthropogenic or natural. Sometimes, there is no clear-cut distinction between the forces causing displacement as shortage of resources may cause human conflict or vice versa. In such cases, it is referred to as mixed displacement. Most global databases study both collectively and often focus on international/cross-boundary displacements.

Some commonly accepted definitions of the abovementioned terms are as follows (Migration Data Portal, 2020):

- A *migrant* is a person who "has moved across an international border or within a State away from his/her habitual place of residence" (IOM, 2019). Here, it is significant to note that the United Nations' definition of migrant doesn't include IDPs, refugees, or asylum seekers (i.e., it only identifies voluntary movement as migration).
- According to International Organization for Migration (IOM) Glossary on Migration, 2019, *forced migration* is "a migratory movement which, although the drivers can be diverse, involves force, compulsion or coercion." It is not a legal term and is widely used to denote refugees, displaced persons, and victims of human trafficking.
- The United Nations (1951) defined any person who crosses international border due to "fear of being persecuted for reasons of race, religion, nationality, membership of a particular social group or political opinion" as a *refugee* or *asylum seeker*.
- The United Nations Commission on Human Rights (1998) defined *internally displaced persons* as "persons or group of persons who have been forced or obliged to flee or to leave their homes or places of habitual residence, in particular as a result of or in order to avoid the effects of armed conflict, situations of generalized violence, violations of human rights or natural or human-made disasters, and who

Table 9.2 Migration and displacement based on distance and source of motivation

Involuntary		Voluntary
Within national boundary	Internal migrants	Internally displaced people
Across national boundary	International migrants	Refugees or asylum seekers

have not crossed an internationally recognized State border."

- The IOM defines *mixed migration* as "a movement in which a number of people are traveling together, generally in an irregular manner, using the same routes and means of transport, but for different reasons."
- According to Bank and Fröhlich (2018), "*disaster-induced migration* denotes population movements in the wake of global environmental change, including both fast-onset events such as floods, storms, or fires, and slow-onset events such as droughts, land degradation, and sea-level rise." (Table 9.2)

Majority of the global studies are mainly concentrated on refugee and conflict-induced displacement; very little has been written upon displacement induced by disasters and climate change, particularly internal displacement. Moreover, this has not been surveyed frequently and systematically which makes an accurate comparison of trends or change in pattern difficult to achieve.

General Trends in Forced Migration Due to Climate Change and Disasters

- *Temporal:* It becomes clear from the given image (Fig. 9.3) that there has been a rapid rise in forced migration from 2009 to 2018, with more than 50% of global forced migrants being internally displaced. This number has progressively increased from approximately 26 million in 2009 to 42 million in 2018. In 2019, 33.4 million new displacements were recorded, 24.9 million of which were due to disasters. Highest numbers of these disaster-induced displacements were seen in India, the Philippines, Bangladesh, China, and the USA, i.e., five million, 4.1 million, 4.1 million, four million, and 916,000, respectively. It is important to note that four of these countries are a part of the Global South (IDMC, 2020) (Fig. 9.2).
- *Spatial:* According to the UNHCR (2019), IDPs are mainly concentrated in the Global South (showing a South to South movement trend), with many having experienced multiple displacements, i.e., multiple uprooting and resettlement to and from their place of origin and place of destination. But internationally displaced people move either to neighboring countries in the Global South or developed nations of the Global North. The forces behind these movements are also varied. For example, Syria experiences conflict-induced displacement, Pacific Ocean islands are most severely affected by rise in sea level (disasters), and India experiences mainly development-induced internal displacement. This becomes clear from Fig. 9.3.

Dynamics of Climate Change, Disasters, and Displacement in the Global South

Climate change is a natural process, but climate scientists across the globe are of the opinion that the rate of climate change has increased as a result of human activities, and this change has increased the probability of displacement due to associated disasters by 60% compared to that in the 1970s. Climate change, as a driver of mobility, has both short-term and long-term variables. Short-term or fast-onset events are natural disasters like floods, storms, heat and cold waves, and droughts whereas long-term or slow-onset events include changes in temperature and precipitation variables over the years, changes in terms of rise or fall in averages, and changes in frequency and intensity of weather events (Bedarff & Jackobeit, 2017; and Abel et al., 2019). This makes the study of dynamics between displacement and climate change very complex (Fig. 9.4 showing

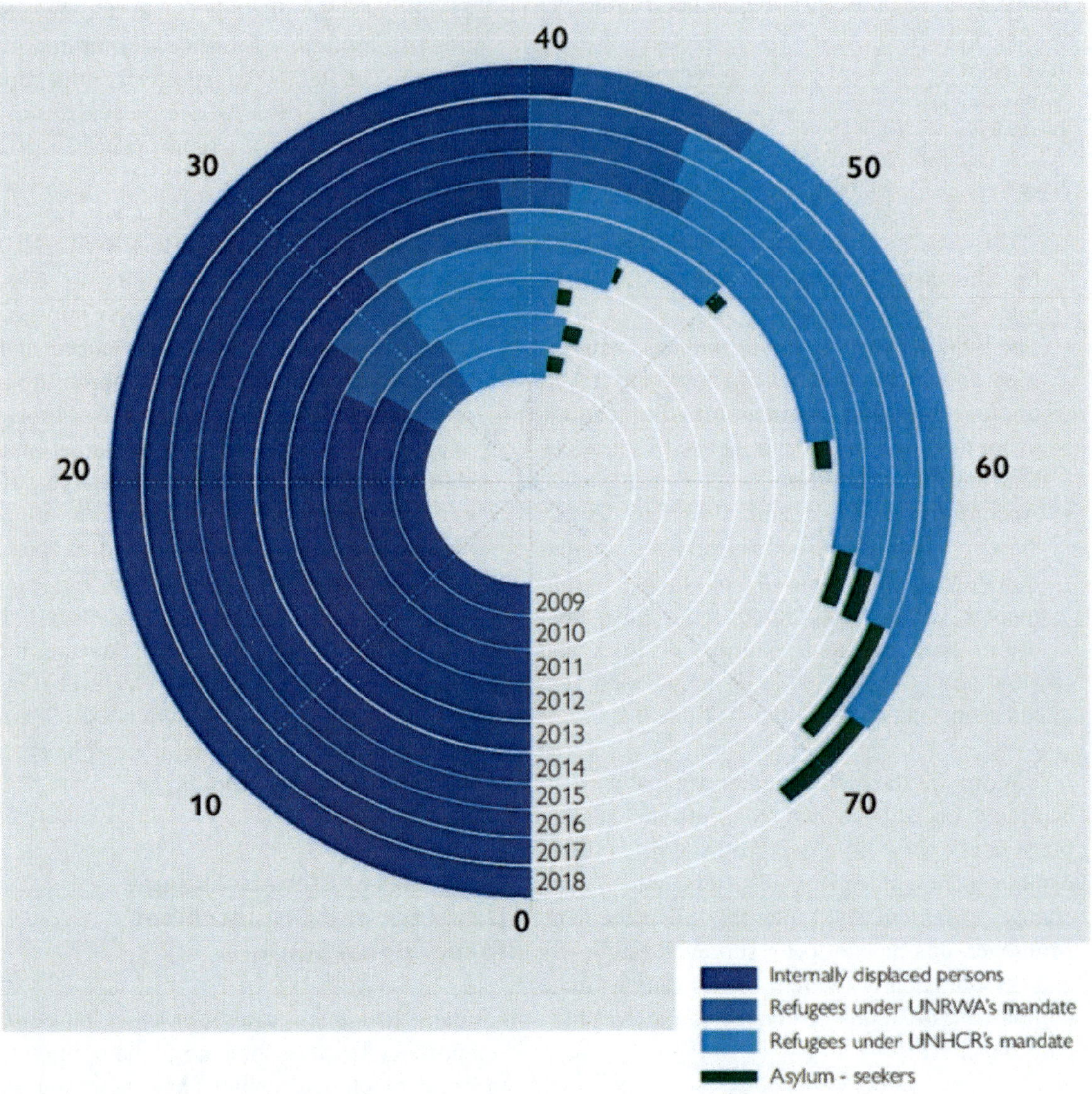

Fig. 9.2 Global displacement trends (2009–2018). (Source: UNHCR, 2019 (Global Trends in Forced Displacement in 2018))

variables of climate change- and disaster-induced displacement in 2019).

Natural disasters have traditionally been associated with sudden, forced, short distance, and temporary displacements, but with climate change-induced slow-onset changes like rise in sea level, salt water intrusion, coastal erosion, increased rainfall, and temperature unpredictability, gradual, long-term, long-distance displacements have also been on a rise in the last 50 years. From over 20 million, climate change-induced displacements between 2008 and 2018, future projections by 2050 vary between 25 and 500 million. In spite of the variability of numbers, it is certain that climate change-induced disasters will become a major displacement force in the coming times (Gemenne, 2011; IPCC, 2012; and Wilkinson et al., 2016).

Although it is a threat to the globe as a whole, in particular, climate change has become an existential threat to the Global South as majority of its countries are still developing, have high den-

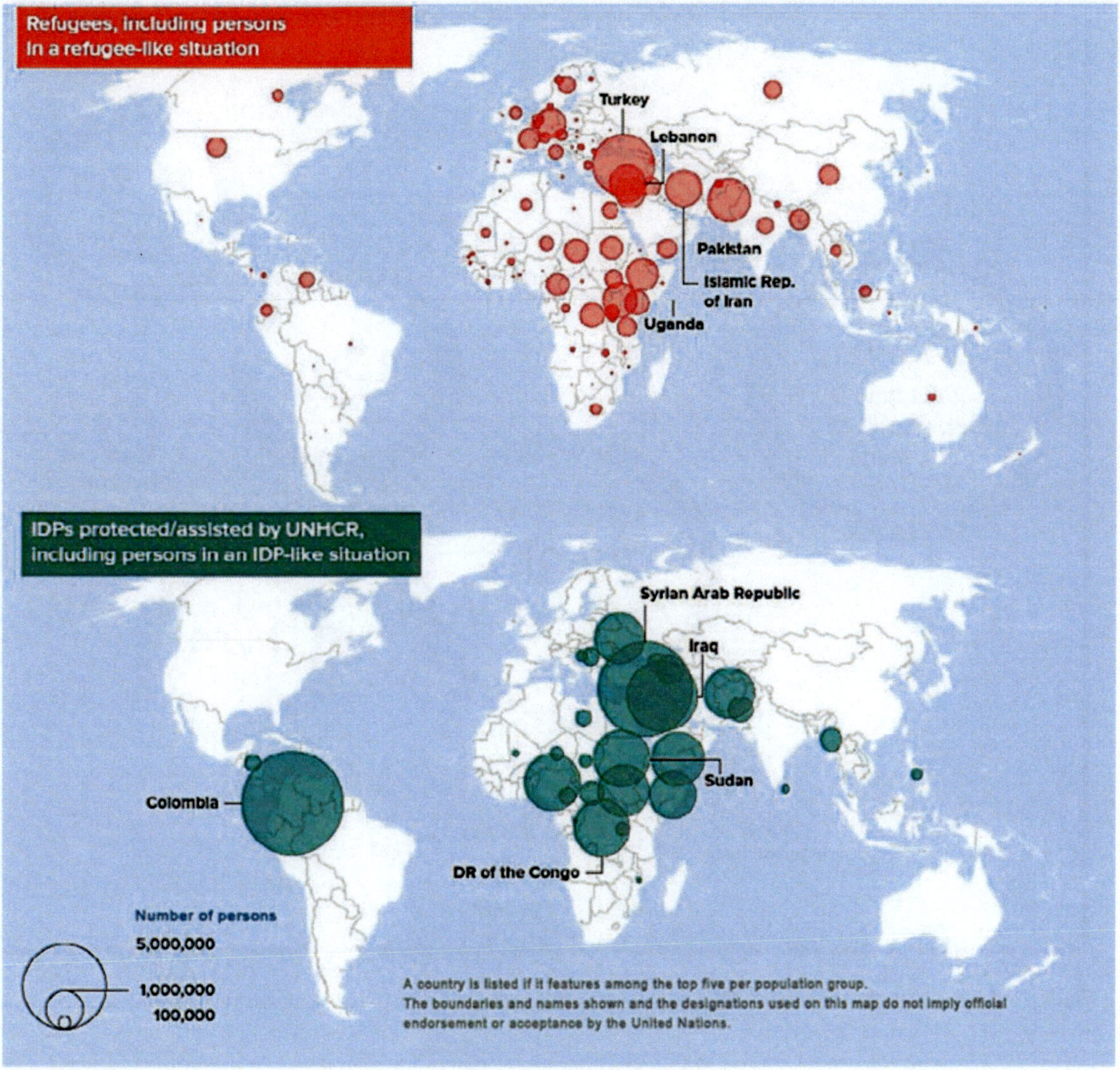

Fig. 9.3 Flow of forced migration. (Source: Bank and Fröhlich (2018))

sity of population, and are located in vulnerable and low-lying areas. Moreover, these nations have a low level of economic development which further enhances their exposure and vulnerability to post-disaster challenges. It is not merely a problem for the future but a present reality that is altering the very nature of global weather events and disasters and causing extreme devastation. Its impacts and mechanisms vary from place to place based on geography; and related losses and impacts are directly influenced topography, location, socio-political structure, and economic development. However, it is obvious that the Global South, owing to the lag in development and research in area of climate change, has an inherent vulnerability to the associated disasters. In this context, this paper reviews major disasters and displacements induced by climate change (i.e., droughts, floods, tsunami, sea level rise) by reflecting upon case studies of different countries and regions in the Global South.

Drought and Heat Stress

Droughts have been increasing in severity, the reason of which cannot be attributed to natural variability alone. Changes in rainfall, runoff, and rise in temperature (which affects the groundwater table, soil moisture, etc.) have changed the

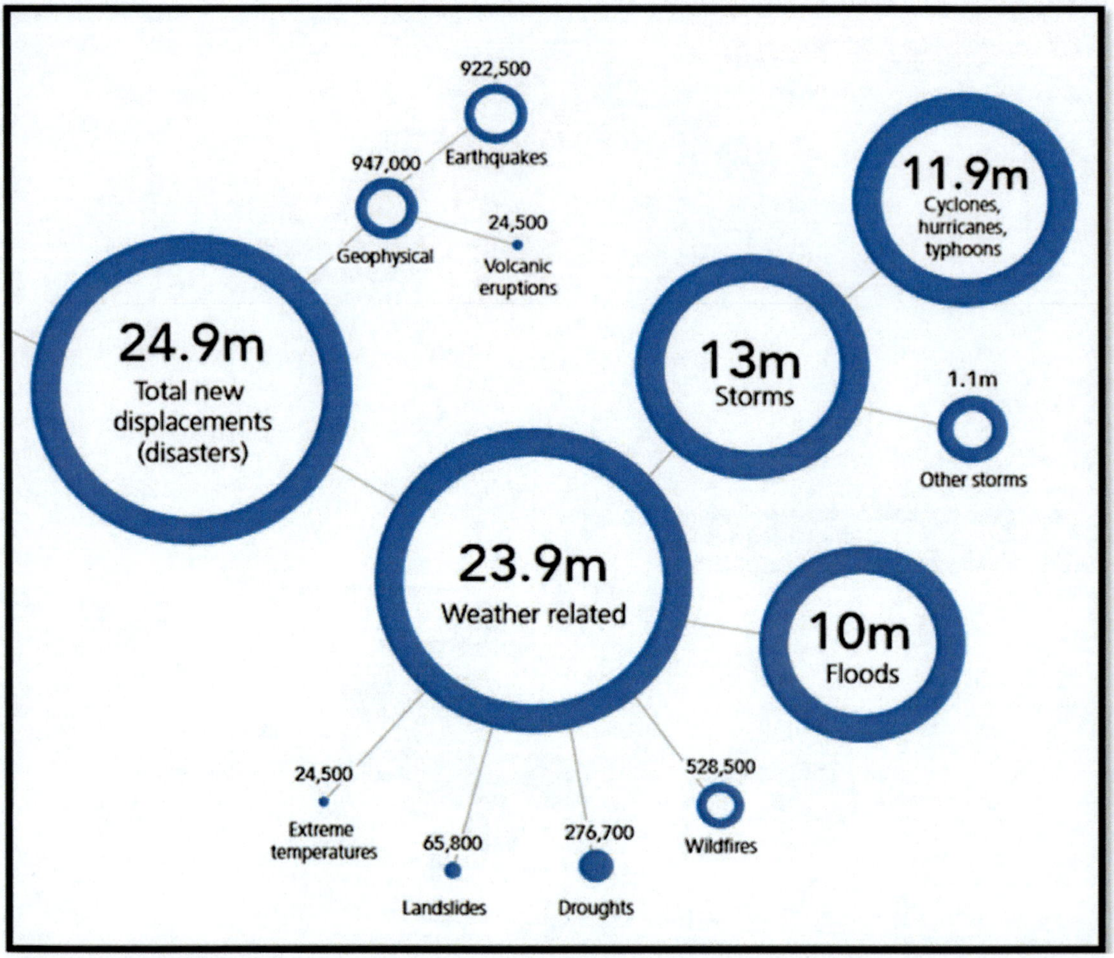

Fig. 9.4 Break-up of global disaster-induced displacement factors in 2019. (Source: GRID-IDMC, 2019)

characteristics of droughts, which have become more frequent and intense. Due to complexity in identifying different types of droughts and its slow-onset nature, it becomes difficult to monitor displacements triggered by it. Also, it is difficult to discern between voluntary and involuntary movement caused due to drought. For example, in 2016, a drought year, Telangana experienced 1.4 million displacements, but as this region experiences seasonal migration, it became difficult to categorize this in absence of a strict monitoring policy, in spite of the number of migrants being a lot more than usual (GRID-IDMC, 2019). Droughts experienced in Syria, Somalia, Iraq, East Mediterranean, and Turkey from 2007 to 2017 have shown increased inter-linkages with climate change. Data suggests with 98% certainty that these droughts were drier than the last 500 years and are linked with climate change. Moreover, majority of forced migrants from Iraq and Afghanistan point toward shortage of water and increased dry conditions as the reason for their displacement and inability to return home. For example, the result of four consecutive years of below average rainfall and high average temperatures in North-West Afghanistan caused severe drought in 2018 and forced over 370,000 people to migrate, more than 71% of which have no intention of returning even if government assistance is granted (Gleick, 2017; and GRID-IDMC, 2019). Pointing out a direct relationship between intention to move and ultimate displacement, a sample survey on people of Malaysia, Indonesia, and the Philippines concluded that slow-onset events like gradual rise in average temperature and increase in heat stress that can lead to drought, food shortage, etc. are responsible for gradual displacement of people (as more than 60% of people showed a desire to move, with nearly 98% claiming the reason as heat stress). Majority of this movement is internal (to a more suitable region) but the more affluent prefer to move across the borders. Although the immediate cause of movement is heat stress, other factors like desire for better living standard, etc. also play a role, and in this context, a better methodology to quantify the same needs to be formulated (Zander et al., 2019).

Floods and Cyclonic Storms

Cyclonic storms occur in tropical and subtropical regions, and their linkages with climate change and rising temperature have been studied by various scientific bodies. The IPCC (2012) has pointed out the increasing severity and intensity of cyclones, hurricanes, and typhoons as the result of increasing sea surface temperature and rise in sea level. In 2016 alone, 13 million people were displaced as a result of the damage and losses caused by it. Countries like India and Bangladesh experiencing 3–5 cyclones every year face maximum damages in spite of good early warning systems due to low preparedness level to post-disaster response. Majority of people thus displaced are sent to temporary shelters and have to settle ultimately in poor, informal settlements, thus enhancing the adverse impact of internal displacement (Heslin et al., 2019). This became evident with nearly 2.7 million displacements in India in 2018 after recurring floods (Kerala) and three cyclonic storms (Titli, Phethai, and Gaja) which aggravated the losses of displaced and increased their vulnerability and poverty. Government of the Republic of Suriname (2015) has projected the possibility of displacement of the country's entire population by 2020 based on its own climate change models and has pointed out its dilemma of deciding whether to focus on building adaptation in the current location or to prepare for eventual forced displacement (Thomas & Benjamin, 2017). In addition to cyclonic storms, increased flooding caused by changes in rainfall patterns has become a trigger for displacement in the Global South. Floods in Niger and Benue rivers of Nigeria displaced 600,000 people from 34 out of its 36 states (i.e., 80% of the country) in 2018. This is in addition to the already displaced two million people as a result of ongoing conflict with Boko Haram and struggle for access to resources. Poor urban planning and unplanned development along the exposed river banks and flood plains have made this African nation vulnerable to frequent floods and forced migration (GRID-IDMC, 2019). The quantitative household survey on intention, mobility, adaptive capacity, and exposure to disaster risks in coastal cities of Bangladesh formulated a displacement model till 2100 (based on coastal flooding and denudation risks), which pointed toward a directly proportional relationship between disaster stress, mobility, and climate change (Adams & Kay, 2019).

Tsunami

In 2007 and 2009, tsunami destroyed parts of Solomon islands and Samoa, respectively, as a result of which, thousands of people were forced to move both temporarily and permanently to either relocate uphill or to settle inland (Tabe, 2019). The 2004 tsunami led to government-funded displacement of 20,000 people from smaller to larger islands in the Pacific Ocean. Although less than half of those displaced returned to their places of origin within 2 years, the rest couldn't due to damage to islands and habitability conditions.

Sea Level Rise

There has been an average rise of 3.2 mm of sea level every year due to rise in global temperature and melting of ice caps (IPCC, 2012). Millions of people living in coastal regions and their settlements are at the most risk particularly those located at low elevated zones and densely populated coastal cities. Small Island Developing States (SIDS) in the Pacific and Indian Ocean face a high degree of threat from climate change and sea level rise due to their low elevation level. For example, people of Papua New Guinea's Carteret Islands have been forced to displace to

Bougainville Island (Connell, 2016; McLeman, 2018). Both short- and long-term disasters have induced displacement in the region since historical times, particularly in atoll islands like Tuvalu, Fiji, Solomon Islands, Marshall Islands, etc. where settlements are constructed within a few meter distance from the coast and thus, floods, tsunami, storm surges, etc. force population to displace. The general trend has been of internal displacement within larger islands due to short-term rapid disasters like floods, tsunami, hurricanes, etc., whereas long-term disasters like sea level rise create food, habitation, and health challenges and force people to seek shelter across national borders. In the longer run, these islands are expected to submerge completely as a result of sea level rise, and their entire populations will have to be relocated (Bank & Fröhlich, 2018). Additional challenge of excessive rainfall, increased CO2 emissions causing ocean acidification, etc. have also led to displacement to the extent that 94% of the households in Kiribati have experienced natural disasters, with 80% of them being rise in sea level (Heslin et al., 2019). Another example is Bangladesh which is expected to lose 17% of its cultivable land, and 30 million of its population are expected to be displaced as a result of 1 meter rise in sea level according to IPCC report, 2012. Being a densely populated nation with a large coastline, it gets hit by cyclones, storm surges, excessive rainfall, floods, etc. on a regular basis. According to Refugee and Migrants Movement Research Unit, over 16 million people are expected to be displaced in Bangladesh by 2050 as a result of climate-induced disasters (Siddiqui & Mahmood, 2015; and Khanam, 2018).

Disaster Mitigation and Emergency Response in the Global South

As is the case with most of the natural disasters, disasters associated with climate change cannot be stopped. The only solution to reducing their adverse impacts is to either slow down the speed of climate change or to increase adaptability to the changes induced by it. Countries in the Global South have not been major contributors to increase in global greenhouse emissions or rapid industrial development, but they are the most affected by their adverse impacts. On the one hand, these nations are raising their voice against this injustice in the United Nations, e.g., Alliance of Small Island States' (AOSIS) "1.5°C to Stay Alive" campaign (Thomas & Benjamin, 2017), and trying to pressurize the Global North to reduce their emissions; on the other hand, they have started preparing for the opposite scenario too. Most of these nations have formulated national disaster policies and are building frameworks to minimize disaster impacts. Following three types of actions are being taken at national level to prepare and respond to climate change and disasters.

Mitigation Measures

Countries in the Global South have become aware of the looming danger of climate change and disasters to their land, people, and economy. This has brought to light the need to strengthen individual and infrastructure capacities. To bring this to fruition, they have made attempts to slow down the rate of climate change by shifting toward sustainable development practices like developing renewable energy infrastructure, increased carbon tax, using energy-efficient appliances, developing public transport systems, etc.

Adaptation Measures

As mitigation of climate change is a long-term process with uncertainty in terms of returns, most disaster-affected nations see it fit to focus more on adaptation and response measures. Despite having different economic, political, and geographical structures, the main adaptation measures (to reduce vulnerability to climate change impacts) of all nations focus on ensuring food and water security (by using drought-resistant seeds, micro irrigation practices, preventing salt water intrusion, raising groundwater table), protecting livelihoods and health of people (diversification of agricultural practices, better vaccination and training of health professionals, awareness generation), and minimizing risks associated with extreme weather events (by strengthening emergency warning and response systems, providing projections and forecasts, better mapping of vulnerability and exposure, etc.).

Response Measures

The following measures were seen in action both during and post disasters:

- *Planned community-based migration* – Island nations of the Pacific Ocean, Afghanistan, and the Philippines have started using relocation and resettlement within and across nations in case of disasters as an adaptation and response practice. The Kiribati motto of "Migrate with Dignity" has been used for displacement of population from the Cook Islands and Samoa Islands to New Zealand and Kiribati's population displacement to Fiji in the recent times (Thomas & Benjamin, 2017).
- *Climate change action plan and fund* – Kenya, Bangladesh, India, Burkina Faso, etc. have their own national disaster response action plans and policies and have also established funds to ensure availability of finance in case a disaster strikes. This ensures increased level of political commitment to invest and take action for the cause of climate change.
- *Integration of geo-intelligence in climate change and disaster mitigation* – In addition to general use of satellites, UAVs, GIS, and artificial intelligence for hazard mapping, resource location, monitoring action plans, etc., technological integration between institutional and individual levels has been a turning point in disaster response measures in the Global South. For example, the Philippines has been using Facebook and other social media data analysis for mapping displacement patterns and used the same for resettlement practices during flood events since 2016.

Despite the formulation of the abovementioned Disaster Management and Recovery plans in majority of the countries, response to disasters is often carried out on a case-to-case basis and no single plan focuses on dealing with the threat of dual or multiple disasters. This shortcoming was witnessed in early 2020 when Cyclone Amphan hit the Bay of Bengal along with the ongoing COVID pandemic and the looming danger of unpredictable monsoons.

To reduce the burden of disasters on displaced people, the "Nansen Initiative-Protective Agenda" highlights the following *priority areas* for future action:

- Collection of data related to cross-border and internal displacement, and based on that building a mechanism to analyze gender, age, and region-specific data, in context of planned or unplanned displacement as a strategy to deal with climate change and its impacts
- Enhancing intergovernmental and interagency cooperation to reduce gaps in present datasets and to provide clarity on disaster trends and patterns
- Building national and international frameworks to ensure dignity of movement, life, and human rights to the displaced
- Establishing systems in place for easier and dignified stay, return, and settlement of displaced people, particularly women, children, elderly, and indigenous sections
- Strengthening disaster response systems at regional, national, and local level by enhancing technical capacity, revising laws and policies, etc. to build resilience and integration of the same with the Sendai Framework for Disaster Risk Reduction (The Nansen Initiative, 2015)

Climate change is an issue whose impacts, frequency, and intensity are supposed to increase with time. Thus, strategies need to be built as soon as possible to not only combat climate change but as far as possible prevent and minimize forced migration caused by it.

Conclusion and Recommendations

This paper has reviewed literature and case studies on displacement caused due to climate-induced disasters in the Global South and provided a brief overview of different issues faced by countries in different regions and adaptive practices followed by them. To summarize, the following generalizations can be made:

- Displacement includes cross-boundary and intra-boundary movements caused by conflict, disasters, development, etc. and is a narrower term than migration. However, there is a huge gap in data availability of specific cause, effect, and location-related movements which make it difficult to study trends and make a comparative assessment.
- Climate change is a major driver to displacement and migration, dominantly affecting the poor and vulnerable. However, the linkages among climate change, disasters, and displacement are nonlinear, i.e., lines blur when slow-onset movements are involved or seasonal and repeated displacements are considered.
- People generally move back after a sudden-onset disaster but slow-onset disasters lead to potentially permanent planned movements.
- Climate change is not the sole factor responsible for migration, but it influences other variables to a great degree, which in turn, affect habitability and survival conditions in a region and lead to displacement. For example, conflict, drought, and displacement linkages in Syria.
- Disaster and vulnerability risk assessment often takes into account a point of origin in totality (population, geography, socioeconomic structure, etc.) but forgets to take into account migration patterns and demographic changes.
- Most adaptive strategies focus on post-disaster settlement and point of origin, without stressing about the people who aren't able to move or to the long-term impacts on point of destination.

As the intensity and frequency of disaster and displacement events keeps increasing, the biggest challenge in finding a solution is the lack of reliable, accurate data. To study disasters and displacements in a holistic manner, the following steps need to be taken:

- A universally accepted definition and methodology of data collection needs to be formulated to increase interoperability of data and for reliable assessment and strategy building.
- Creation of a common, open international data source is required which is complementary with big data and artificial intelligence. Also, new technologies like mobile phones and satellites should be integrated with existing data collection methods.
- Data to be collected and categorized based on gender, region, frequency of data collection, disaster type, adaptability and vulnerability capacity, and in-flows and out-flows of population.
- Disaster resilience capacity, sustainability, and resource burden on destination need to be studied (in addition to point of origin) to reduce undue stress on infrastructure and resources and relocation risks.
- Indigenous and traditional knowledge base needs to be utilized by learning from their practices and experiences. Also, it is necessary to directly involve locals in policy making to improve sustainability and resilience.
- Comparative study of multiple nations having similar issues, using the same methodology, or applying a successful adaptive technique from one region to another can be made for better understanding. Also, movement tracking systems for individual countries required along the lines of the UNHCR and Danish Refugee Council tracking and monitoring systems need to be established.
- Need for integration of climate change adaptation policies and disaster risk management

programs for making better policy decisions and action plans based on lessons from inter-regional, global, and trans-national experiences and cooperation.

- Need for an international agreement of all UN members to protect environmental refugees along the lines of African Union's Kampala Convention. Additionally, IDPs also need to be provided with certain securities and guarantees (legal basis) to maintain a basic living standard as these are often the most ignored category of migrants.

Here, it becomes necessary to point out that disaster-induced migration is a temporary solution to climate change. This is because the focus on relocation can take attention away from need of change in present resource use and development practices that have led to acceleration in the rate of global warming- and climate change-related extreme events. Thus, the priority should be to minimize emissions and mitigate climate change impacts. Although various programs, frameworks, etc. are in force to deal with climate change and disasters, the Sustainable Development Goals do not address these two issues in conjunction but in isolation from one another. The increased incidents of disasters and displacement thus provide an opportunity to fill this loophole and work for better coordinated action at global, regional, national, local, and individual level.

References

Abel, G., et al. (2019). Climate, conflict, and forced migration. *Global Environmental Change, 54*, 239–249. Retrieved 21 May 2020 from. https://doi.org/10.1016/j.gloenvcha.2018.12.003

Adams, H., & Kay, S. (2019). Migration as a human affair: Integrating individual stress thresholds into quantitative models of climate migration. *Environmental Science and Policy, 93*, 129–138. Retrieved 17 June 2020 from. https://doi.org/10.1016/j.envsci.2018.10.015

Bank, A., & Fröhlich, C. (2018). Forced migration in the global south: reorienting the debate. Retrieved 12 June 2020 from ISSN 1862-3581.

Bedarff, H., & Jackobeit, C. (2017). Climate change, migration and displacement: The underestimated disaster. *Greenpeace*. Retrieved 26 June 2020, from https://www.climate-diplomacy.org/publications/climate-change-migration-and-displacement-underestimated-disaster

Connell, J. (2016). Last days in the Carteret Islands? climate change, livelihoods and migration on coral atolls. *Asia Pacific Viewpoint, 57*, 3–15. https://doi.org/10.1111/apv.2016.57. issue-1.

Gleick, P. (2017). Climate, water, and conflict: commentary on Selby et al 2017. *Political Geography, 60*, 248–250. Retrieved 23 June 2020 from. https://doi.org/10.1016/j.polgeo.2017.06.009

Heslin, A. et al. (2019). Displacement and resettlement: understanding the role of climate change in contemporary migration. https://doi.org/10.1007/978-3-319-72026-5_10.

Huggel, C. et al (2015). How useful and reliable are disaster databases in the context of climate and global change? A comparative case study analysis in Peru. *Natural Hazards and Earth Systems Sciences*. https://doi.org/10.5194/nhess-15-475-2015

Hussain, M. (2011). *Human geography* (4th ed.). Rawat Publications.

Internal Displacement Monitoring Centre. (2019). Global Report on Internal Displacement (GRID), 2020. Retrieved on 28 April 2020 from https://www.internal-displacement.org/sites/default/files/publications/documents/2020-IDMC-GRID.pdf

IIDMC-GRID. (2020). Global Report on Internal Displacement 2020, retrived on 28 April 2020 from https://www.internaldisplacement.org/global-report/grid2020/

Kaul, I. (2013). The Rise of Global South: Implications for the provisioning of global public goods. *United Nations Development Program*. Retrieved 12 June 2020, from http://hdr.undp.org/en/content/rise-global-south

Khanam, A. (2018). On the link between climate change and forced migration: the impact of climate change in the arctic and global south an analysis in the context of Bangladesh. *Arctic Yearbook, 2018*. Retrieved on 29 April 2020 from https://arcticyearbook.com/images/yearbook/2018/Briefing_Notes/2_AY2018_Khanam_BN.pdf

McLeman, R. (2018). Migration and displacement risks due to mean sea level rise. *Bulletin of the Atomic Scientists, 74*, 148–154. https://doi.org/10.1080/00963402.2018.1461951

Migration Data Portal. (2020) Types of migration. Retrieved on 3 July 2020 from https://migrationdataportal.org/themes/forced-migration-or-displacement#footnoteref1_cgcj7fd

Ponserre, S. and Ginnetti, J. (2019). Disaster displacement: A global review. *International Displacement Monitoring Centre*. Retrieved 21 April 2020 from https://www.internal-displacement.org/sites/default/files/publications/documents/201905-disaster-displacement-global-review-2008-2018.pdf.

Siddiqui, T. and Mahmood, A. (2015). Impact of migration on poverty and local development in Bangladesh. Retrieved 5 July 2020, from http://www.migratingoutofpoverty.org/files/file.php?name=wp2-impact-of-migration-on-poverty-and-development.pdf&site=354

Suliman, S. (2016). Mobility and the kinetic politics of migration and development. *Review of International Studies, 42*, 702–723. https://doi.org/10.1017/S0260210516000048

Tabe, T. (2019). Climate change migration and displacement: Learning from past relocations in the Pacific. Social Sciences, MDPI. Vol. 8, pp. 18.Retrieved 13 June 2020 from https://ideas.repec.org/a/gam/jscscx/v8y2019i7p218-d250058.html

The Nansen Initiative-Agenda for the protection of cross-border displaced persons in the context of disasters and climate change. (2015) Retrieved 10 July 2020 from https://nanseninitiative.org/wp-content/uploads/2015/02/PROTECTION-AGENDA-VOLUME-1.pdf

Thomas, A., & Benjamin, L. (2017). Policies and mechanisms to address climate-induced migration and displacement in Pacific and Caribbean small island developing states. *International Journal of Climate Change Strategies and Management, 10*, 86–104. https://doi.org/10.1108/IJCCSM-03-2017-0055

UNHCR. (2016). Retrieved on 22 June 2020, from https://www.unhcr.org/news/latest/2016/3/56e95c676/refugees-migrants-frequently-asked-questions-faqs.html

Wilkinson, E. et al. (2016). Climate change, migration and the 2030 Agenda for sustainable development. *Swiss Agency for Development and Cooperation*. Retrieved on 12 April 2020 from https://doc.rero.ch/record/308986/files/29-5._ODI_Climatechange

Zander, K., et al. (2019). Human mobility intentions in response to heat in urban South East Asia. *Global Environmental Change, 56*, 18–28. Retrieved 5 July 2020 from. https://doi.org/10.1016/j.glsoenvcha.2019.03.004

10 Climate Change, Disaster and Adaptations: Human Responses to Ecological Changes

Richa Singh, Kirpa Ram, Chandrashekhar Yadav, and Azizur Rahman Siddiqui

Introduction

Climate change is long-term change in the earth's overall temperature, humidity, precipitation pattern, etc. It is a complex global environmental challenge faced by humanity these days which has led to crop failure, disturbed natural ecosystem, scarcity of freshwater supplies and health issues with some extreme climatic conditions like rise in earth's surface temperature and sea level (Beniston, 2003). The consequences of climate change have impacts on regional as well as global scales which can be witnessed in various earth's climate systems since the beginning of earth's formation. However, due to certain man-made activities, climate change shoots up and causes warming of 0.1 °C per decade observed over the last 50 years (Ramanathan et al., 1985).

After the industrial revolution, the human-caused climate drivers have been increasing, and their effect dominates natural climate drivers. Burning of fossil fuels and increased sophistication have disbalanced the natural world. The human-induced environmental crisis is the biggest problem of these times, and this shoots up the climate changes. The world population has already crossed the planetary boundary. According to the Intergovernmental Panel on Climate Change (IPCC, 2000; IPCC, 2001), the increase in global mean temperature is largely due to anthropogenic activities such as industrialisation which may likely increase global mean temperature between 1.4 and 5.8 °C by 2100. This exceptional change would result in severe impacts on the hydrological system, agriculture, vegetation cover, wind pattern, sea level, melting glaciers and permafrost, ecosystem disturbances and other related processes. The report further highlights that coastal areas and developing countries, including India, will be more prone to these changes as they have a high emission rate and comparatively high percentage of population having lower economic conditions.

One of the possible reasons behind the present-day climate scenario is failure of policy to reduce greenhouse gas emission and uncontrolled developments without sustainability goals. The policies being made by the international agencies such as the United Nations Framework Convention on Climate Change (UNFCCC) and the Kyoto Protocol are not implemented properly and thus, are insufficient to deal with climate change effectively. This challenge can be addressed with sustainability goals like water conservation, dependence on renewable and efficient energy sources, forest and biodiversity con-

R. Singh · K. Ram (✉)
Institute of Environment and Sustainable Development, Banaras Hindu University, Varanasi, UP, India
e-mail: ram.iesd@bhu.ac.in

C. Yadav · A. R. Siddiqui
Department of Geography, University of Allahabad, Allahabad, UP, India

A. R. Siddiqui, A. Sahay (eds.), *Climate Change, Disaster and Adaptations*, Sustainable

servation, etc. Due to the poor economic conditions and lack of awareness among the citizens, developing nations like India are likely to face more climate challenges and opt migrations as adaptation strategies.

This chapter summarises the basics about climate changes (e.g. natural vs human-induced climate change) and consequences of climate change which induces extreme events such as cloud burst leading flood, landslides, avalanches, etc. Some long-term changes like increase in earth temperature leading to melting of ice cover, sea level rise causing to coastal area flood, etc. are also discussed. We also elaborate on the impact on cropping pattern, biodiversity loss as well as human survival and poor economic condition of the country due to climate change.

Natural Versus Human-Induced Climate Change

Earth is continuously receiving solar radiation in which ~29% is radiated back into space, 23% absorbed in the atmosphere and the remaining 48% absorbed at the surface (Wald & Basics, 2018). This radiation energy is absorbed by the land, water and other components of atmosphere and then reradiates back in the form of heat. This heat energy warms the atmosphere. The mixing of gases traps some energy and obstructs them escaping directly into space. Due to this, the atmospheric blanket gets warm and maintains optimum temperature to support life called as greenhouse gas (GHG) effect. Figure 10.1 represents the driver of climate change and their effects.

Earth surface emits radiation in the form of molecules having various vibration frequencies, some having in the range of infrared radiation. When this IR radiation interacts with the GHGs, the molecular vibrations are enhanced because there is a match between the wavelength of the light and the vibrational frequency of the molecule. This makes vibration more vigorous and in turn heating the surrounding air (Harren & Cristescu, 2019). These molecules also emit IR radiation in all directions, some of which reaches earth's surface and causes the greenhouse effect (Earle, n.d.).

According to the law of thermodynamics, the total energy absorbed by the earth should be equal to the amount radiated back to space. But in actual conditions, it does not happen; some amount of solar radiation remains entrapped within the atmosphere. Therefore, the difference between the total incoming and outgoing radiation is known as planet's radiative forcing (Iacono et al., 2008). This climate forcing factor will change the climate system. When the radiation absorbed is higher than the radiation emitted, the planet warms up and is called positive radiative forcing; on the other hand, when outgoing radiation is greater than the incoming radiation, the planet cools and is called negative radiative forcing.

The sun is the main component which impacts longest-term natural forcing. Life and metabolic activities have evolved and changed the atmospheric condition and amount of CO_2 and CH_4 – greenhouse gases (GHGs) – over geological time scale. Both these GHGs contribute to the temperature balance, and its percentage makes the planet cool enough to make it habitable. But increased amount of CO_2 in the atmosphere from fossil fuel burning and increased carbon footprints push climate forcing (Bennington, 2009). This CO_2 traps heat from the atmosphere and radiates it back which causes warming.

During volcanic eruption, lava and exploding rock fragments evolved with particulate matter and gases. Sulphur dioxide and CO_2 are the most important among them. Sulphur dioxide is an aerosol that reflects incoming solar radiation and has a net cooling effect (Earle, n.d.). Its half-life is short and doesn't typically contribute to longer-term climate change (Charlson et al., 1992). However, volcanic CO_2 emissions can contribute to climate warming. It is widely believed that the catastrophic end-Permian extinction (at 250 Ma) resulted from warming initiated due to eruption of the massive Siberian Traps over a period of at least a million years (Liu et al., 2020).

The increased warming changes vegetation patterns; contributes to the melting of snow, ice and permafrost; causes sea level rise, ocean acidi-

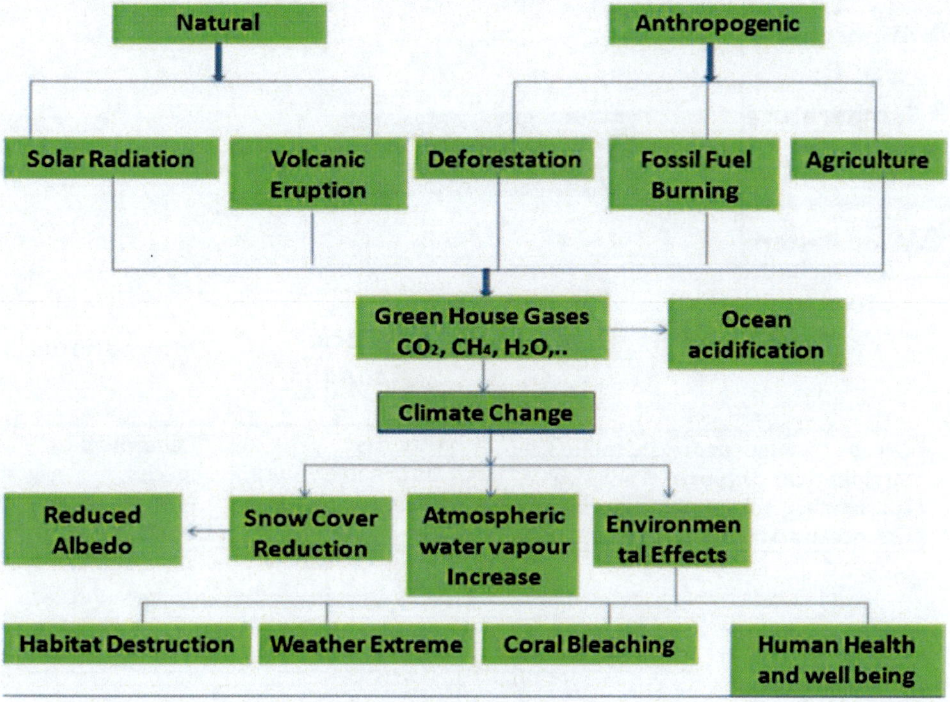

Fig. 10.1 Climate change driver and its effects. (Source: prepared by the authors)

fication which reduces the solubility of CO_2 in seawater, coral bleaching and toxic habitat; and has a number of other minor effects that are not noticed significantly (Waldbusser & Salisbury, 2014). Most of these changes contribute to more warming. Melting of snow and glaciers acts as strong positive feedback because frozen soil contains trapped organic matter that is converted to CO_2 and CH_4 (Khvorostyanov et al., 2008). On the other hand, increased CO2 can cause more vegetation growth and thus, reduces the warming effect (negative feedback) (Foley et al., 2005). Recent studies suggest that glaciers and permafrost are melting at faster rates which contribute to positive feedback and warming climate (Adger & Winkels, 2014a, b).

Climate Change Impacts

Impact of Climate Change on Agriculture Sector

The agriculture sector is very sensitive to adverse effects of climate change because it directly affects the output because disturbance in weather patterns like cold wave, heat storm and irregular monsoon can hamper the yield (Masud et al., 2017; IPCC, 2007). It results in the decline of socio-economic condition of farmers. Therefore farmers of developing nations like India are more vulnerable and facing many difficulties with climate change because they are more or less dependent on monsoon for irrigation, etc., which put extra pressure on them (Verchot et al., 2007; IPCC, 2007). The impacts of climate change are more severe to the small-scale farmers than the large-scale because climate change drastically reduces agricultural productivity which adversely affect the rural per capita income and poverty levels (Esham & Garforth, 2013). The impact of climate change on ocean and other ecosystems is presented in Fig. 10.2. According to Redfern et al. (2012), climate change induces a decline in world rice production which led to increase the poverty level of the farmers. The agriculture sector suffers from the following: (i) reduced agricultural productivity; (ii) food scarcity; (iii) increased temperature of air and water; (iv) economic crisis; (v) reduced power generation; and

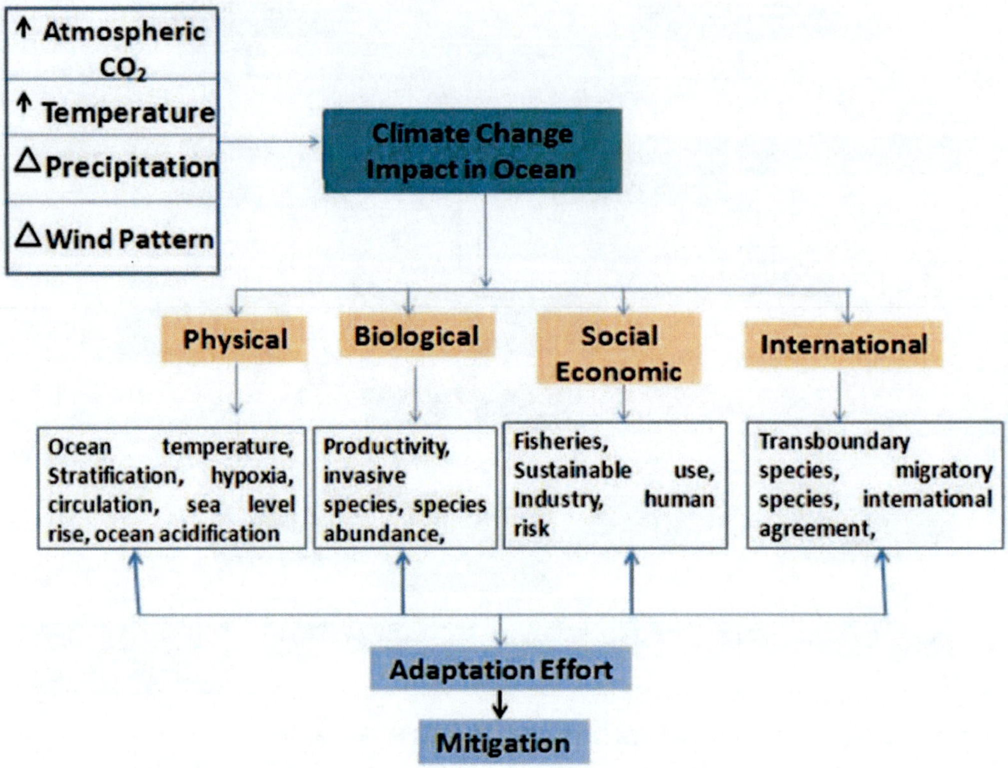

Fig. 10.2 Climate change impact study on ocean. (Source: Prepared by the authors)

(vi) change in cropping pattern due to change in precipitation, temperature, cloud cover and increased carbon dioxide concentration (Baker & Allen, 1993).

Impact of Climate Change over Polar Regions

One of the worst scenarios of climate change can be seen in the Arctic where the glaciers are melting at an alarming rate resulting in the shrinking of cryosphere cover. According to Comiso et al. (2008) on 14 September 2007, the extent and area of the ice cover were 4.1×10^6 km^2 and 3.6×10^6 km^2, respectively, which are 24% and 27% less than the previous record of 5.4×10^6 km^2 and 4.9×10^6 km^2 which was on 21 September 2005. This data is 37% and 38% less than the climatological averages (Comiso et al., 2008). The year 2018 was recorded as the warmest (Cheng et al., 2019). The temperature of air above the Arctic Ocean in September 2018 was 1.5 °C higher than the Arctic land (Kumar et al., 2020). The Arctic sea generally melts during summer and again frozen during winter, but now its freezing speed is lesser than the melting because summer exists for more time than the winter and due to which permafrost not getting enough time to freeze again properly (Abraham et al., 2013). The devastating melting of permafrost results in the increased temperature of the earth as the energy stored in the form of ice or snow comes out and there is additional warming. The Arctic is warming roughly twice as fast as the rest of the world (Taylor et al., 2017).

In the Arctic, polar bears are most threatened. The female bears face lots of challenges in building their den in autumn and feeding their cubs in spring on the sea ice surfaces because of decreased density of sea ice (Kovacs et al., 2011). Their main prey sources are seals which are also suffering from climate changes as they have no ice land to raise their young ones. Due to unavailability of sufficient foods, they are feeding on lesser amounts than they previously did, and consequently their population is declining day by day. It is estimated that its population would decline by 30% by the middle of this century (Derocher, 2004).

The devastating effects of climate change can be clearly seen on the global food supplies, water scarcity, conflict, pandemics and global instability. These all speed up the migration of the human population to suitable place.

Consequences and Adaptation to Climate Change

Adaptation is a slow but continuous strategy of all living towards their changing environmental factors posed by climate change to secure their survival, enhance resilience and perpetuate their further generation on this earth (Adger et al., 2009). Figure 10.3 shows adaptation strategies and technological intervention towards climate change. It is a two-step process in which first is to perceive climate change and associated risks and the second is to minimise the adverse effects. According to the UNDP (2005), climate change adaptation strategies are a broad programme of action dealing with the impacts of climate change. At present time, adaptation strategies are quite important for agriculture sectors. If we talk about the Indian farmer, the first image that comes to our mind is of a poor and weak person. An Indian farmer has very little idea about climate change and they don't know how to cope with it. Therefore farmers should be well trained with essential recognition of climate change and able to do needful adjustment in cropping patterns to minimise the negative impacts of climate change (Masud et al., 2017). To implement adaptation methods and techniques in agriculture, a clear understanding of climate change is necessary for farmers (Somda et al., 2017).

These are the responses which enhance the resilience of vulnerable systems and thereby reduce risk of climate variability on humans and other components of the ecosystem (Adger & Winkels, 2014a,b). Generally adaptation takes

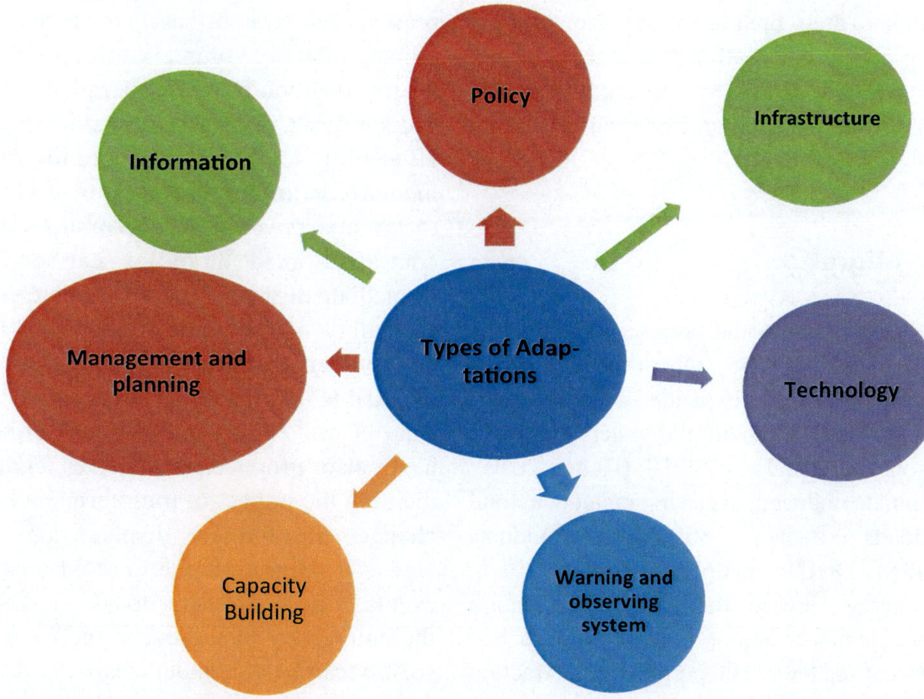

Fig. 10.3 Adaptation towards climate change. (Source: Prepared by the authors)

place when resources shrink or not easily available. According to the Gaia hypothesis developed by James Lovelock in the 1960s, "organisms evolve in the ways that contribute to ensure that their environment remains habitable". Adaptation is a risk management policy. Although many uncertainties exist about the potential consequences of future climate change, existing evidence suggests that the climate is changing and will continue to change (McMichael et al., 2006). This will result in both beneficial and adverse effects on human health, ecosystems and economic systems that are sensitive to changes in climate. Investments in adaptation are warranted to reduce the vulnerability of systems to climate change and to exploit opportunities that may increase social wellbeing.

Genetic Adaptation by Sea Nomads

A study suggests some seafaring people may have evolved over thousands of years to push the limits of typical dive responses even further. Genetic changes have allowed one population in Southeast Asia to grow plus-size spleens that may enhance their breath-holding capabilities, according to an international research team's analysis (Ilardo et al., 2018). The study suggests human genetic diversity and adaptation to the underwater environment.

Human Migration

Human displacement could be a serious outcome of environmental change (Oliver-Smith, 2012). Cataclysmic events have expanded over the previous century, with ~370 natural disasters displaced 38 million individuals in 2010 (Rong, 2010). These climate induced disasters include a total of 182 floods, influenced 180 million individuals and killed 8100 people (Rong, 2010). Approximately 10% of the world's population resides beside coasts among that are 10 m or less above current sea level. The perplexing interaction of rehashed extraordinary climate occasions and ongoing biophysical measures, for example, disintegration and climate-induced ocean level rise, may for all time dislodge the occupants of numerous waterfront networks, especially in low-lying island, dying downstream deltas and zones of dynamic beachfront disintegration (McGranahan et al., 2007). In this specific situation, network movement incorporates the remaking of occupations just as the reconstructing of lodging and public foundation in an area, away from weak danger inclined seaside and riverine territories. Sometimes, it is impossible for some species to incorporate adaptation in them according to the changing environment caused due to climate change. Therefore, they migrate to suitable places. Migration is a livelihood strategy (Fig. 10.4). It can be either voluntary (as a risk management and livelihood diversification strategy) or forced (where livelihoods are no longer plausible due to extreme conditions) (Liao et al., 2015).

In the context of long-term environmental change, both immediate events and long-term processes can lead to voluntary and forced migration depending on the nature of the disaster and the livelihood strategy. Figure 10.5 presents ecosystem-based responses towards climate change. Voluntary migration can occur where livelihoods are temporarily disrupted by sudden-onset disasters (climate events) or are affected by long-term deterioration of environmental conditions (climate processes) leading to gradual poverty (Krishnamurthy, 2012). In contrast, forced migration can occur where immediate disasters (events) threaten the physical safety of populations or where long-term environmental changes (processes) lead to unfeasible livelihoods (Krishnamurthy, 2012; Liao et al., 2015). International organisations have also produced extensive reports documenting the impact of long-term environmental change on human displacement patterns. However, migration due to rapid-onset disaster events is perhaps easier to identify because of the emergency of the event and its impacts on society can be concluded easily.

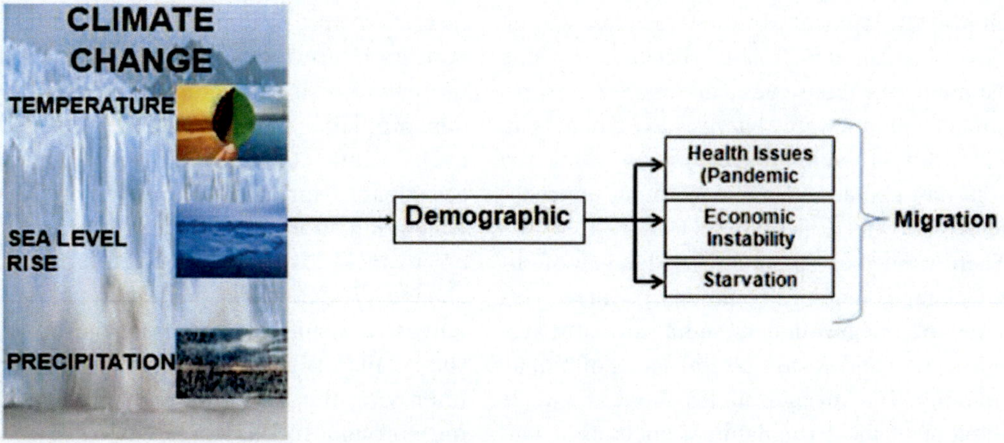

Fig. 10.4 Impact of climate change on demography. (Source: Prepared by the authors)

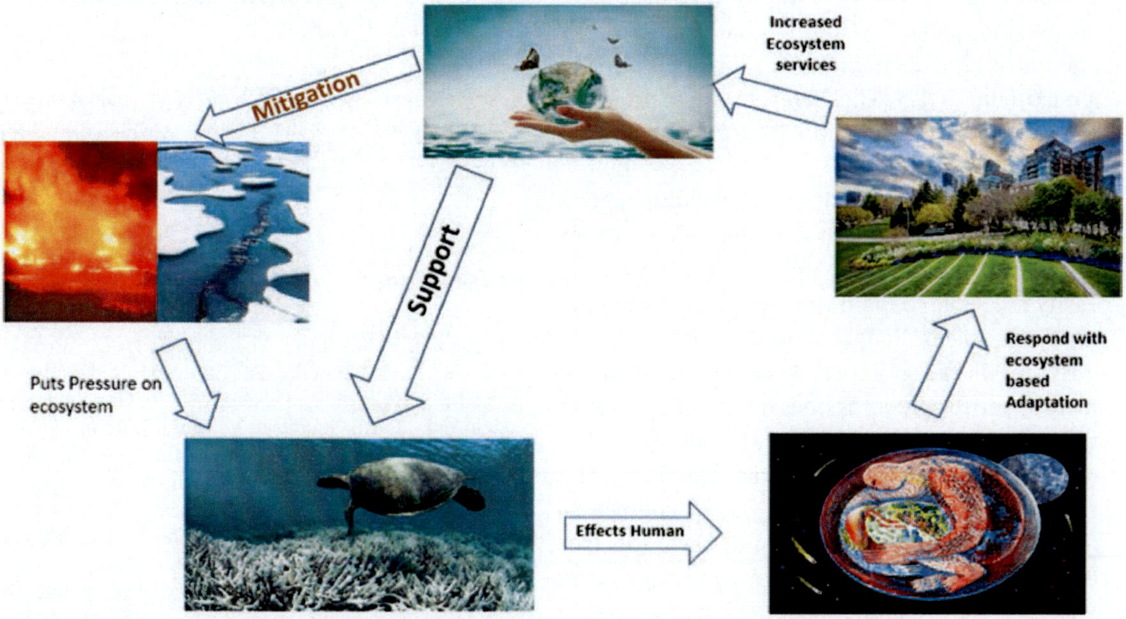

Fig. 10.5 Ecosystem-based responses towards climate change. (Source: Prepared by the authors)

Conclusions and Recommendation for the Future

Resilience towards climate risk is the best mode of securing life. Therefore it is important to evaluate the policy implication of resilience. The main aim of the climate policy is to reduce the climate risk and convert the adversity into opportunities like rain water harvesting in areas of torrential rain (Rabe et al., 2005). According to the Intergovernmental Panel on Climate Change (IPCC, 1996) Second Assessment, "With the growth in atmospheric concentrations of greenhouse gases, interference with the climate system will grow in magnitude, and the likelihood of adverse impacts from climate change that could be judged dangerous will become greater" (IPCC, 1996). Sometimes nations are unaware of the probable climatic changes and fail to invest on the adaptation so it results on the very adverse impacts, severe consequences, economical and social instability and overall downfall of the country's development (Smith & Lenhart, 1996). The world is facing a climate crisis. China is one of the main greenhouse gas emitters but the largest investor in renewable energy also. There is an urgent need of being responsible towards the present-day climate change and to act upon climate breakdown (Kolk et al., 2008).

There is need to fill the knowledge gap of climate change among common people. This requires the integral understanding of earth system processes and socio-cultural status of human community. The strong policies should be made to implement the sustainability goals as it will increase the better survival opportunities with less emission rate and strong socio-economic condition and capacity building. According to recent study done by Guldberg et al. (2018), we have only 12 years to limit the global warming to a maximum of 1.5 °C. Therefore, most probably we are the last generation to act upon climate change.

Government should focus on providing renewable energy to the common people with various schemes. Renewable energy is now becoming easily available to them with lesser cost and one-time investment. People should prefer not to use dirty fossil fuels. There is an urgent need of investment for green business and control pollution of major cities. Along with that, there is a need for development of technology and a system for 100% shift on renewable energy sources so that the people and nature could thrive together (Omer, 2010). This will surely require huge funds and awareness among people, but this would surely benefit nature and organisms here.

To fight against the vulnerability of climate change, we need to focus on building new and effective policy framework and appropriate adaptation strategies. However there are certain limitations of adaptation such as lack of understanding of climate change among common people (Gifford et al., 2011), insufficient funding and least knowledge technologies among common people (Jones & Boyd, 2011) (IPCC, 2007). Adaptation strategies are vital to tackle hostile impacts of climate change on nations overall development; therefore, it is necessary to understand the attitude and perceptions of farmers towards climate change and adaptation barriers (Karimi et al., 2018). The inability of farmers to resist climate change may have serious problems like displacement, ailment and even death (Bhatta & Aggarwal, 2016). Sustainable agricultural practices are a way of adaptation strategy which can be used by farmers to tackle the detrimental effect of climate change. The farmers are facing several challenges to adaptation therefore the policymaker should plan to minimise the adverse impact by improving the irrigation and weather forecasting through technological advancement, cost reduction of farm input, providing information regarding their farmland and crop pattern in it and providing agricultural subsidies and access to agricultural market.

Acknowledgement KR thankfully acknowledges research grant DST under Strategic Programmes, Large Initiatives and Coordinated Action Enabler (SPLICE) climate change programme (DST/CCP/Aerosol/87/2017).

References

Abraham, J. P., Baringer, M., Bindoff, N. L., Boyer, T., Cheng, L. J., Church, J. A., Conroy, J. L., Domingues, C. M., Fasullo, J. T., Gilson, J., Goni, G., Good, S. A., Gorman, J. M., Gouretski, V., Ishii, M., Johnson, G. C., Kizu, S., Lyman, J. M., Macdonald, A. M., … Willis, J. K. (2013). A review of Global Ocean temperature observations: Implications for ocean heat content estimates and climate change. *Reviews of Geophysics, 51*(3), 450–483.

Adger, W. N., Dessai, S., Goulden, M., Hulme, M., Lorenzoni, I., Nelson, D. R., Naess, L. O., Wolf, J., & Wreford, A. (2009). Are there social limits to adaptation to climate change? *Climatic Change, 93*(3–4), 335–354.

Adger, W. N., & Winkels, A. (2014a). Vulnerability, poverty and sustaining Well-being. In *Handbook of sustainable development* (2nd ed., pp. 206–216). Edward Elgar Publishing Ltd.

Adger, W. N., & Winkels, A. (2014b). Vulnerability, poverty and sustaining well-being. In *Handbook of sustainable development* (2nd ed., pp. 206–216). Edward Elgar Publishing Ltd..

Baker, J. T., & Allen, L. H. (1993). Contrasting crop species responses to CO2 and temperature: Rice, soybean and citrus. *Vegetatio, 104–105*(1), 239–260.

Beniston, M. (2003). *Climatic change in mountain regions: A review of possible impacts* (pp. 5–31). Springer.

Bennington, J. Bret. 2009. The carbon cycle and climate change.

Bhatta, G. D., & Aggarwal, P. K. (2016). Coping with weather adversity and adaptation to climatic variability: A cross-country study of smallholder farmers in South Asia. *Climate and Development, 8*(2), 145–157.

Charlson, R. J., Schwartz, S. E., Hales, J. M., Cess, R. D., Coakley, J. A., Hansen, J. E., & Hofmann, D. J. (1992). Climate forcing by anthropogenic aerosols. *Science, 255*(5043), 423–430.

Cheng, L., Zhu, J., Abraham, J., Trenberth, K. E., Fasullo, J. T., Zhang, B., Fujiang, Y., Wan, L., Chen, X., & Song, X. (2019). 2018 continues record Global Ocean warming. *Advances in Atmospheric Sciences, 36*(3), 249–252.

Comiso, J. C., Parkinson, C. L., Gersten, R., & Stock, L. (2008). Accelerated decline in the Arctic Sea ice cover. *Geophysical Research Letters, 35*(1).

Derocher, A. E. (2004). Polar bears in a warming climate. *Integrative and Comparative Biology, 44*(2), 163–176.

Earle, Steven. (n.d.) Physical geology. Retrieved from https://opentextbc.ca/geology/chapter/19-1-what-makes-the-climate-change/

Esham, M., & Garforth, C. (2013). Agricultural adaptation to climate change: Insights from a farming Community in Sri Lanka. *Mitigation and Adaptation Strategies for Global Change, 18*(5), 535–549.

Foley, J. A., DeFries, R., Asner, G. P., Barford, C., Bonan, G., Carpenter, S. R., Stuart Chapin, F., Coe, M. T., Daily, G. C., Gibbs, H. K., Helkowski, J. H., Holloway, T., Howard, E. A., Kucharik, C. J., Monfreda, C., Patz, J. A., Colin Prentice, I., Ramankutty, N., & Snyder, P. K. (2005). Global consequences of land use. *Science, 309*(5734), 570–574.

Gifford, R., Kormos, C., & McIntyre, A. (2011). Behavioral dimensions of climate change: Drivers, responses, barriers, and interventions. *Wiley Interdisciplinary Reviews: Climate Change, 2*(6), 801–827.

Harren, F. J. M., & Cristescu, S. M. (2019). Photoacoustic spectroscopy in trace gas monitoring. In *Encyclopedia of Analytical Chemistry* (pp. 1–29). Wiley.

Hoegh-Guldberg, O., Jacob, D., Bindi, M., Brown, S., Camilloni, I., Diedhiou, A., Djalante, R., Ebi, K., Engelbrecht, F., Guiot, J., Hijioka, Y., Mehrotra, S., Payne, A., Seneviratne, S. I., Thomas, A., Warren, R., Zhou, G., Halim, S. A., Achlatis, M., … Robert, B. (2018). Zougmoré. In *Chapter 3: impacts of 1.5°C global warming on natural and human systems*. IPCC Secretariat.

Iacono, M. J., Delamere, J. S., Mlawer, E. J., Shephard, M. W., Clough, S. A., & Collins, W. D. (2008). Radiative forcing by long-lived greenhouse gases: Calculations with the AER radiative transfer models. *Journal of Geophysical Research Atmospheres, 113*(13).

Ilardo, M. A., Moltke, I., Korneliussen, T. S., Cheng, J., Stern, A. J., Racimo, F., de Barros, P., Damgaard, M. S., Seguin-Orlando, A., Rasmussen, S., van den Munckhof, I. C. L., ter Horst, R., Joosten, L. A. B., Netea, M. G., Salingkat, S., Nielsen, R., & Willerslev, E. (2018). Physiological and genetic adaptations to diving in sea nomads. *Cell, 173*(3), 569–580.e15.

IPCC. (1996). *Climate change 1995, second assessment report of the IPCC*. Cambridge University Press.

IPCC. (2000). *Land-use, land-use change and forestry. Special report of the intergovernmental panel on climate change* (p. 375). Cambridge University Press, UK.

IPCC. (2001). *Climate change 2001: Impacts, adaptation and vulnerability. Report of the working group II* (p. 967). Cambridge University Press, UK.

IPCC Intergovernmental Panel on Climate Change. (2007). *Climate change 2007: The physical science basis. Contribution of working group I to the fourth assessment report of the intergovernmental panel on climate change*. Cambridge University Press.

Jones, L., & Boyd, E. (2011). Exploring social barriers to adaptation: Insights from Western Nepal. *Global Environmental Change, 21*(4), 1262–1274.

Karimi, V., Karami, E., & Keshavarz, M. (2018). Climate change and agriculture: Impacts and adaptive responses in Iran. *Journal of Integrative Agriculture, 17*(1), 1–15.

Khvorostyanov, D. V., Krinner, G., Ciais, P., Heimann, M., & Zimov, S. A. (2008). Vulnerability of permafrost carbon to global warming. Part I: Model description and role of heat generated by organic matter decomposition. *Tellus, Series B: Chemical and Physical Meteorology, 60 B*(2), 250–264.

Kolk, A., Levy, D., & Pinkse, J. (2008). Corporate responses in an emerging climate regime: The institutionalization and commensuration of carbon disclosure. *European Accounting Review, 17*(4), 719–745.

Kovacs, K. M., Lydersen, C., Overland, J. E., & Moore, S. E. (2011). Impacts of changing sea-ice conditions on Arctic Marine Mammals. *Marine Biodiversity, 41*(1), 181–194.

Krishnamurthy, P. K. (2012). Disaster induced migration: Assessing the impact of extreme weather events on livelihoods. *Environmental Hazards, 11*(2), 96–111.

Kumar, A., Yadav, J., & Mohan, R. (2020). Global warming leading to alarming recession of the Arctic Sea-Ice Cover: Insights from remote sensing observations and model reanalysis. *Heliyon, 6*(7), e04355.

Liao, C., Barrett, C., & Kassam, K.-A. (2015). Does diversification improve livelihoods? Pastoral households in Xinjiang, China. *Development and Change, 46*(6), 1302–1330.

Liu, Z., Selby, D., Zhang, H., & Shen, S. (2020). Evidence for volcanism and weathering during the Permian-triassic mass extinction from Meishan (South China) osmium isotope record. Palaeogeography, Palaeoclimatology, Palaeoecology. p. 553.

Masud, M. M., Azam, M. N., Mohiuddin, M., Banna, H., Rulia Akhtar, A. S. A., & Ferdou. Alam, and Halima Begum. (2017). Adaptation barriers and strategies towards climate change: Challenges in the agricultural sector. *Journal of Cleaner Production, 156*, 698–706.

McGranahan, G., Balk, D., & Anderson, B. (2007). The rising tide: Assessing the risks of climate change and human settlements in low elevation coastal zones. *Environment and Urbanization, 19*(1), 17–37.

McMichael, A. J., Woodruff, R. E., & Hales, S. (2006). Climate change and human health: Present and future risks. *Lancet, 367*(9513), 859–869.

Oliver-Smith, A. (2012). Debating environmental migration: Society, nature and population displacement in climate change. *Journal of International Development, 24*(8), 1058–1070.

Omer, A. M. (2010). *Environmental and socio-economic aspects of possible development in renewable energy use* (Vol. 2). Academic Journals.

Rabe, B. G., Roman, M., & Dobelis, A. N. (2005). State competition as a source driving climate change mitigation. *New York University Environmental Law Journal, 14*.

Ramanathan, V., Cicerone, R. J., Singh, H. B., & Kiehl, J. T. (1985). Trace gas trends and their potential role in climate change. *Journal of Geophysical Research, 90*(D3), 5547–5566.

Redfern, S. K., Azzu, N., & Binamira, J. S. (2012). Rice in Southeast Asia: facing risks and vulnerabilities to respond to climate change. *Build Resilience Adapt Climate Change Agri Sector, 23*(295), 1–14.

Rong, F. (2010). Understanding developing country stances on Post-2012 climate change negotiations: Comparative analysis of Brazil, China, India, Mexico, and South Africa. *Energy Policy, 38*(8), 4582–4591.

Smith, J. B., & Lenhart, S. S. (1996). Climate change adaptation policy options. *Climate Research, 6*(2), 193–201.

Somda, J., Zougmoré, R., Sawadogo, I., Bationo, B. A., Buah, S., & Abasse, T. (2017). Adaptation processes in agriculture and food security: Insights from evaluating behavioral changes in West Africa. In *Evaluating Climate Change Action for Sustainable Development* (pp. 255–269). Springer International Publishing.

Taylor, P., Maslowski, W., Perlwitz, J., & Wuebbles, D. (2017). *Arctic changes and their effects on Alaska and the Rest ofthe United States.* Publications, Agencies and Staff of the U.S. Department of Commerce.

UNDP. (2005). In: Lim B, Spanger-Siegfred E (eds) Adaptation policy framework for climate change: developing policies strategies and measures. Cambridge University Press, Cambridge.

Verchot, L. V., Van Noordwijk, M., Kandji, S., Tomich, T., Ong, C., Albrecht, A., Mackensen, J., Cynthia Bantilan, K. V., & Anupama, and Cheryl Palm. (2007). Climate change: Linking adaptation and mitigation through agroforestry. *Mitigation and Adaptation Strategies for Global Change, 12*(5), 901–918.

Wald, Lucien and Lucien Wald Basics. 2018. *Basics in solar radiation at earth surface.*

Waldbusser, G. G., & Salisbury, J. E. (2014). Ocean acidification in the coastal zone from an Organism's perspective: Multiple system parameters, frequency domains, and habitats. *Annual Review of Marine Science, 6*(1), 221–247.

11 Assessing Impact of Climate Variability on Potential Agricultural Land Suitability in Nalanda District, Bihar

Shashi Sekhar, Shyamal Dutta, Mehebub Sahana, G. Areendran, and Krishna Raj

Introduction

Agriculture plays a very prominent role in the Indian economy and contributes up to 18% to GDP that provided up to 50% of their means of livelihood (Madhusudhan, 2015). Agriculture is one among the important sectors of the economy of Bihar state that employs up to 77% of the whole workforce and is accountable for generating 35% of the state's domestic product (Census India, 2011a,b). Out of the entire population, 88% of the state's poor were residing in the rural areas. The Nalanda district is among the prosperous districts of Bihar in agriculture and agricultural products. Here, agriculture is the backbone of their livelihood with many of the population engaged in agricultural practices. The important crops that are accustomed to grow within the district are paddy, pulses, wheat, maize, potato, fruits, and vegetables. A number of problems arose in agricultural practices, and the lack of consolidation ends up in misuse of land and resources (Kumar, 1986). With its predominantly agricultural character, the district is split into the hills in the southern part and lowland in the northern part. The whole district is surrounded by rivers and streams, which contain little or no water during the summer and winter seasons but are filled from bank to bank at the time of heavy rainfall. In the global context, agriculture has been reported to be very prone to agricultural pollution and global climate change. And in due course of time, inadequate human activities, intensive farming, and improper management of the natural resources have highly contributed to land degradation in agricultural areas in many parts of the nation. Management of natural resources is a crucial aspect of agricultural sustainability and ensuring food security within the reporting district. Agricultural land suitability analysis becomes impressive in working out the crop's suitability in a region with exceedingly limited resources (Sarkar et al., 2014; Sajjad & Nasreen, 2016; Singh et al., 2015). This study is an attempt to spot the present land utilization pattern of Nalanda district of Bihar. Agricultural land suitability may refer to "the fitness of a specific land type for specified quite use" (FAO, 1985). In the context of the present scenario, the population of the world is growing dramatically. In this regard, the farmer's communities are fac-

S. Sekhar
PG Department of Environmental Sciences, A. N. College, Patna (affiliated to Magadh University), Bihar, India

S. Dutta
Independent Researcher, West Bengal, India

M. Sahana (✉)
School of Environment, Education and Development, University of Manchester, Manchester, UK

G. Areendran · K. Raj
IGCMC, New Delhi, India
e-mail: gareendran@wwfindia.net; kraj@wwfindia.net

A. R. Siddiqui, A. Sahay (eds.), *Climate Change, Disaster and Adaptations*, Sustainable

ing many problems and apparently need to produce more and more crops to satisfy the increasing demand for food and their products. On the wayward, there is an increasing global concern toward environmental aspects, food, and human health on the use of higher amount of chemical fertilizers, pesticides, insecticides, etc. (Feizizadeh & Blaschke, 2013). These effects could unquestionably be liable for the reduction of soil fertility and their ability to nourish crop growth, which consequently could immensely impact agricultural production. One among the intense threats to the agricultural sector is erosion, which is continuously alarming in many parts of our country, and additionally, our study area is affected by such a problem. Hence, priority should be kept on choosing suitable land for growing of crops, and it is our main concern in this research work.

The crops requirements, land characteristics, and soil characteristics are the major determining factors of suitability towards any agricultural-land analysis procedure (Kumar et al., 2018). Agricultural soils are among the predominant factors for the assembly of crops and are also very helpful within the main tenancy of unpolluted air and water, reduction to the emission of greenhouse gases, preserving biodiversity, and ultimately ensuring food security (Mazahreh et al., 2019). Matching such criterion on any farmland will certainly meet the specified goal. Excessive use of chemical fertilizers, immoderate irrigation, and damage to vegetation roots were fairly often and well versed for runoff, waterlogging and salinization of soil, loss of infertility, nutrient deficiency, pollutants discharge, and sedimentation to surface similarly as groundwater bodies, etc. (Ennaji et al., 2018). Aside from the soil and land characteristics, local/global market position, socioeconomic condition, and technological implementation are additionally driving factors that may influence the productivity of crops. Accuracy and reliability of cropland suitability analysis help within the identification of which area is suited to the cultivation or not and also provide information where it has to improve the present paradigm (Jamil et al., 2018). The assembly of quality crops is the key to precise farming practices, which is economically efficient and socially acceptable (Dengiz et al., 2015). In order to properly utilize accessible resources for agricultural practices, there's much-needed evaluation of land resources that helps our food providers comprehend every aspect alongside the policymakers to implement new methods to urge and obviate the present situation (Chandio et al., 2013). The analysis of agricultural land is applied in such the simplest way that it helps farmers, local communities, and anxious officialdom to require authentic decision toward the optimum production to the world.

Land suitability analysis is a multidisciplinary approach that incorporates the impact of the many areas consisting soil science, horticulture, economics, science and technology, management, etc. In the present study, the land suitability analysis, there are basically two approaches: current situation suitability and potential suitability. Current situation suitability evaluates the current conditions with none major improvement within the particular area. This approach evaluates land on a broader scale mainly depending on the farmer's experiences and institutional intuitive prior knowledge. On the opposite hand, potential suitability is predicated on the units of land within the future condition or upcoming data after the desired some improvement paradigm has got to be invoked where necessary (Bozdağ et al., 2016; FAO, 1983; Sys et al., 1993). Based on the characterization of soil, rainfall, slope and elevation, and temperature distribution, the Bihar state is riven into four major agroclimate zones. The Nalanda district falls under the agroclimate zone 3-B (southern-west) in conjunction with the 10 more districts of Bihar state. Distribution of soil resources during this zone except "Tal lands" and "Diara lands" wasn't properly well drained, and soil's texture varies from medium textured to heavy textured alongside soil pH is moderately acidic to slightly alkaline. The soil of the district is moderate to poorer in nutrient values regarding nitrogen and is not properly enriched in phosphorous and potassium contents (Department of

Agriculture, Govt. of Bihar, b). Uninterruptedly inadequate utilization of the agricultural land in the past few decades has encountered much more destruction than the provided resources (FAO, 1976). Agriculture in the entire state is severely affected by the natural calamities, south Bihar districts are struggling with the absolute climatic change in the form of droughts, and north Bihar districts are continuously fighting with the recurrent floods. Hence, the proper evolution of agricultural land use pattern is important to unravel recent or upcoming problems (Bozdağ et al., 2016). The number of literature and research work has been conscientious to the analysis toward the agricultural land using multicriteria decision analysis (MCDA), which is extremely helpful in such context (Elsheikh et al., 2013). Rasheed and Venugopal (2009) used agroecological characteristics for accessing cropland suitability. Bandyopadhyay et al. (2009) uses parameters like soil depth, soil texture, soil properties, organic matter content, erosion, slope, elevation, approximate to road, and land use land cover to investigate the suitability of land for agricultural purposes (Akpoti et al., 2019). Remote sensing and GIS within the context of multicriteria decision-making (MCDA) analysis provide information from different data, which parameters were necessary for identification of appropriate spatial pattern for future land use (Bunruamkaew & Murayama, 2012). For that reason, GIS-based MCDA AHP technique is employed, data from different parameters together with mapping criteria for suitability analysis toward the agricultural land in Nalanda district (Dalavi et al. 2015). Analytical hierarchical process (AHP) is a very common and widespread used method, which is introduced by Thomas L. Saaty (2008). The AHP introduced by Thomas Saaty (1980) has emerged as a preferred decision-making technique for solving multicriterion problems which emphasizes on the additive weightage model (Ayehu et al. 2015; Malczewski, 1999; Feizizadeh et al. 2013; Cinelli et al., 2014). AHP is an effective technique in addressing complex decision-making tool, which involves in identification and weighting criteria selection, analyzing the information and assigning a rank to that, and executing decision-making process in multicriterion decision analysis (Hossain et al., 2007). Additionally, the AHP analysis states the compatibility of the choice maker's evaluations, which help in the reduction of prejudice into the method of decision-making aspects. Multicriteria decision analysis (MCDA) within the context of remote sensing and GIS might be very useful and effective tools in the aspect of agricultural land use planning and management (Bailey et al. 2010; Goswami et al., 2012; Adeniyi, 1993; Malczewski, 2006; Shearer & Xiang, 2009; Jamil et al., 2018; Qiu et al., 2014; Kalogirou, 2002; Saaty, 2004). Within the period of time of technological advancement in GIS, the AHP method was used for agriculture and horticulture (Böhme et al., 2011). It is the very first introduction to a GIS application by Bagchi and Rao (1992). Aside from that, this method was initially developed to outside GIS environment through the number of analytical resources. AHP model is incredibly effective and simple to implement to unravel multicriteria decision problems (Guo & He, 1999). The steps of the AHP criterion are intended by Saaty and further explained by Mau-Crimmins et al. (2005). Improving agricultural performance and related rural nonfarming activity is vital for improving livelihoods, and curbing poverty is one among the aims of this research work. One of the objectives of this work is to investigate problems within the changing climatic patterns to enhance rural livelihood by the implementation of latest technologies and techniques with prime concentration on the combination of farming system in the way to increase income, standards of living, livelihood enhancement, and food security in an optimistic way. Geospatial techniques are very useful tools within the identification of the suitable sites for the agricultural practices based on different criteria like soil characteristics, the topography of the region, geology, drainage density, river basin, and transporting system of any particular area (Duc, 2006; Pramanik, 2016). Crop yield is principally counting on many factors like soil type, its available nutrients status, rainfall, management practices, and other inputs also. Therefore, it's mandated for efficient planning with the use of the latest technology together with the correct understanding of agroclimatic conditions (Bera et al., 2017).

Land suitability analysis involves arriving at important decisions at different levels ranging from the selection of criteria, analysis on major land use type paradigms, an arrangement of the criterion supporting above-said method, and determination of suitability limits for every class of the parameters that are employed in this study area. This study is the step toward understandings to the present problems and provides better suggestions to the agricultural planning authorities, which is the demand of the time. The reliability of accessible information, timely implementation on the extent, and distribution and kind of land use pattern aware decision-makers to require appropriate steps toward sustainable farming. The present literature relies on the land use analysis on the above-said district that was three decades old, so it's much needed to understand the changes occurring within the course due of time with the assistance of newest technologies and supply some mechanism to enhance the farming practices. Multicriteria decision analysis using different parameters such as soil, climate, and other environmental constituents plays a significant role in defining suitable areas for agricultural land evaluation (Baniya, 2008). This study assumed that the technique used is acceptable to integrate soil, climate, and other environmental constituents in the viewpoint of GIS context. Moreover, the weighting factor method produced important information, which will be useful for forthcoming studies on land suitability analysis. Satellite images provide a powerful tool for the identification of crop suitability analysis. Hence, crop suitability/land suitability analysis becomes an indispensable tool for the higher cognitive process that will be beneficial to the farmers. Additionally, it also helps the government in forming new policies to implement new schemes to enhance farmer's productivity and livelihood. This research work is an integration sort of GIS and MCDA to the assessment of land suitability for crops.

Database and Methods

Study Area

Nalanda district is among the 38 districts of Bihar state. Nalanda district can be geographically accessed through 85.4788° E longitude and 25.2622° N latitude. Agriculture is the backbone of this district in which almost 60% of the population engaged in farming practices. The Nalanda district occupies the area of 2324.678 sq. km (898 sq. miles), which is equivalent to Cornwall Island of Canada. Most of the land in this district falls into Indo-Gangetic plain, which is very fertile in nature. In the extreme south end, the *Rajgir* hillsides, and a small hillock in the district headquarter of Bihar Sharif, Nalanda district consists of 20 blocks, and Bihar Sharif is the administrative headquarter. *Phalgu*, *Mohane*, *Jirayan,* and *Kumbhari* are the major rivers of this district. Rice, wheat, maize, potato, and pulse are the main crops of these districts. The net sown area on the Nalanda district is about 119,792 ha, production 381,094 (M. P), and yield 3181 kg/ha (Department of Agriculture, Govt of Bihar, a) (Fig. 11.1).

The latest trend of population growth in Nalanda is 21.18% as per the 2001–2011 census data. According to the 2011 Census of India, the population of the district was 2,872,523 of which 84.09% of the population lives in rural areas and only 15.91% of the remaining population resides in the urban areas (Census-India, 2011a,b). The average annual temperature ranges from 14 °C to 44 °C, whereas the average annual precipitation is about 120 cm (903.5 mm). The typical soil distribution type ranges from sandy to clay loam. The main sources of irrigation in this district are canals, tube well, etc. Development in the agriculture field will not only improve the quality of life of the people residing in rural areas but also helps in promoting their livelihood dependency. However, it will also prevent the local migration of the laborer in search of their jobs in the industrial cities, which is very often in this study area. Apart from that, tourism is also considerable due to UNESCO's declaration as to the Nalanda World Heritage Site for the ruins of ancient Nalanda University, which attracts the tourist all

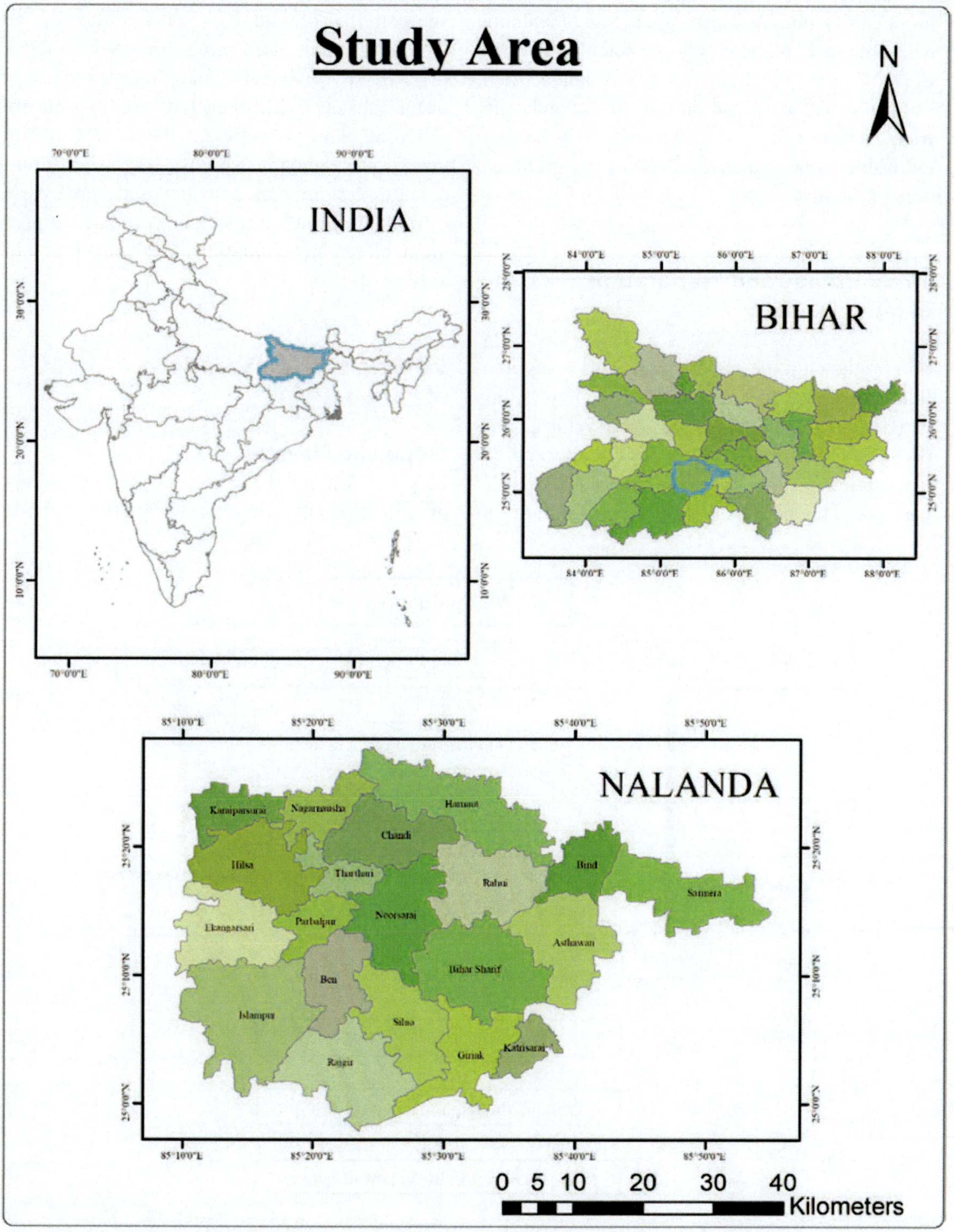

Fig. 11.1 Location of the study area

around the globe (Archaeological Site of Nalanda Mahavihara at Nalanda (a), Archaeological Site of Nalanda Mahavihara at Nalanda (b)). Following the land use pattern of Nalanda district, it is found that 181,130 ha of land belongs to cultivable area, which is 77% to geographical area of the district.

Methodology and Preparation of Input Dataset

After reviewing the various literature and guidelines mentioned for land suitability analysis, this work uses suitability analysis to develop appropriate and potential zonation of agricultural practices based on the set of parameters used. We have taken the parameters like soil texture, drainage pattern, climatic data, and satellite data and then further processed using ArcGIS and ERDAS IMAGINE software for mapping purposes. GIS raster datasets used for each dataset is used from different sources to obtain the land suitability maps for Nalanda district. Based on their importance and significance in this study area, seven different criteria are selected for the accomplishment of the crop suitability analysis (Fig. 11.2, Table 11.1).

Selection and Description of Parameters

Slope and Elevation

Slope is among the basic parameters of topographic elements for agricultural land suitability

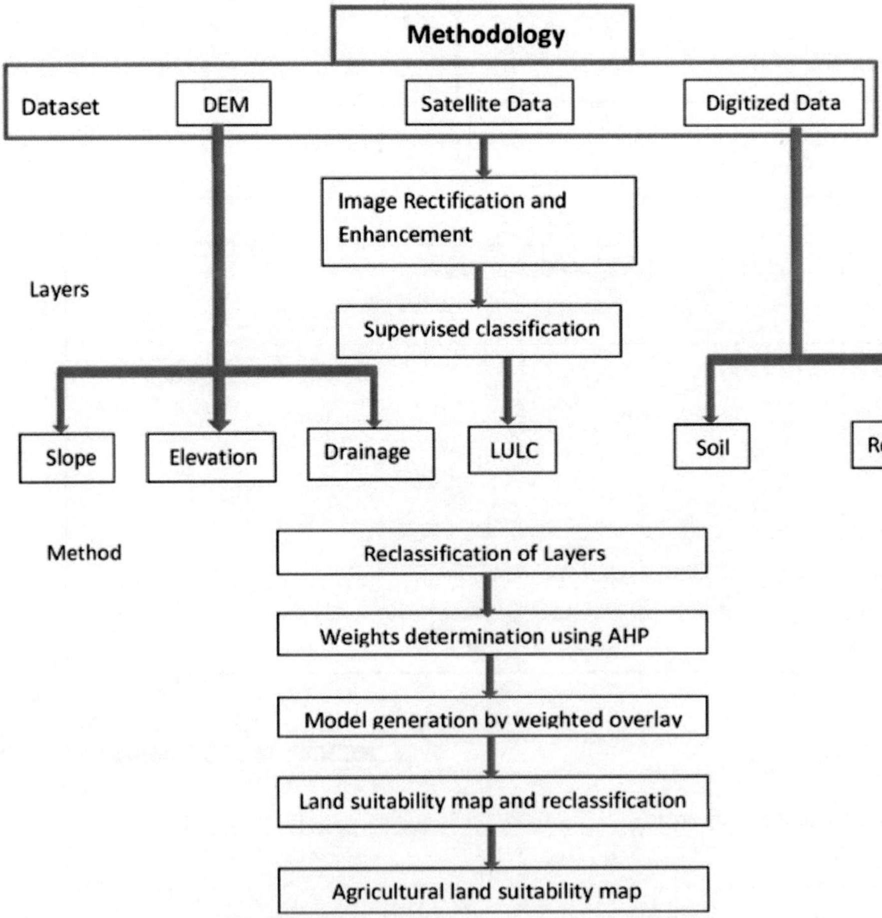

Fig. 11.2 Methodological framework of present study

Table 11.1 Datasets used in the study

Description of the datasets	
Satellite data	*Remarks*
Sentinel	LULC (tiles, T45RUJ, T45RUH, 29/12/2018)
Shuttle radar topography mission (DEM)	Slope degree
	Elevation
	Drainage and river basin
	Drainage density
Website	*Remarks*
USGS Earth Explorer	Satellite data (sentinel)
Indian Meteorological Department	Rainfall map
Central Ground Water Board	Soil type map
DACNET	Soil pH map
DIVA-GIS	Road network map
Census of India	Population dynamics and socioeconomic status of the district
United Nation Food and Agriculture Organization (FAO)	Assess global food security aspects
Department of Agriculture, Govt. of Bihar	Assess district profile and factors
Customized Rainfall Information System (CRIS)	Annual average rainfall

Source: Prepared by the authors

analysis. The shuttle radar topography mission (SRTM) data sources are used globally for DEM data elevation information. The slope map of Nalanda district is obtained with the help of SRTM-DEM (30m) spatial resolution data acquired from USGS Earth Explorer. Further, the slope is a layer obtained from the spatial function of ArcGIS 10.1 software using Spatial Analyst Toolbox. Generally, with the increased amount of slope, the agricultural suitability is reduced gradually as the water-holding capacity is decreased (Figs. 11.3 and 11.4).

Soil Texture

As we all know, the soil is the most dominant and very essential component present on the planet Earth. Soil is considered as a natural body, which developed under the natural forces acting on it and environmental medium too for plant growth. The physical and chemical properties of the soil determine the selection and distribution of the crops over any region. The soil texture map is obtained from the IIRS website for the Nalanda district. The thematic layer for soil texture parameters obtained converts based on their value fields in ArcGIS 10.1 and is further classified into seven classes (Ayehu & Besufekad, 2015). In the spatial context, it is deep or shallow, rich or poor, but everywhere, it consists of organic matters, minerals, air, and water. Therefore, soil texture identification and mitigation are an important aspect of this research (Fig. 11.5).

Soil pH

Soil pH is an appropriate indicator of soil's suitability for plant growth. The soil pH map for Bihar state is extracted from the cereal systems initiatives for south Asia. Geo-referenced shape file creation and geocoding were done for the Nalanda district. Then the obtained map is further classified into eight classes ranging from 6 to 9 (Fig. 11.6).

Rainfall

This district comes under the moderate rainfall zone and major source of irrigation through the canals and streams. The rainfall data of this district is extracted by the IMD station and plotted to the map. The district is divided into five categories ranging from 940 mm to 995 mm for the year 2018. There is a positive established relation between rainfall amount and agricultural production of any region, which is applicable in this study area too (Fig. 11.7, Table 11.2).

Drainage and River Basin

Drainage pattern ensures the appropriate soil aeration and reduces soil nutrient loss because of runoff in the form of erosion. Total area in the context of drainage is best suited for agricultural practices except the hilly region and some plain areas. The three major river basins are Mohane, Jirayan, and Kumbhar. High drainage density ensures optimum supply of water through irrigation practices, which ultimately help the land to become highly suitable for agriculture (Fig. 11.8).

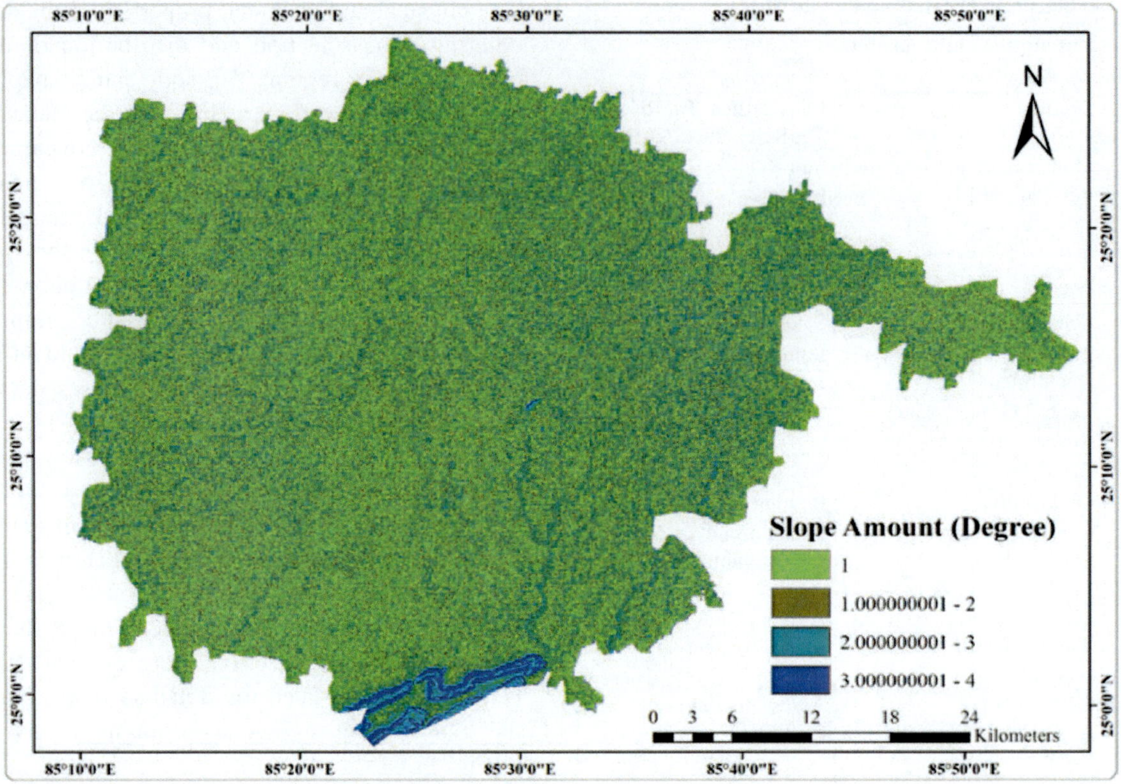

Fig. 11.3 Slope amount (in degree) map of the study area

Drainage Density

It is defined as channel length per unit watershed area in km per sq. km. It can be obtained by the total length of all streams and rivers in a drainage basin divided by the total area of the drainage basin. Drainage density plays an important role in surface runoff processes, concentration, and sedimentation load with respect to water balance in a drainage basin. Basically, the permeability of rocks has greater influence on the presence of drainage density. Clay and shale are the examples of the impermeable rocks, which produce higher surface runoff. In addition to this, humid regions have greater drainage density texture than the semiarid region(Fig. 11.9).

Distance to Major Road

The road network is obtained from the Nalanda district website and further classified into three major categories such as major district road (MDR), national highway (NH), and state highway (SH) (Fig. 11.10).

Road network plays very crucial role in transporting the cultivated crops into the nearest local markets.

Land Use Land Cover

The societal growth is the function of social and economic development by the use of available land resources. The LULC maps play very crucial and significant role in the utilization aspects of resources available in any particular area. In addition, LULC maps were important key elements in management, planning, and monitoring at regional, local, and national levels. This type of paradigm provides brief information in relation to the land utilization patterns, which helps in the formation of policies and development programs. In this study, sentinel satellite image data for the Nalanda district is obtained from USGS Earth Explorer agency. The diverse land use land cover maps were generated using sentinel data using image classification techniques through ERDAS IMAGINE software. Supervised classification

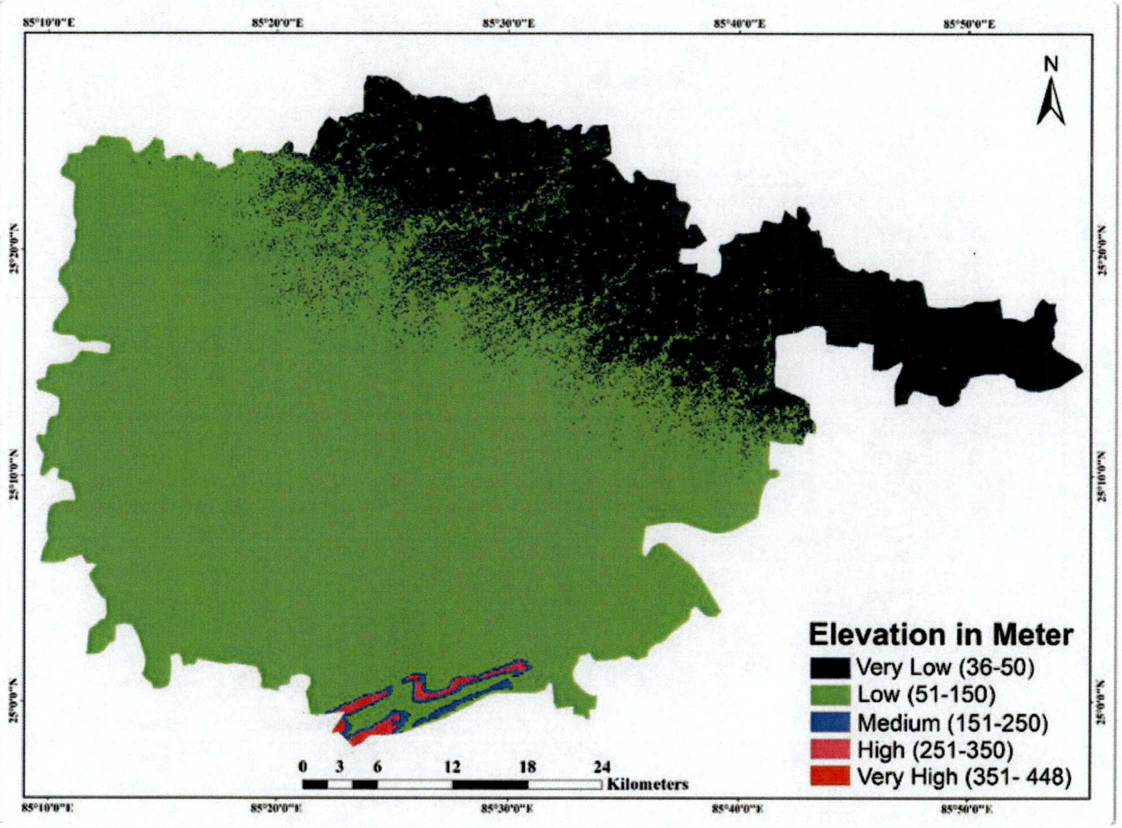

Fig. 11.4 Elevation map of the study area

technique is used in this study to obtain detailed LULC map (Fig. 11.11).

Weightage Estimation Criteria

Thomas L. Saaty (1980) developed the pairwise comparison technique called the AHP model for decision-making, which was based on expert's opinions for assigning the weightage over the various parameters. In the decision-making procedure, each factors/parameters are compared with other factors/parameters with respect to its importance and influence on the basis of Saaty scale, which comprises of rank from 1 to 9 (Yohannes & Soromessa, 2018). Origination of the AHP model starts with the first step to construct the hierarchy of the parameters. In this study area, we have taken seven parameters and grouped them into clusters and further interdependence of parameters for each cluster in design. The importance of each parameter in the form of ranks to each class has been evaluated based on expert opinion acquired from the designed questionnaires (Saaty, 2004) (Balasubramanian, 2017). For an instance, 1 is given for the equally importance and 9 for extremely importance to the parameters. Finally, priority weightage can be calculated using pair-wise comparison matrix and eigenvector values using this formula:

$$\lambda_{\max} = \sum_{j=1}^{n} aij \frac{Wi}{Wj} \tag{11.1}$$

where w_i is the sum of pair-wise comparison, and n is the size of matrix.

The consistency ratio (CR) is computed to check the consistency level of comparison. The sum of weightage criteria should be equal to 1. The CR ratio of 0.1 or less value is being considered as reasonable level of consistency (Saaty, 1980) (Table 11.3).

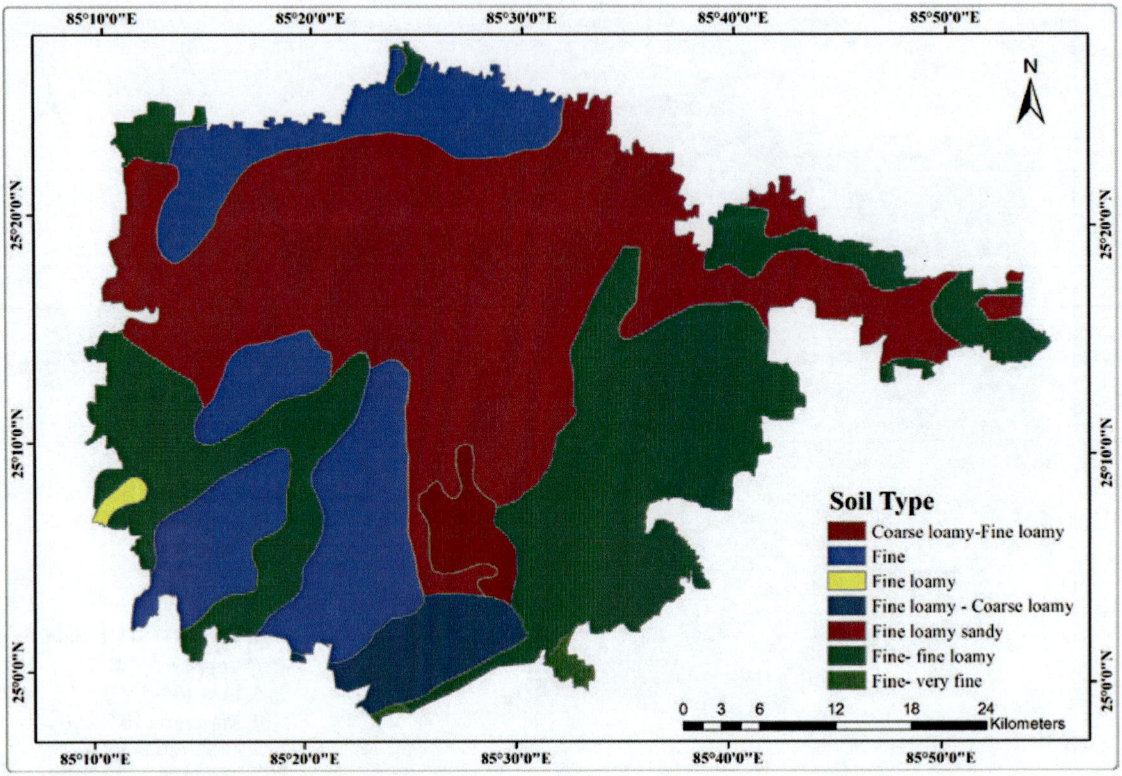

Fig. 11.5 Soil type map of the study area

The consistency index (CI) may be computed through this formula:

$$CI = \frac{\lambda_{\max} - n}{n - 1} \tag{11.2}$$

where $\lambda_{\max}$ is the highest eigenvalue of pairwise comparison matrix, and n is the number of classes.

Then consistency ratio is computed as

$$\text{CR} = \frac{\text{CI}}{\text{RI}} \tag{11.3}$$

where RI is average value or ratio index of CI for a random value based on Saaty's Table (1980).

Weightage Aggregation

In this study area, weightage overlay analysis method was used to aggregate the criteria and weighted criteria for land suitability analysis. The obtained map is further classified into specified number of criteria, which is our final suitability map for the Nalanda district. The formula is given below (Table 11.4):

$$S = \sum_{i=0}^{n} W \times X \tag{11.4}$$

where S indicates suitability, W indicates weight of factor, and X indicates to rank of factor.

Results and Discussion

Delimitation of fertile areas is a most important process toward the successful operation of the development of a sustainable agriculture. Land suitability analysis was performed by using the AHP method through assigning different weights to all parameters.

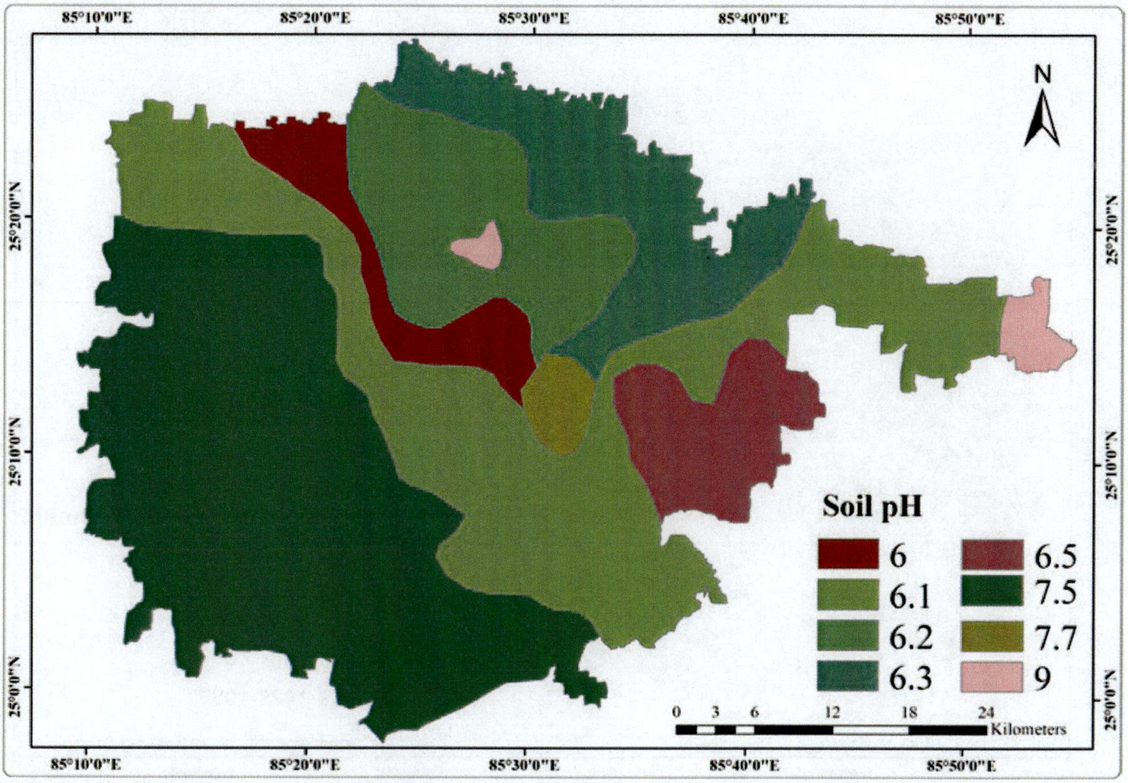

Fig. 11.6 Soil pH map of the study area

Spatial Outlooks of Different Input Parameters

Overall, the topographic elevation is indicated to flat surface, which is reasonably good for developing agricultural practices. Slope is slightly high in the southern part of the study area, which belongs to hilly areas under Rajgir hills. Most areas of the district are flat, which is reasonably well for developing agricultural practices. The highest degree of slope lies in the hilly areas of the Rajgir region named as "Rajgir hill," which is not appropriate for growing crops because of surface runoff of the rainfall water. Following the rainfall pattern, it is clear that there is an increasing trend of rainfall from northwest to southeast direction. Only some parts adjoining to Rajgir hills received annual rainfall above 1000 mm. High drainage density is found in the central part of the study area, which is covered by three major river basins, namely, *Mohane*, *Jirayan,* and *Kumbhar* Rivers. The rest of the region belongs to less drainage density zone in the district. Making assemblage with the slope and topography with rainfall and drainage facility over the study area, there are different soil types in the whole region with varying chemical condition mainly in terms of soil pH. Most of the areas have suitable pH range except some part of Chandi, Bihar Sharif, and Sarmera blocks, which is in extreme red color. The maximum area is covered by fine, loamy, sandy soil, which is surrounded by fine-fine, sandy, and fine soil in Nalanda district. Soil pH is high (7.5 and above) in the western part of the district and in varying condition in some pockets all over the region. From land use perspectives, seven LULC classes are identified in terms of forest, open forest, riverbed, standing crops, built-up, water bodies, and mature crop. Standing crops and matured crop cumulatively cover above 80% of the study area. Almost 10% of the land are used for built-up activities, which are allocated in the central and western part of the study area (Table 11.5).

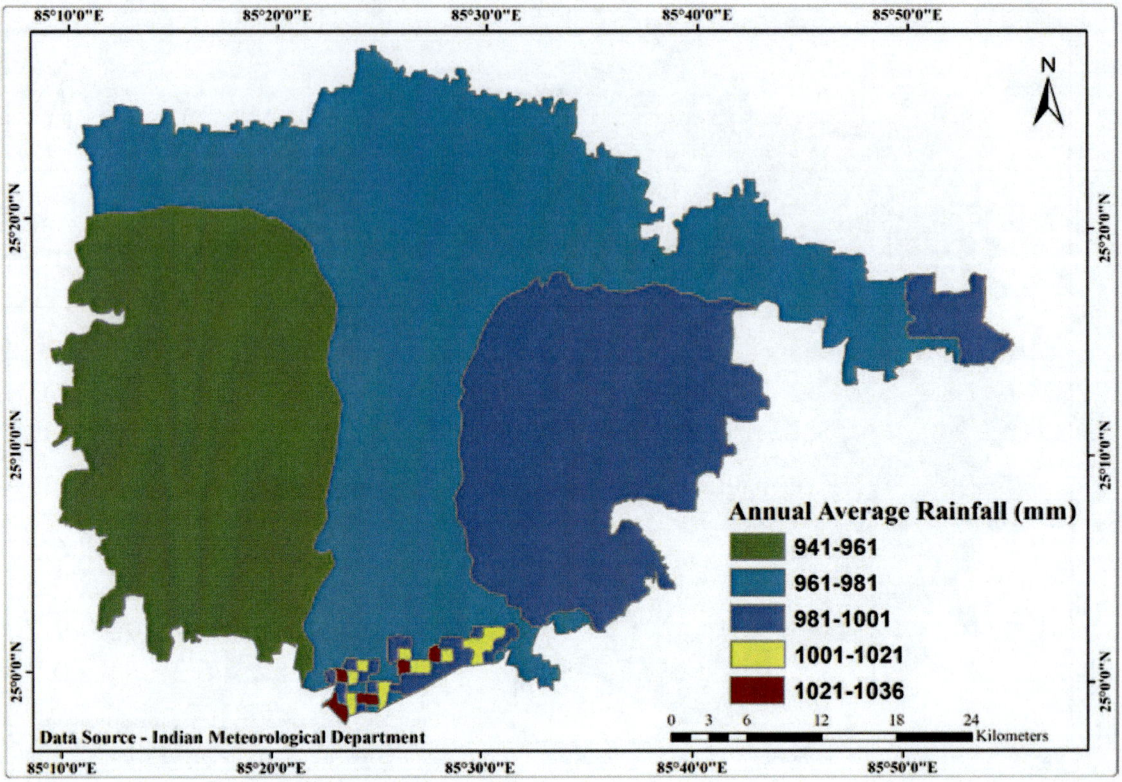

Fig. 11.7 Annual average rainfall map of the study area

Table 11.2 Annual average rainfall in Nalanda district

Year	R/F	%DEP
2014	87.8083	12
2015	55.6583	−26.25
2016	86.675	−26.75
2017	58.3917	−67.75
2018	53.2583	−58.583

Source: Customized Rainfall Information System (CRIS)
Note: (1) The district rainfall in millimeters (R/F) shown below is the arithmetic averages of rainfall of stations under the district
(2) % dep. are the departures of rainfall from the long period averages of rainfall for the district

Agricultural Land Suitability

All the factors found significant in the present study are chalked out from the Spearman rho Rank test (99% significant level). Agricultural suitability of this district has been calculated from slope, rainfall, soil type, drainage density, road density, and LULC. Following the selection of all the six indices, each of them has been categorized based on their respective characteristics, and class weights were assigned (Table 11.4) in a systematic and scientific way that will lead us toward greater accuracy level. Thereafter, AHP has been applied for weight assignment of different themes, and pair-wise comparison matrix has been used for normalizing the weights (Table 11.6) (Saaty 1990). Considering the goodness as a statistical technique for spatially circulated variables, AHP has been used with great precision. Following assignments of weights, six layers containing six different themes have been put into the GIS environment, and weighted sum overlay has been executed to unearth the spatial variation in agricultural suitability. Spatial output in terms of suitability status map exhibits the facts that there was a variation in agricultural suitability, and its coverage was not uniform over the study area. Considering all these parameters with their respective weightages, agricultural land suitability has been evaluated in this district, among which 1144.6 sq. km (49%) and 787.3 sq.

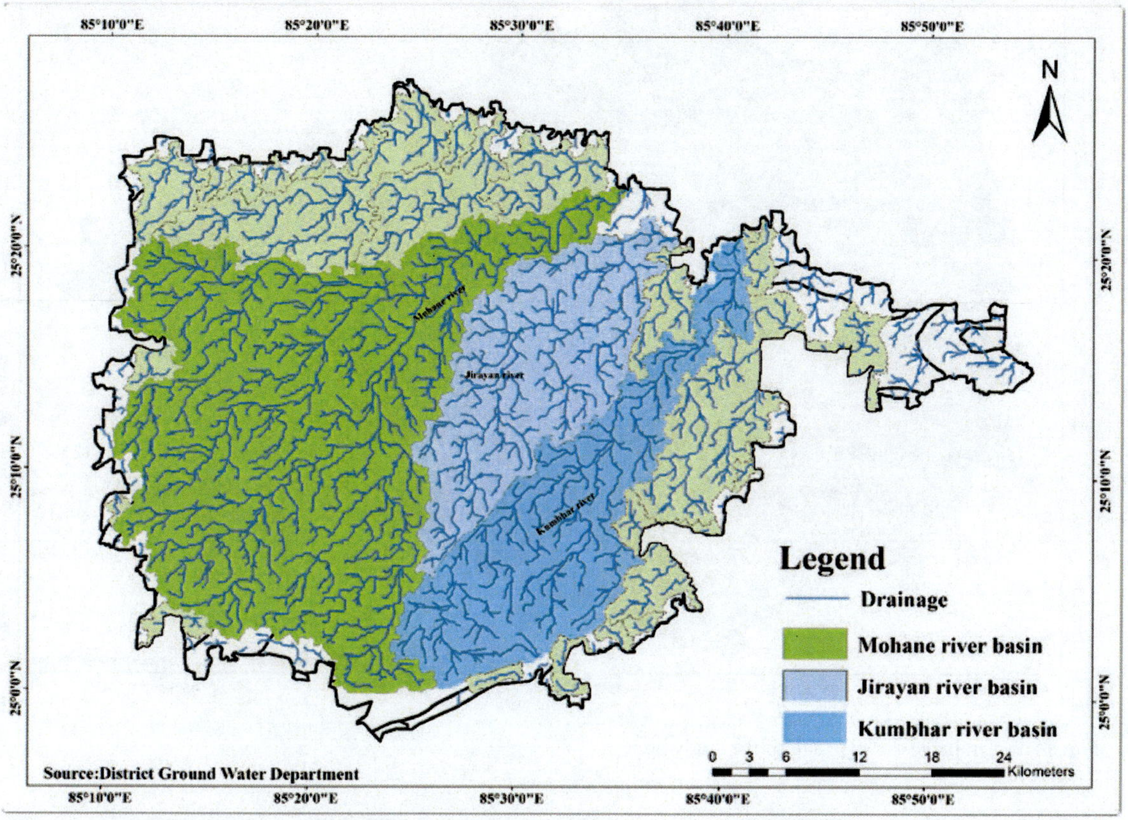

Fig. 11.8 Drainage and river basin map of the study area

km (34%) area mainly in the western and central part of the district covering the location of *Chandi, Tharthari, Parbalpur, Ben, Islampur, and Hilsa* belongs to highly suitable zone and moderately suitable zone, respectively. Marginally suitable zone along with vulnerable zones is found in the eastern part adjoining to *Silao, Bihar Sharif, Bind, Asthawan, and Giriak* locations. This region is mainly covered by sparse vegetation along with built-up land comparison to the rest of the part of the district (Table 11.6).

Any assessment on determination of soil quality comprises a wide-ranging evaluation process in directive to recognize the appropriate agricultural land location. The identified suitable soils must contain the key nutrients for basic plant nutrition in order to reduce organic, mineral, chemical inputs to minimize economic cost and environmental damages. The land in the Nalanda district mainly consists of Kewal kind soil. Kewal soil consists of hard clay and which is very suitable for rice cultivation. This kind of soil has the capacity to retain moisture and is most suitable for rabbi crops. On uplands near the farmstead, "doras" soil is also found that is generally used for growing potatoes and vegetables. The stage of groundwater development of the study area is only 64.9%, which implies that there is a huge scope for further groundwater development. However, the artificial groundwater recharge techniques including rainwater harvesting should be taken to augment the groundwater reserve. Exploitation of groundwater in regions of the study area can be done through both shallow and deep tube wells. Generally, small and marginal farmers frequently use shallow tube wells, and farmers' cooperative can choose for high-discharge deep tube wells. The organic farming techniques were also successfully adopted after being introduced in 2010, and the villages of

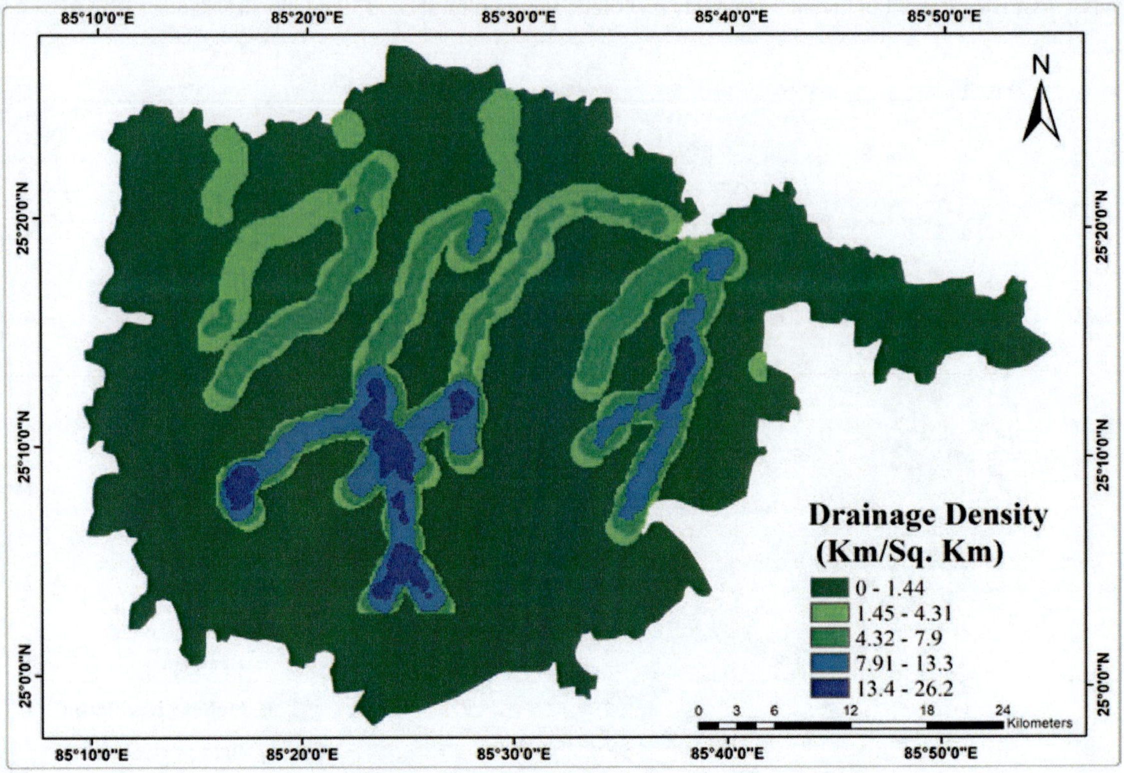

Fig. 11.9 Drainage density map of the study area

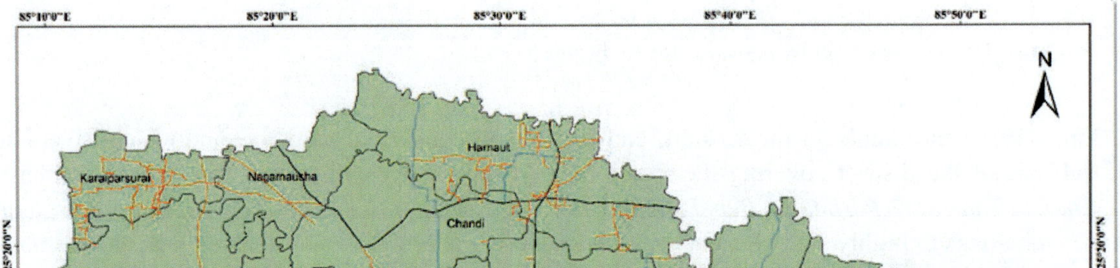

Fig. 11.10 Distance to major road map of the study area

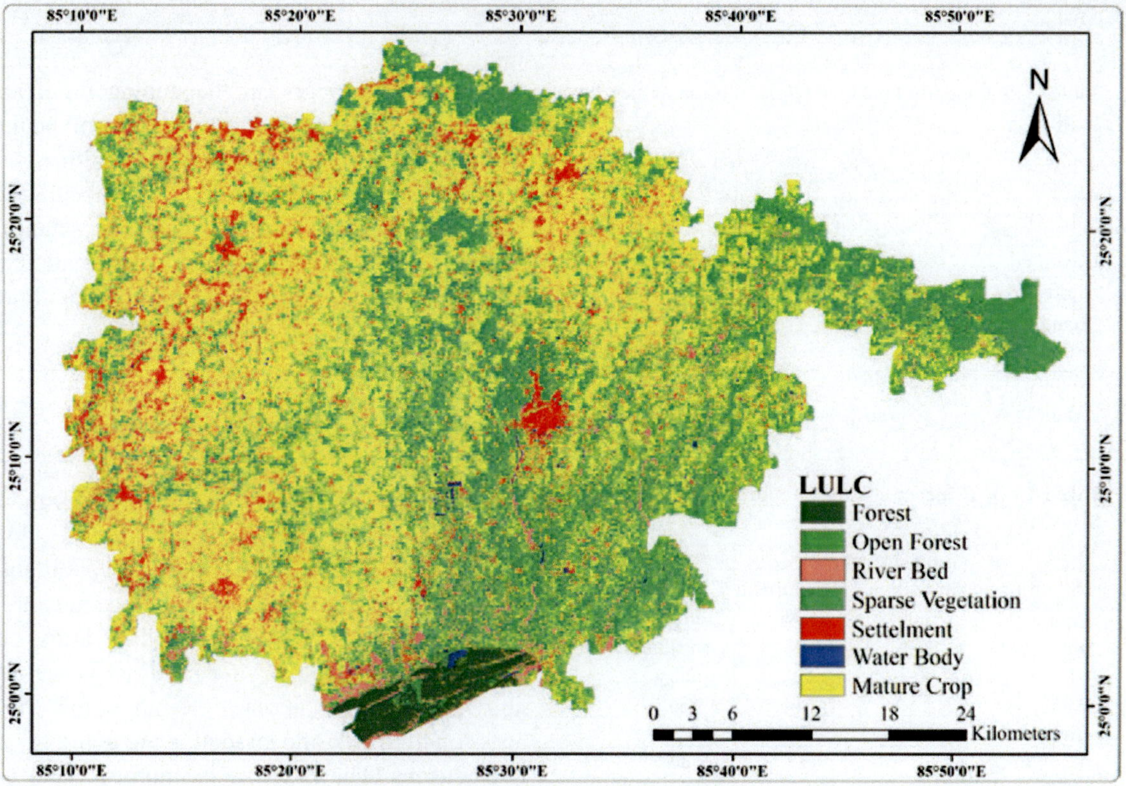

Fig. 11.11 Land use land cover map of the study area

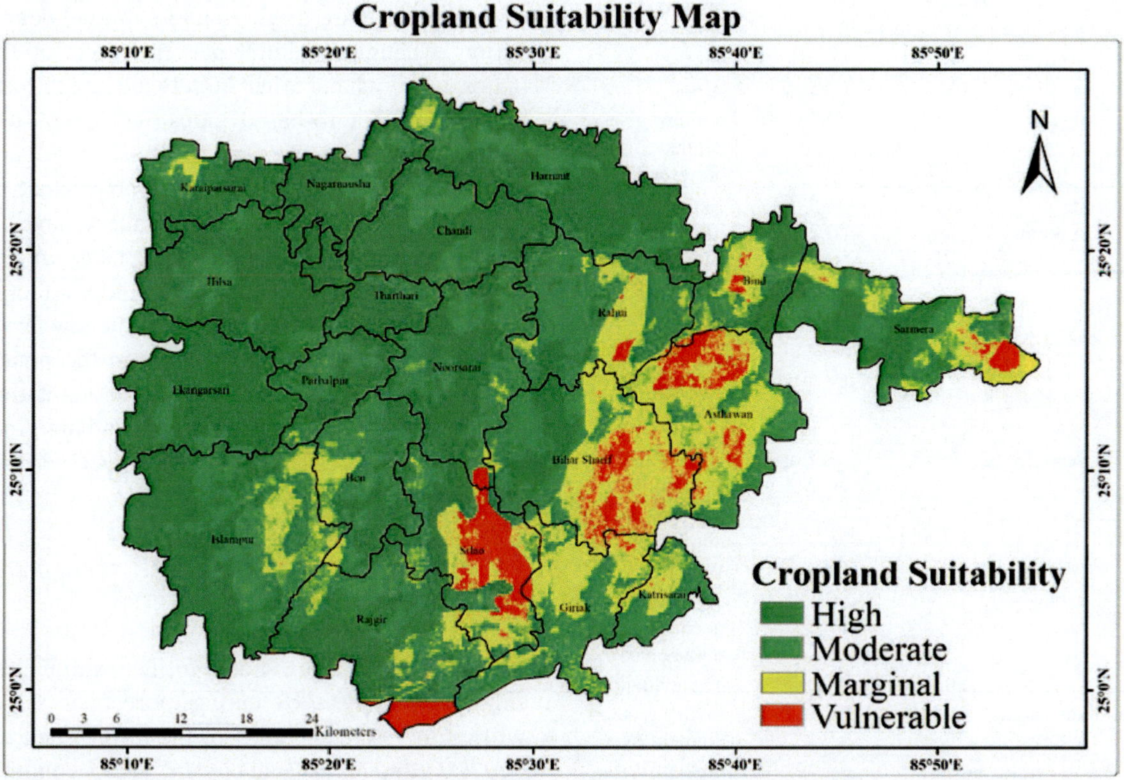

Fig. 11.12 Cropland suitability map of the study area

Table 11.5 Land use land covers distribution in the study area

Classes	Area in hectares	Total % distribution
Forest	3885.86	1.67
Open forest	803.95	0.35
Riverbed	7666.55	3.29
Built-up	21985.80	9.46
Standing crops	68634.80	29.52
Mature crops	128746.00	55.38
Water body	744.81	0.32

Table 11.3 Scale of relative importance (according to Saaty, 1977, 1980)

Definition	Intensity of importance	Reciprocal	Explanation
Equal importance	1	1/1 or 1.00	Two activities contribute equally.
Weak, slight importance	2	1/2 or 0.500	When compromise needed between two
Moderate importance	3	1/3 or 0.333	Experience and judgment moderately favor one element over another.
Moderate plus	4	1/4 or 0.250	When compromise needed between two
Strong importance	5	1/5 or 0.200	Experience and judgment strongly favor one element over another.
Strong plus	6	1/6 or 0.167	When compromise needed between two
Very strong	7	1/7 or 0.143	One element is favored very strongly over another.
Very very strong	8	1/8 or 0.125	When compromise needed between two
Extreme importance	9	1/9 or 0.111	The evidence favoring one element over another is of the highest possible order.

Source: Saaty, 1980

Nalanda and farmers are depending on it and contributing part in the way to sustain soil health, environmental stability, and human health.

Hence, land suitability like AHP techniques were showing good results by using weightage of the indicators and subindicators to estimate their influence in the selective rain-fed agricultural farming areas (Fig. 11.12).

Conclusion

It is estimated that from the overall assessment, 49% of the land area is highly suitable, 34% moderately suitable, 11% marginally suitable, and 6% high risk of degradation of the study area. According to their mutual suitability ratios, i.e., highly, moderately, marginal, and risk or vulnerable suitable areas, it was determined that those areas on high risk of degradation are mainly from two reasons. The first one is improper land use results in land degradation as well as decline in agricultural productivity, and the other is excessive use of chemical fertilizers and changing rainfall pattern are driving factors to land degradation. Changing rainfall pattern is one of the major concerns that affect largely the agricultural practices and agro-based industries. Crop suitability analysis involves major decisions at various levels to build up an efficient crop production system. Additionally, decision-makers have to resolve the land degradation problems by implementing new farming techniques to those affected areas. Adopting organic farming by the few farmers in the district set new records in the potato production. Such type of agricultural adaptation ultimately acts as a path to the small and marginal farmers to beyond the threshold level without affecting the environment.

Recommendation

Research in the future should further examine the suitability analysis by taking more factors that will improve the accuracy of the result, such as-CACO3 content, organic content, socioeconomic condition of the farmers, etc. The extensive study

Table 11.4 AHP matrix table used in the study area

Layers	Soil type	Road	Rainfall	Line density	Slope	LULC
Soil type	1	0.5	1.111	0.5	3.003	2
Road	2	1	0.5	0.5	3.003	2
Rainfall	0.9	2	1	2	2	2
Line density	2	2	0.5	1	3.333	2
Slope	0.333	0.333	0.5	0.3	1	0.5
LULC	0.5	0.5	0.5	0.5	2	1
C.R. = 0.058						

Source: Prepared by the authors

Table 11.6 Area under different land suitability classes

Value	Area (in Km^2)
Highly suitable	1144.699
Moderately suitable	787.3326
Marginally suitable	269.4996
Vulnerable	136.926

Source: Prepared by the authors

should be done in the due course of time using GIS and its proponents because such study has the ability to show greater availability of information from different sources to map land suitability and help in many levels to management authorities. In addition, the work should be on intensifying the analysis to the other part of the Bihar state and examining the impact of climate changes in all conditions in the region on the suitability maps of different parameters, which should be central to a better understanding of the impact on agricultural products and the sustainable use of land resources at the various local and regional levels.

References

Adeniyi, P. O. (1993). Integration of remote sensing and GIS for agricultural resource management in Nigeria. *EARSeL. Advances in Remote Sensing, 2*(3), 6–21.

Akpoti, K., Kabo-bah, A. T., & Zwart, S. J. (2019). Agricultural land suitability analysis: State-of-the-art and outlooks for integration of climate change analysis. *Agricultural Systems, 173*, 172–208.

Archaeological Site of Nalanda Mahavihara at Nalanda, Bihar (a). https://whc.uneso.org/en/list/1502] Accessed on 24 March 2019.

Archaeological Site of Nalanda Mahavihara at Nalanda, Bihar (b), Creative Commons Attribution-ShareAlike 3.0 IGO. License cc-BY-SA IGO 3.0. Accessed on 12 April 2019.

Ayehu, G. T., & Besufekad, S. A. (2015). Land suitability analysis for rice production: A GIS based multi-criteria decision approach. *American Journal of Geographic Information System, 4*(3), 95–104.

Bagchi, P., & Rao, R. P. (1992). Decision making in mergers: An application of the analytic hierarchy process. *Managerial and Decision Economics, 13*(2), 91–99.

Balasubramanian, A. (2017). Bihar-At a Glance. https://doi.org/10.13140/RG.2.2.32377.77929.

Bandyopadhyay, S., Jaiswal, R. K., Hegde, V. S., & Jayaraman, V. (2009). Assessment of land suitability potentials for agriculture using a remote sensing and GIS based approach. *International Journal of Remote Sensing, 30*(4), 879–895.

Baniya, N. (2008). Land suitability evaluation using GIS for vegetable crops in Kathmandu valley/Nepal.

Bera, S., Ahmad, M., & Suman, S. (2017). Land suitability analysis for agricultural crop using remote sensing and GIS-A case study of Purulia District. *IJSRD-International Journal for Scientific Research & Development, 5*(06), 999–1024.

Böhme, M., Pinker, I., Arsanti, I., Chuong, H. V., Baniya, N., & Shevchenko, Y. (2011). 14. Improving productivity in horticultural crops. *Horticulture and Livelihood Security*, 203.

Bozdağ, A., Yavuz, F., & Günay, A. S. (2016). AHP and GIS based land suitability analysis for Cihanbeyli (Turkey) County. *Environmental Earth Sciences, 75*(9), 813.

Bunruamkaew, K., & Murayama, Y. (2012). Land use and natural resources planning for sustainable ecotourism using GIS in Surat Thani, Thailand. *Sustainability, 4*(3), 412–429.

Census-India. (2011a). Census of India. Office of the director of census operations, Bihar, Patna. Online http://censusindia.gov.in/2011provresults/data_files/bihar/Provisional%20Population%20Totals%202011-Bihar.pdf. Accessed on 26 May 2019.

Census-India. (2011b). Census of India. Office of the director of census operations, Bihar, Patna [Online-www.census2011.co.in]. Accessed on 12 May 2019.

Chandio, I. A., Abd Nasir, B. M., WanYusof, K. B., Talpur, M. A. H., Balogun, A. L., & Lawal, D. U. (2013). GIS-based analytic hierarchy process as a multicriteria decision analysis instrument: A review. *Arabian Journal of Geosciences, 6*(8), 3059–3066.

Cinelli, M., Coles, S. R., & Kirwan, K. (2014). Analysis of the potentials of multi criteria decision analysis methods to conduct sustainability assessment.

Dengiz, O., Özyazici, M. A., & Sağlam, M. (2015). Multi-criteria assessment and geostatistical approach for determination of rice growing suitability sites in Gokirmak catchment. *Paddy and Water Environment, 13*(1), 1–10.

Department of agriculture, Govt of Bihar [online-krishi.bih.nic.in/statistics/rice-2017-18]. Accessed on: 26 April 2019.

Department of agriculture, Govt. of Bihar [online-www.Krishi.bih.nic.in]. Accessed on 20 May 2019.

Duc, T. T. (2006, November). Using GIS and AHP technique for land-use suitability analysis. In *International symposium on geoinformatics for spatial infrastructure development in earth and allied sciences* (pp. 1–6).

Elsheikh, R., Shariff, A. R. B. M., Amiri, F., Ahmad, N. B., Balasundram, S. K., & Soom, M. A. M. (2013). Agriculture Land Suitability Evaluator (ALSE): A decision and planning support tool for tropical and subtropical crops. *Computers and Electronics in Agriculture, 93*, 98–110.

Ennaji, W., Barakat, A., El Baghdadi, M., Oumenskou, H., Aadraoui, M., Karroum, L. A., & Hilali, A. (2018). GIS-based multi-criteria land suitability analysis for sustainable agriculture in the northeast area of Tadla plain (Morocco). *Journal of Earth System Science, 127*(6), 79.

FAO (1976). A framework for land evaluation. FAO soils bulletin 52, FAO, Rome, 79p. [Outlines the basic principles of the FAO approach to land evaluation and land use planning].

FAO (1983). Guidelines: Land Evaluation for Rain fed Agriculture. FAO Soils Bulletin, 52, FAO, Rome, 237 p. [An overview of practical applications of the FAO approach to the evaluation of land for rain fed cropping; provides also data on crop growth requirements for a wide range of crops].

FAO (1985). Guidelines: Land Evaluation for Irrigated Agriculture. FAO Soils Bulletin 55, FAO, Rome, 290 p. Provides the evaluation methodology and gives crop growth requirements for irrigated crops.

Feizizadeh, B., & Blaschke, T. (2013). Land suitability analysis for Tabriz County, Iran: A multi-criteria evaluation approach using GIS. *Journal of Environmental Planning and Management, 56*(1), 1–23.

Goswami, S. B., Matin, S., Aruna, S., & Bairagi, G. D. (2012). A review: The application of remote sensing, GIS and GPS in precision agriculture. *Journal of Advanced Technology and Engineering Research, 2*, 50–54.

Guo, L. S., & He, Y. S. (1999). Integrated multi-criterion decision model: A case study for the allocation of facilities in Chinese agriculture. *Journal of Agricultural Engineering Research, 73*(1), 87–94.

Hossain, M. S., Chowdhury, S. R., Das, N. G., & Rahaman, M. M. (2007). Multi-criteria evaluation approach to GIS-based land-suitability classification for tilapia farming in Bangladesh. *Aquaculture International, 15*(6), 425–443.

Jamil, M., Sahana, M., & Sajjad, H. (2018). Crop suitability analysis in the Bijnor District, UP, using geospatial tools and fuzzy analytical hierarchy process. *Agricultural Research, 7*(4), 506–522.

Kalogirou, S. (2002). Expert systems and GIS: An application of land suitability evaluation. *Computers, Environment and Urban Systems, 26*(2–3), 89–112.

Kumar, J. (1986). *Land use analysis: A case study of Nalanda district*. Inter India Publications.

Kumar, N., Singh, S. K., Mishra, V. N., Obi Reddy, G. P., Bajpai, R. K., & Saxena, R. R. (2018). Soil suitability evaluation for cotton using analytical hierarchic process. *International Journal of Chemical Studies, 6*(4), 1570–1576.

Madhusudhan, L. (2015). Agriculture role on Indian economy. *Business and Economics Journal, 6*(4), 1.

Malczewski, J. (1999). *GIS and multicriteria decision analysis*. John Wiley & Sons.

Malczewski, J. (2006). Ordered weighted averaging with fuzzy quantifiers: GIS-based multicriteria evaluation for land-use suitability analysis. *International Journal of Applied Earth Observation and Geoinformation, 8*(4), 270–277.

Mau-Crimmins, T., de Steiguer, J. E., & Dennis, D. (2005). AHP as a means for improving public participation: A pre–post experiment with university students. *Forest Policy and Economics, 7*(4), 501–514.

Mazahreh, S., Bsoul, M., & Hamoor, D. A. (2019). GIS approach for assessment of land suitability for different land use alternatives in semi-arid environment in Jordan: Case study (Al Gadeer Alabyad-Mafraq). *Information Processing in Agriculture, 6*(1), 91–108.

Pramanik, M. K. (2016). Site suitability analysis for agricultural land use of Darjeeling district using AHP and GIS techniques. *Modeling Earth Systems and Environment, 2*(2), 56.

Qiu, F., Chastain, B., Zhou, Y., Zhang, C., & Sridharan, H. (2014). Modeling land suitability/capability using fuzzy evaluation. *GeoJournal, 79*(2), 167–182.

Rasheed, S., & Venugopal, K. (2009). Land suitability assessment for selected crops in Vellore district based on agro-ecological characterisation. *Journal of the Indian Society of Remote Sensing, 37*(4), 615–629.

Saaty, T. L. (2004). Decision making—The analytic hierarchy and network processes (AHP/ANP). *Journal of Systems Science and Systems Engineering, 13*(1), 1–35.

Saaty, T. L. (2008). Decision making with the analytic hierarchy process. *International Journal of Services Sciences, 1*(1), 83–98.

Sajjad, H., & Nasreen, I. (2016). Assessing farm-level agricultural sustainability using site-specific indicators and sustainable livelihood security index: Evidence from Vaishali district, India. *Community Development, 47*(5), 602–619.

Sarkar, A., Ghosh, A., & Banik, P. (2014). Multi-criteria land evaluation for suitability analysis of wheat: A case study of a watershed in eastern plateau region, India. *Geo-spatial Information Science, 17*(2), 119–128.

Shearer, K. S., & Xiang, W. N. (2009). Representing multiple voices in landscape planning: a land suitability assessment study for a park land-banking program in Concord, North Carolina, USA. *Landscape and Urban Planning, 93*(2), 111–122.

Singh, S. K., Mishra, S., & Rai. (2015). A. K. A Geographical Review of Land Use Pattern in Mirzapur District.

Sys, C., Van Ranst, E., Debaveye, J., & Beernaert, F. (1993). Land evaluation. Part III: Crop requirements. Agricultural Publications n° 7, GADC, Brussels, Belgium, 1993, 191 p.

Yohannes, H., & Soromessa, T. (2018). Land suitability assessment for major crops by using GIS-based multi-criteria approach in Andit Tid watershed, Ethiopia. *Cogent Food & Agriculture, 4*(1), 1470481.

12 Adaptation of the Agri-Based Society to Environmental Changes in Thar Desert

Amal Kar

Introduction

The Thar, or the Great Indian Sand Desert, is a distinctive but integral part of the arid western part of Indian subcontinent, covering about 290000 sq. km area in the western part of Rajasthan state and adjoining parts of Haryana state in India, as well as the contiguous parts of Sind and Punjab provinces in Pakistan. The eastern limit of the desert lies along the moisture availability index (Im) of minus 66.67, which runs approximately along the western foothills of the degraded but the world's oldest surviving mountain, the Aravalli hill ranges. In the north, the desert boundary runs roughly along the wide underfit valley of the Ghaggar River and the old sand dunes to its north, while the western boundary runs along the Indus River up to Sukkur and then along the Nara River up to the Great Rann of Kachchh, which forms the desert's southern boundary (Fig. 12.1). The mean annual rainfall varies from about 500 mm along the eastern margin (coefficient of variation, cv, about 35%) to about 100 mm along the western margin (cv about 80%). More than 85% of it is received from SW monsoon during June to September (in ~28 days along the eastern fringe to <10 days in the westernmost part). The mean potential evapotranspiration (PET) is ~1700 mm, (~1500 in the east and ~2000 in the west). Thus, there is a distinct gradient in the rainfall and wetness from east to west. By contrast, the wind strength is the highest in the southwest and declines toward the east and the northeast. As a result of such spatial variability in climate, fluvial activities in the desert gradually decline from east to west. Organized stream network is found mainly in the southeastern part of the desert. Aeolian processes are more efficient in the western part, especially due to stronger wind in a dry sandy environment. Aeolian landforms constitute the most dominant landscape features.

For an agriculture-based rural community to survive in such a harsh environment, the realities to be faced include the erratic behavior of the monsoon rainfall, extremes of temperature, recurrent drought of moderate to severe intensities, scarcity of water over long distances, challenges of food and fodder production from the dominantly sandy and impoverished soils, and dust and sandstorms during summer months that either blow away the top soil or bury under moving sand the productive croplands, infrastructures, etc.

To understand the changing paradigm of adaptation to the desert's environment and the challenges it throws up, we propose to provide first an overview of the region's landscape vulnerability as inherited from its evolutionary history. This will be followed by an analysis of human

A. Kar (✉)
Formerly of Central Arid Zone Research Institute, Jodhpur, Rajasthan, India

A. R. Siddiqui, A. Sahay (eds.), *Climate Change, Disaster and Adaptations*, Sustainable

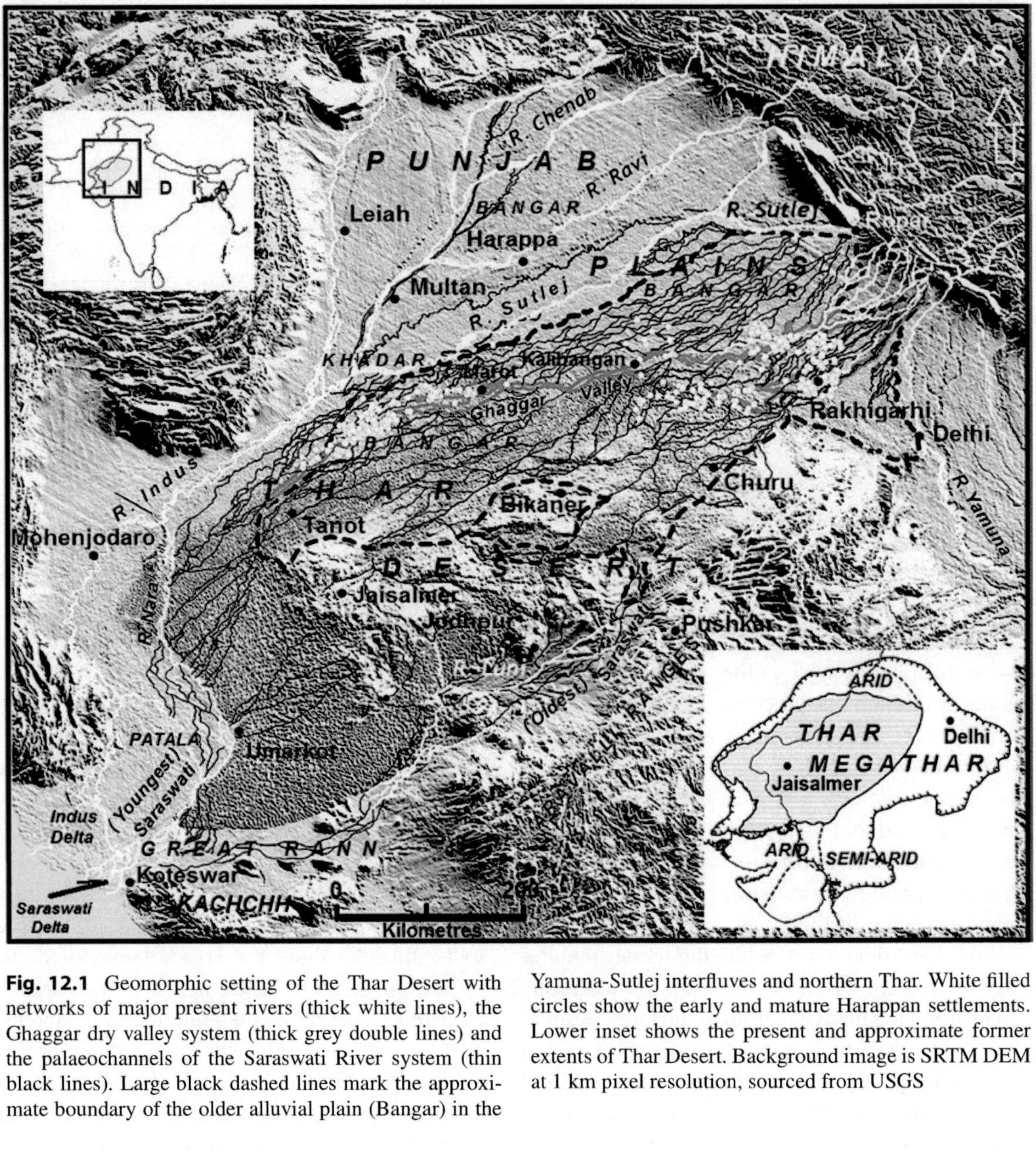

Fig. 12.1 Geomorphic setting of the Thar Desert with networks of major present rivers (thick white lines), the Ghaggar dry valley system (thick grey double lines) and the palaeochannels of the Saraswati River system (thin black lines). Large black dashed lines mark the approximate boundary of the older alluvial plain (Bangar) in the Yamuna-Sutlej interfluves and northern Thar. White filled circles show the early and mature Harappan settlements. Lower inset shows the present and approximate former extents of Thar Desert. Background image is SRTM DEM at 1 km pixel resolution, sourced from USGS

enterprises on the landscape, beginning with the stage of developing skills for adaptation to the environment and for maintaining a symbiotic relationship with the nature to the stage of mastering the technological skills to exploit and manipulate the natural resources for economic gains but not taking adequate care for the vulnerable land system. We then discuss how nature's response to the exploitative enterprises has started to harm the very economies for which the ecosystem health was ignored and how the stakeholders are trying to learn from the losses and what the researchers think about dovetailing the technological innovations with traditional wisdom for sustainable solutions to the emerging problems.

Landforms and Land Use

Landscape Development

Studies so far have revealed that the Aravalli hill ranges in the east of the desert were formed as a fold mountain some 1000 million years (Ma) ago during the Mesoproterozoic era of the Precambrian age and along the western edge of an ancient Rodinia supercontinent, somewhere near the South Pole (Ashwal et al., 2013). Detached from that continent since then, and drifting northward over the next millions of years after being welded to the Indian shield, this old landform assemblage gradually expanded through volcanic activities and subaerial denudation process to its west, which led to the formation of a vast pediplaned surface with Cambrian sandstone and limestone beds and then a sequence of sedimentary beds of the Mesozoic and Tertiary periods further west. These geological formations provided the basement for Quaternary landform development. The oldest surviving Quaternary deposits over large parts of the desert are a mixture of sand and gravel with calcreted aeolian sand sheet deposits (Sinharoy et al., 1998; Wadhawan, 2018; Kar & Kumar, 2020), which suggest complementary roles of both the fluvial and the aeolian processes rather than exclusive role of either of the two (Kar, 2014a). Chronostratigraphic studies on deep sedimentary sequences have revealed alternate phases of wet and dry climate throughout the late-Quaternary period, which led to the relative dominance of fluvial and aeolian processes over time. Broadly, the major wet phases with significant fluvial activities took place at about 127 thousand years before present (i.e., 127 ka), 80–90 ka, 55 ka, 30 ka, and 7–6 ka, while the major aeolian phases occurred at about 155 ka, 115–100 ka, 85–75 ka, 65–60 ka, 55–50 ka, 30–25 ka, 14–7 ka, and 5.0–3.5 ka. The latest phase of sand mobility in the desert started around 0.3 ka, when the rates of dune mobility and sand accretion began to surpass the geological rates due to increasing human pressure (Kar et al., 2001, 2004; Singhvi & Kar, 2004). Contrary to traditional views, the periods of maximum aridity in our desert were high during the transition from a drier to a wetter climate (i.e., from glacial to interglacial phase) when the SW monsoon winds gained strength and started to move northward across the intertropical convergence zone (ITCZ), and the time-averaged rainfall was yet to peak. Remnants of sand dunes beyond the Thar mark the boundary of a Mega-Thar, when the desert extended up to the vicinity of Ludhiana, Delhi, Agra, Gwalior, and Vadodara (Goudie et al., 1973; Kar, 2014a; Fig. 12.1, inset).

The wettest peak of monsoon rains during the late Quaternary period was at about 127 ka, when the desert had an integrated drainage system from the Himalayas. Visual interpretation of the first-generation Landsat MSS and TM images revealed the approximate flow path of two major Himalayan streams, the Sutlej and the Tons-Yamuna, through the presently dune-infested areas of the desert, which were named as the Proto-Saraswati and the Proto-Drishadvati (Ghose et al., 1979; Kar & Ghose, 1984). Recently, the author carried out digital processing of several satellite sensor data, which helped him to map numerous giant valleys and palaeochannels through the Punjab Plains and the Thar Desert, several linearly arranged evaporite beds along them, and a high-level older alluvial plain (the Bangar) that was built by the system (Kar, 2020; Fig. 12.1). Due to avulsion in the Himalayan foothills, a slow uparching of the Yamuna-Sutlej Interfluves (YSI), and secular variations in climate, the two rivers gradually shifted away (Fig. 12.2). By about 80 ka, the Sutlej opened a new west-flowing channel along the presently underfit valley of the Ghaggar-Hakra-Nara River to flow into the Arabian Sea. The Tons-Yamuna explored a westward course through another valley that was subsequently occupied by the Chautang River. Till about 25 ka, the oscillating Sutlej used to feed the Ghaggar-Hakra-Nara valley at different points, after which an extremely dry glacial climate, the Last Glacial Maximum (LGM, 25–18 ka), severely restricted the stream discharge (Singh et al., 2017). When the SW monsoon recovered after the LGM, the Ghaggar valley began to receive discharge mainly from the small Siwalik streams (Clift et al., 2012; Singh et al., 2017; Singh & Sinha, 2019). The Sutlej

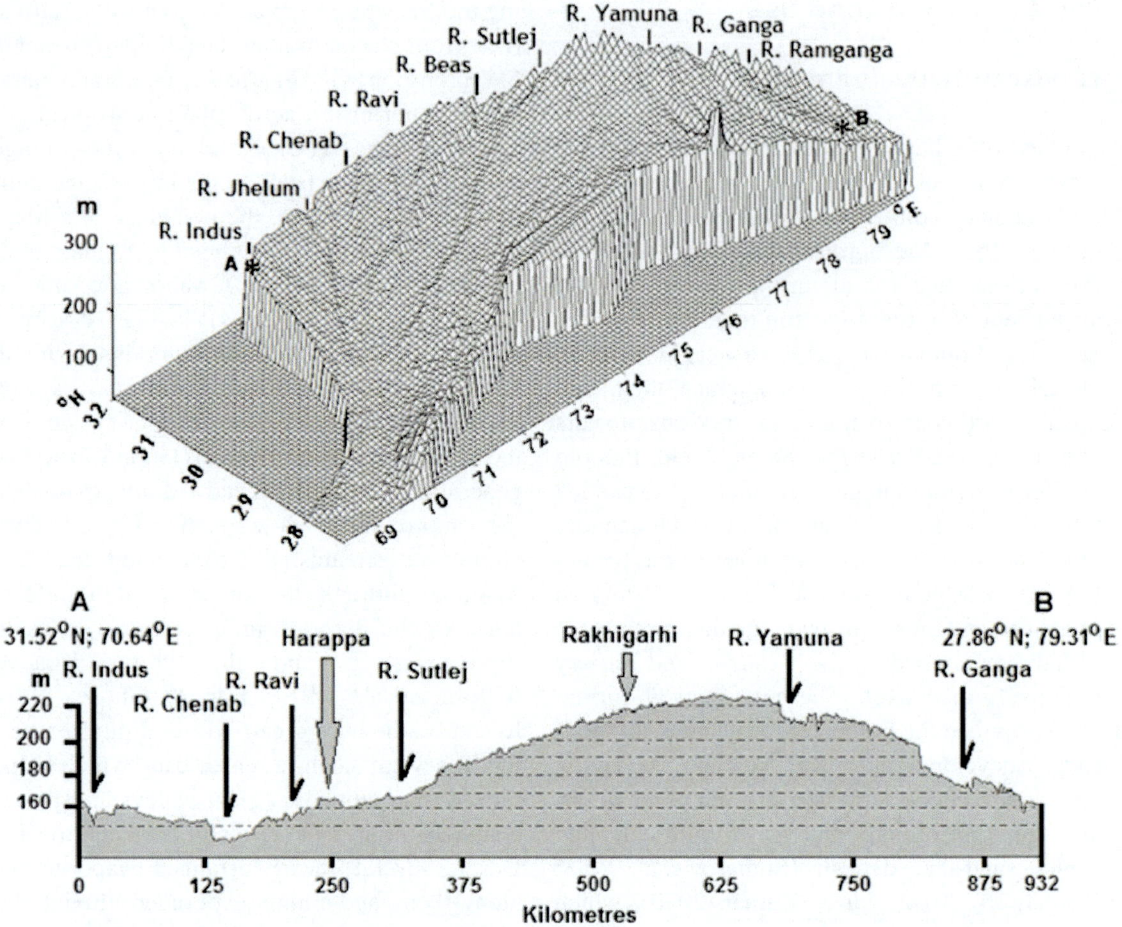

Fig. 12.2. Wire-fence diagram of the relative altitudinal positions of major stream valleys in the sub-Himalayan Punjab Plains till the Ghaggar-Hakra valley and the Panjnad's confluence with the Indus (top), and surface profile across the terrain between points A and B in the wire-fence diagram (below)

had by then abandoned the Ghaggar part of the valley to flow through the Beas valley in the west, and the combined flow used to discharge through the Hakra-Nara segment. Finally, it joined the Indus through the Panjnad in subrecent time. The Tons-Yamuna got pirated by a tributary of the Ganga by about 75 ka and turned eastward. From its apical cone at Ropar to the apical base along the Ghaggar-Hakra valley, the Sutlej shifted east to west by about 500 km, while from its apical cone near Jagadhri to the apical base along the Chautang, the Yamuna shifted by about 350 km.

The Ghaggar valley was identified by James Tod (1832) as the "lost river of the desert." James Rennell (1788) first mapped the river's "supposed" former course to the sea through the western part of the desert, while Survey of India mapped the misfit valley accurately by the 1800s. C. F. Oldham (1874, as anonymous) recognized it as an abandoned course of the Sutlej and of the Saraswati River of the Rig Veda. Although the Sutlej and the Tons-Yamuna have left the desert, their vestiges as thick alluvium of the "Bangar," as evaporites along the dry valleys, and as potable groundwater below the sand have endowed the region with a rich hoard of natural resources. Exploitation of groundwater in the dune-covered areas of north-central Thar has started to show the alignment of some of the deeply buried palaeochannels of the Saraswati system which were interpreted from satellite data (Fig. 12.3). The configuration of the valleys below the thick

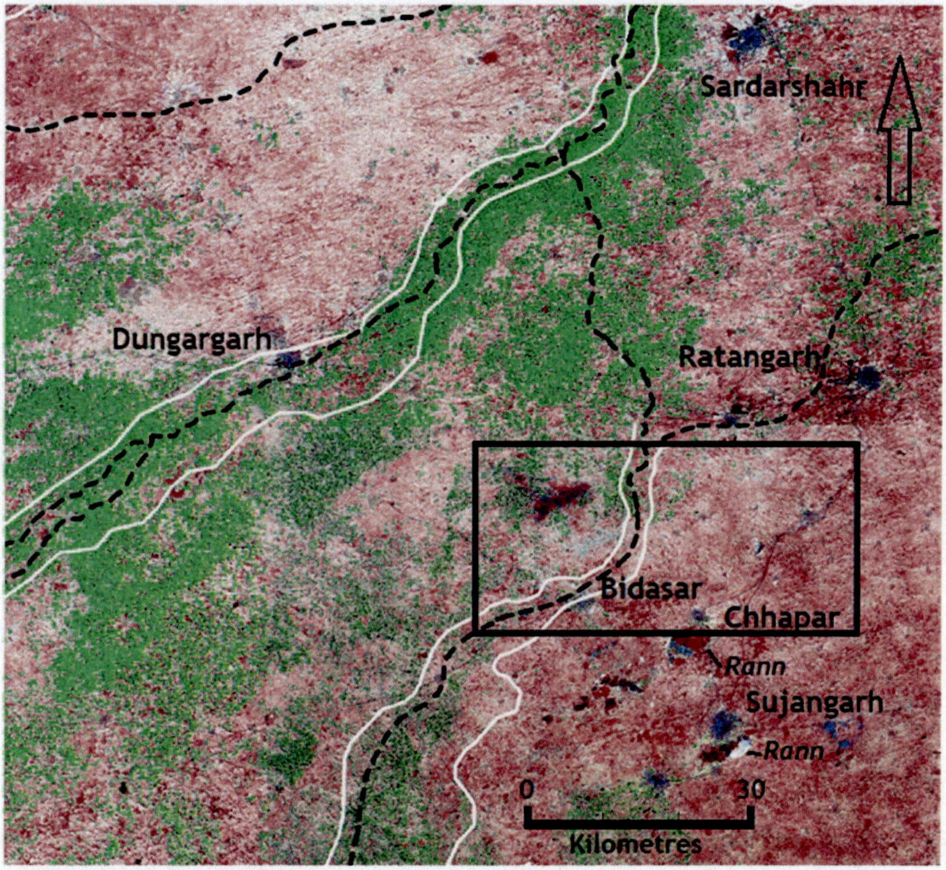

Fig. 12.3. Sentinel-2 FCC mosaic of 30 September 2017 and 29 March 2018 for the Sardarshahr-Sujangarh-Dungargarh area in northern Thar, showing groundwater-irrigated croplands (dark green) along the buried stream valleys (white lines) and palaeochannels (black dashed lines) of the Saraswati River system that were interpreted from digital processing of the SRTM1km and AW3D30m DEM data with ASTER, Landsat ETM+ and Sentinel-2 optical sensor data. Inset within the map shows the area for Fig. 12.4

aeolian sand could be partly deciphered from digital processing of the DEM data (Fig. 12.4; Kar, 2020) and from geophysical depth sounding (Fig. 12.5; Kar and Shukla 2000).

Landforms and Their Use Potentials

The major fluvial landform sequence in the desert is hills and uplands-pediments and pavements-colluvial plains-older alluvial plains-younger alluvial plains-riverbeds. The major aeolian landforms are sand sheets, sandy hummocks, and sand dunes of different types. A number of salt playas (locally called the Ranns) have also been formed in the desert due to insufficient drainage, deflation, and neotectonic activities. The above landform units subsume in them a recurring pattern of soil, water, natural vegetation, etc., which make them suitable for exploitation of certain natural resources, but the units have also their specific inherent vulnerability to pressures and degradation. The hills, the pediments, and the uplands are good for water conservation and rangeland development, while the younger alluvial plains usually host shallow potable groundwater and well-drained, nutrient-rich soils suitable for double cropping. The deep colluvial and older alluvial plains provide a niche environment for cropping, especially along the major palaeochannels, and also host calcrete, gypsum, placer deposits, salt, etc. Most old, tall sand

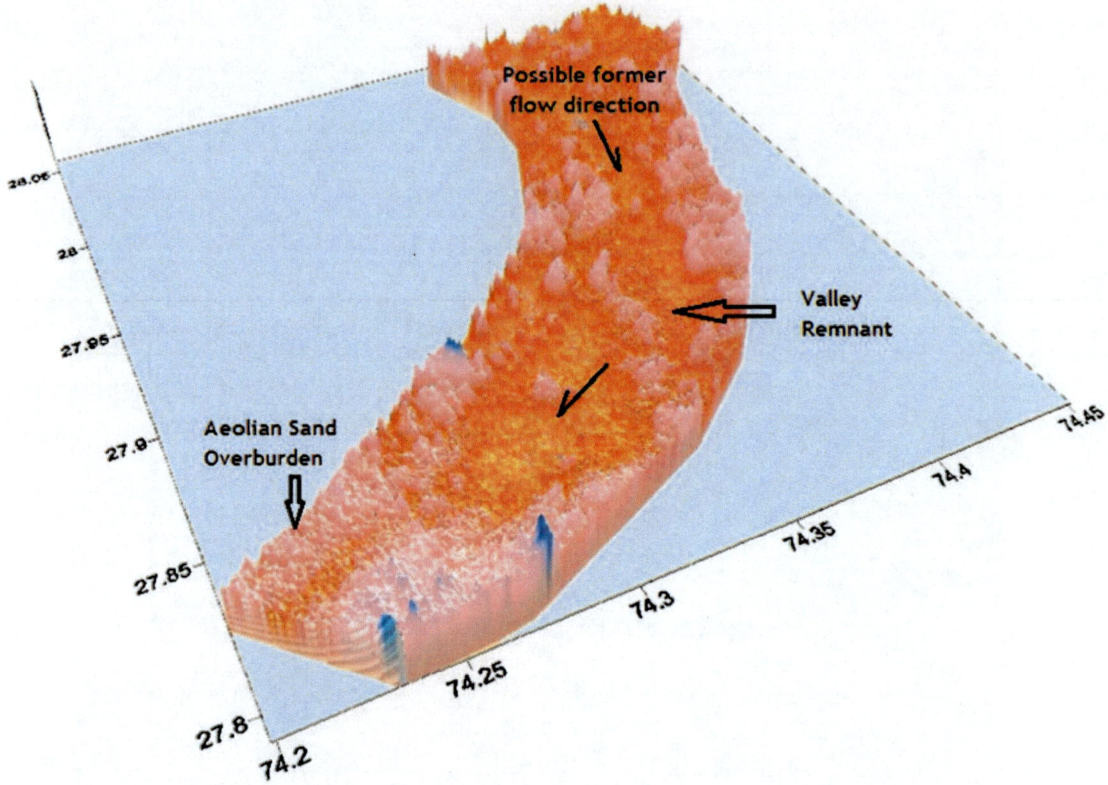

Fig. 12.4. Approximate configuration of a giant stream valley below the tall sand dunes through the vicinity of Dhadhru, Bidasar and Sandwa, as revealed from digital processing of SRTM1km and AW3D30m DEM data

dunes commonly grow a variety of shrubs and grasses, providing rich grazing resources and fuel wood. Many old dunes have also developed an incipient soil profile below the transient surface sand. Often, a mulch or a fine silt layer below the surface hosts colonies of cyanobacteria and other microbial communities, which capture and retain moisture during the rains, and slowly release it to the shallow plant roots during the dry months (Rao et al., 2015). This enormously helps the growth of shrubs and grasses on the dunes, which in turn aid in localized carbon and nutrient cycling through root and litter decay. Because of such traits, the farmers used to grow traditionally gram and other winter crops on these sand dunes without any irrigation. Many such traits are possessed by the different landforms that have been gainfully used by farmers for centuries.

Traditional Agriculture

Agricultural Land Use During the Harappan Period

Agriculture in the Thar Desert and its fringes dates back to the Harappan period. Geoarchaeological studies now reveal that the Harappan civilization flourished in the Ghaggar-Hakra valley from about 6 ka to 3.9 ka (Sarkar et al. 2016; Petrie et al., 2017; Neogi et al., 2019). Our mapping of the Harappan settlements in the context of the terrain reveals that most settlements were situated away from the large active rivers, and on a 10 m or higher surface of the older alluvium, the Bangar (Kar, 2020; Fig. 12.6). The Harappa township was also built away from an active stream, the Ravi, and at least 5 m above the Khadar surface, along an abandoned oxbow lake (Schuldenrein et al., 2007). When the early Harappans (~6.0–4.5 ka) began to populate the

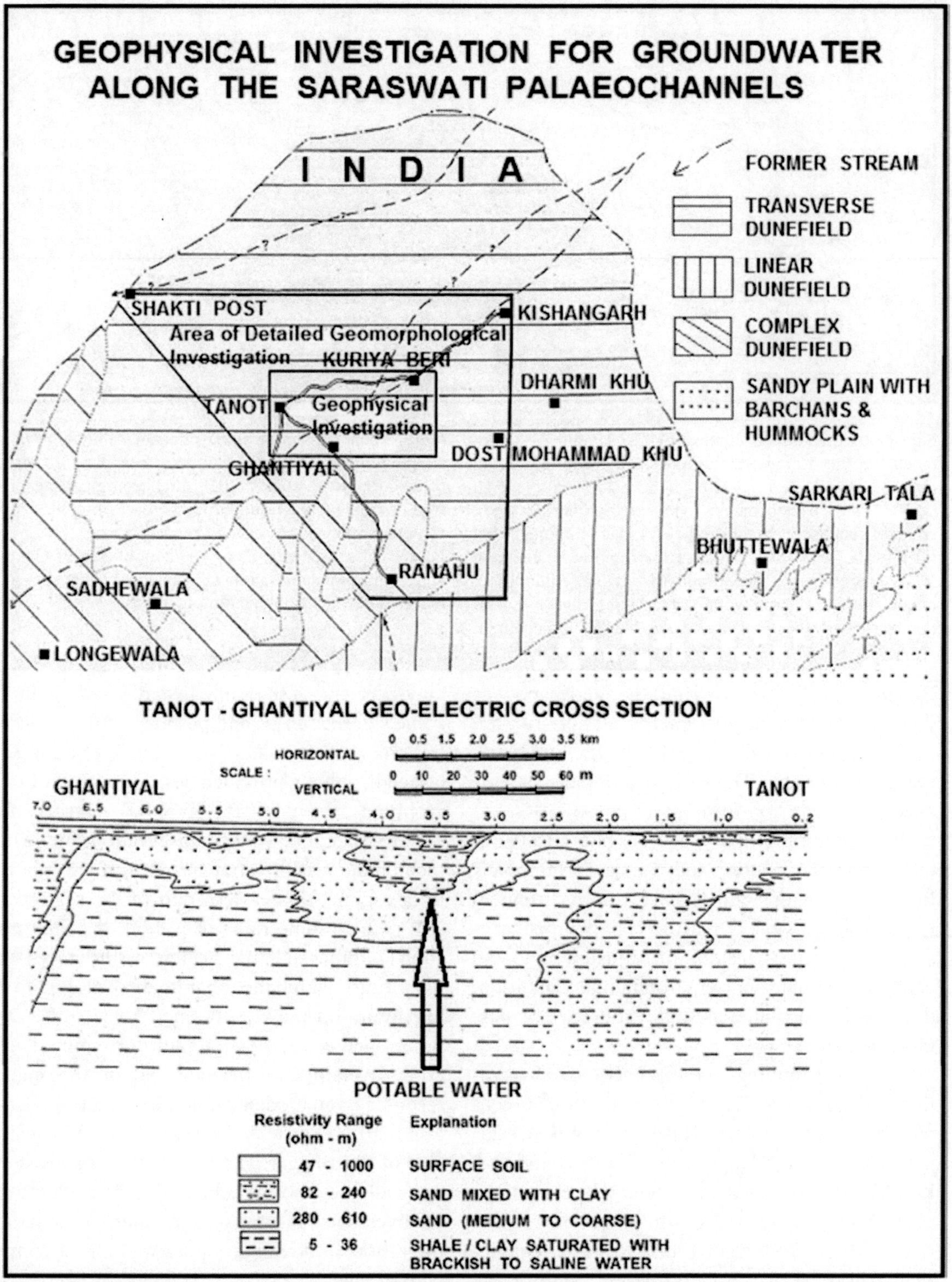

Fig. 12.5. Identification of a potable aquifer through geophysical depth sounding along a Saraswati palaeochannel below the high sand dunes in the western part of Jaisalmer district

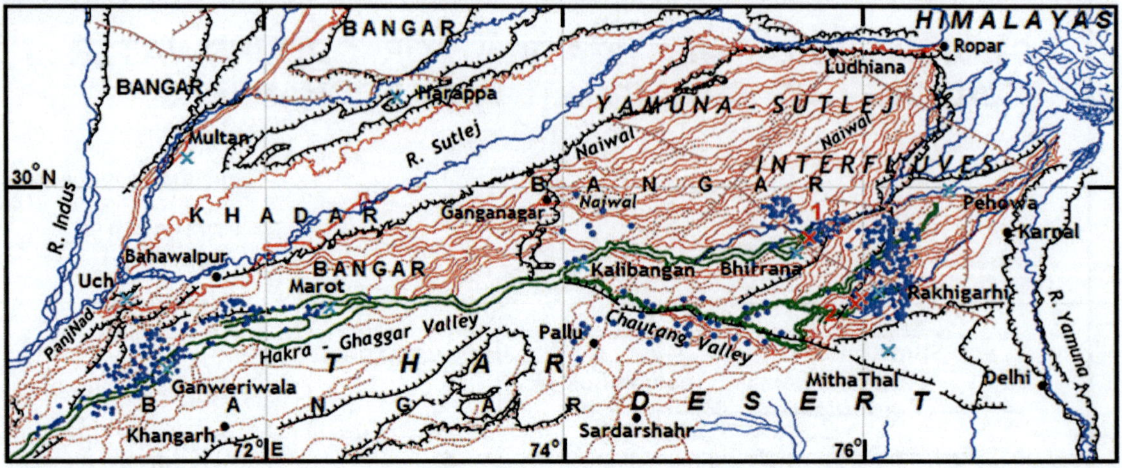

Fig. 12.6. Landscape context of the Harappan settlements in the Yamuna-Sutlej interfluves and adjoining parts of Thar Desert, showing large palaeo-valleys (red double line), palaeochannels (red single dashed line), recently abandoned stream courses (red single line), wide misfit valleys of the Ghaggar-Hakra and the Chautang (dark green double line), modern streams (blue line), major alluvial scarps within and along the Bangar's boundary with the Khadar surface (black stippled line), major lineaments controlling channel behaviour (brown stippled line), early and mature Harappan settlements (blue dot), Neogi et al.'s (2019) archaeological sites, Burj (red cross, 1) and Masudpur (red cross, 2), other ancient settlements (cyan cross), and modern towns (black dot). Paucity of settlements along the central part of Ghaggar valley and abundance along the Chautang valley and some Naiwals have hydrological implications

Bangar area of northern Thar, the landscape was dominated by some low-energy streams within the wide, abandoned valleys left by the Sutlej and the Tons-Yamuna. The early Harappans experienced the transition from a wet phase (10–6 ka) to the beginning of a dry, aeolian phase (5.0–3.5 ka), while the mature and the late Harappans lived in a strong dry phase. Archaeological research has revealed that the Harappan success story is basically a story of adjustment to the changing climate and environmental condition, utilizing the available natural resources in a sustainable manner.

While examining the archaeological sediments at the early Harappan sites of Burj (occupation ~4.8–4.5 ka) and Masudpur (~4.4–4.1 ka) near Rakhigarhi, Neogi et al. (2019) found 8–6 ka old paleosols over the ancient levees of the paleo-Sutlej and the paleo-Tons-Yamuna, respectively. These soils were found to have organic-rich A horizon over a well-drained sandy loam, and there were ample evidences that the early Harappans used these levee soils for multiple cropping during both summer and winter/spring. Down the slope, the soil was found to be more clay-rich and less-drained, and so were fit for some summer crops and pasture, while the valley bottom was often marshy, waterlogged and salt-affected, which provided less opportunities for cropping but more for livestock rearing, fuel wood collection, water conservation, etc. Since the climate was then transiting from a wetter to a drier regime, this eastern part of the region was still getting some rains from both the SW monsoon (summer) and the western disturbance (winter/spring), which the inhabitants exploited to the maximum on the well-drained levee soils. Over time, such success with multi-cropping led to crop diversification, better use of the natural resources, and socioeconomic security (Bates et al., 2018; Petrie & Bates, 2017). The western part of the Bangar was not so well-disposed climatically, and so the inhabitants depended more on livestock rearing, pasture management, and migration, although cropping was also a component of land use.

Such environmental control was also noticed on the distribution pattern of the mature (~4.5–3.9 ka) and the late Harappan (~3.9–3.5 ka) settlements (Pawar et al., 2013; Petrie et al., 2018).

In the Bangar area of the Cholistan in the west, the early Harappan sites were hardly occupied by the mature and the late Harappans (Mughal, 1982) because the inhabitants were always in search of virgin or rested land that could assure them of some extra soil nutrients and good pasture. Traditionally, transhumance was a necessity in this zone to feed the livestock from the low-yielding soils and to maintain soil nutrient balance. Moving over to the Khadar nearby was risky for them as the Sutlej was joined there by two other mighty and oscillating rivers, the Ravi and the Chenab, which rendered the area marshy and extremely flood-prone.

By contrast, most Harappan sites in the east had mature and late Harappan cultural layers right over the early Harappan settlements (Green & Petrie, 2018). This was possibly of a memory-based impression that the successive generations of Harappans carried of the old levees being more productive for cropping and so needed to be exploited without transgressing through settlement construction. This was despite the realities of a gradually weakening SW monsoon that compelled the inhabitants to depend more on the erratic and meager winter/spring rains.

The inherited wisdom on crop diversification and slope- and soil-specific land uses helped the successive generations of Harappans to manage the scarce resources under the worsening climate and to take calculated risk for overall socioeconomic modulation (Madella & Fuller, 2006). In the meantime, diversification of skill to mineral processing and trading had begun to provide additional opportunities of livelihood in the east. As the smelting, manufacturing, and trading of different metals became the new livelihood sources in the foothills of the mineral-rich Aravalli hills, several settlements in the Mitathal-Rakhigarhi sector began to flourish (Prabhakar, 2013). Throughout the Harappan period, the villages and the semi-agrarian settlements in the hinterlands of the cities like Kalibangan, Rakhigarhi, Ganweriwala, Harappa, etc., not only supplied food, fodder, and fiber to those cities, but many also used to serve as the trading and manufacturing hubs and as conduits for metals, precious stones, and manufactured goods from the arid hinterlands to the larger townships (Kenoyer, 1995; Prabhakar, 2013).

Interesting details have also emerged on the crops grown during the Harappan period (Madella & Fuller, 2006; Fuller, 2011; Bates et al., 2018; Neogi et al., 2019). During the kharif season, several varieties of millets used to be grown widely on the sandy plains, with cotton as one of the major intercrops. Perhaps the millets were as popular a food item then as at present. Cotton, especially the traditional "Narma" variety, gradually became a famous export item from the region. Rice used to be grown in the marshy lowlands. During the rabi season, barley, wheat, and pulses were the most favored crops, but their area under production was perhaps much less than that of the millets during the summer. Rice, wheat, and barley were possibly consumed more by the affluent town dwellers. It follows that the dominant land use in the Bangar was kharif cropping. Rabi cropping was limited to very small areas. Interpretation of the Vedic literature suggested that animal husbandry was an integral part of agriculture in that distant past (Roy, 2009), and so land parcels were also allocated for animal grazing.

Land Use During Historical Period

In western Rajasthan, the early written records of land ownership, major occupations, crops grown, and their prices are found in some stone edicts and tablet inscriptions. According to one such edicts during the Pratihara rule (ca. 750–1018 AD; originally established in southeastern part of the region, with capital at Jalor, but expanding to as far as Bengal during its peak), the king was the owner of all the marshy and barren lands, woodlands, jungles, mines, and salt pits. The major occupations were trade and commerce, cropping, and pasture, but presumably all the three were practiced by a joint family to withstand the impacts of recurrent drought on agricultural production. Other edicts describe the cultivation of winter wheat (on conserved moisture), coarse millet, and pulses in the southeastern part (Sharma, 1966). Bharadwaj (1961) suggested that land tenure system in the region was developed from the ancient Aryan system. Cultivable and uncultivated land of any kind belonged to the Crown (Khalsa land), who would provide some land to his chieftains for agricultural development under some taxes or as grant. Gradually, the chieftains received ownership right (Jagirdari) on the developed land (some with irrigation facility) and appointed tenants to cultivate the land. The land rights, however, belonged mostly to the descendants and relatives of the chieftains, but with time, the cultivators also got right to their land (Pattadhari).

Till the middle of the nineteenth century, agriculture in the region was largely subsistence-oriented and tradition-bound. Records from the 1660s to the 1880s reveal that the per capita cropland gradually increased from 0.58 ha to 0.85 ha, and large parts were uncultivated scrublands. As cropping depended almost totally on the fluctuating monsoon rains (except in small patches in the north where irrigation facilities were introduced), there used to be large interannual variability in production. Technologies evolved slowly through traditional wisdom and as per the societal needs, especially to tackle rainfall variability, evapotranspiration loss, and sand control. Rainwater conservation in "tanka" (small underground tank), "jhalra" (large step well), "nadi" (pond), and "khadin" (a system of runoff conservation through embanking of valleys for winter cropping), growing of crops and plants that demanded limited water, animal husbandry, and animal migration were the normal practices (Dhir, 2003). Small village-level pastures (locally called the "oran," "gochar," and "bir") and forests used to be maintained as common property resources (CPRs) to feed the animals and for fuel wood collection, for which funds were used to be collected through donation or taxation (Jodha, 1985).

Keeping the land fallow was an important pillar of the system. It recognized that the sandy soils were poor in soil nutrients and that after a certain period of cropping, the soils would get fatigued and so had to be rested for crop yield in the subsequent seasons. Farmers used to intuitively decide about keeping a parcel of his land under "short fallow" (usually for 1 year) or "long fallow" (for 2–10 years). They also used to keep rotating the fallow areas within their landholding and practice animal grazing in the fallow lands.

Most individual rural households used to practice an integrated farming system (IFS), which involved several optional mixes of crops, trees, shrubs, grasses, and animal rearing, based mainly on the natural resources endowment of the given land and the capacity of the household (Bhati & Joshi, 2007). In the easternmost part with mean annual rainfall of 400–500 mm, the farmers used to prefer arable cropping with crop diversification, but also maintained trees within and along their crop fields for agroforestry, and practiced livestock rearing. Dependable rains during the crop-growth stages used to encourage the farmers to grow multiple crops during the kharif season (e.g., pearl millet, sorghum, mung bean (*Vigna radiata*), sesamum, etc.), not in a mix but in several rows or plots devoted to single crops.

In the 300–400 mm rainfall zone, more erratic rainfall used to compel the farmers to practice mixed cropping, in which a mixture of seeds belonging to cereals, pulses, oilseeds, etc. (e.g., pearl millet, mung bean, moth bean (*Phaseolus aconitifolius*), sesamum, castor) used to be sown or broadcast so that at least some of the crops would survive in case of rainfall deficiency. Agroforestry involved both trees and shrubs, and livestock rearing used to have more importance than in the wetter east. A rudimentary horticulture used to be practiced on some permanent pastures.

In the 100–300 mm rainfall zone, livestock rearing was the most important component, followed by mixed cropping of pearl millet, moth bean, and cluster bean, as also agroforestry in which shrubs and grasses were the dominant components (Bhati et al., 2009). So important was animal husbandry that, till independence, the major agricultural exports from the three princely states of Marwar, Bikaner, and Jaisalmer were camel, bullock, sheep, goat, horse, "ghee" (clarified butter), and high-quality wool (Erskine, 1909). Marwari bullocks used to fetch very high prices since the early centuries AD, especially because they could walk effortlessly on sand and had enough strength and stamina for long-distance transportation between Delhi and Gujarat. Camels of Jaisalmer used to plow about two acres of land per day in a sandy terrain, and so were highly priced. To maintain large stocks of such animal assets, the inhabitants used to maintain large pastures for common use. One such, the Pali Jungle to the north of Ramgarh-Mohangarh in Jaisalmer district, used to cover about 5000 sq. km area and stretch for about 150 km along the present India-Pakistan border. Located in a sandy deflation plain over the Saraswati alluvium, and visited by occasional rains, it used to produce, till the mid-1990s, the best-quality nutritious *Lasiurus sindicus* shrub for cattle across Rajasthan. The land is now converted into irrigated croplands. Another major pasture area was the ~75 km-long Lakhi Jungle to the west of Ganganagar, where several palaeochannels of the Sutlej (the Naiwals) used to meet. It used to supply some of the best grasses in northwest India and was a major horse-breeding area (Francklin, 1805). The area became an irrigated cropland since the beginning of the twentieth century. Fortunately, the small village pastures received legal protection against land use conversion since independence, but their condition deteriorated fast due to the lack of regulatory mechanism and grazing policy. This is despite the fact that the farmers know how the livestock stabilizes their economy during moderate to severe droughts when crops fail totally. In this context, it is worthwhile to note that presently, the mean per ha percent income from irrigated cropping is 27% of the total, while that from bovines and ovines is 31%, horticulture 20%, rain-fed cropping 6%, and household industries 16% (Anonymous., 2011). Thus, animal husbandry still holds the maximum potentials for development, and the western part of the desert is best suited for it.

Croplands in western part increased from about 3% in the 1880s to 9% in the 1940s, while in the northern and eastern parts, they increased from about 10% to 20% (Haynes, 1999). We estimated that during the time of independence, an average household with 8–10 ha land in the eastern part of the desert used to get about 70–80% of his total agricultural income from crop cultivation, while the animal products used to provide him the rest of the income. In the central part, crops used to fetch about 60–70% of the total income, while in the west, crops could hardly fetch more than 30–40% of the total. This reflected a strong environmental control over the components of agricultural land use. The system subsequently received a strong challenge from irrigated farming.

Modern Agriculture

Cropping Through Canal Irrigation

The first major change in the traditional rain-dependent agricultural system was made when a palaeochannel of the ephemeral Chautang River in the Hansi-Hisar-Bhadra tract was canalized under the dictates of Emperor Firoz Shah Tughlaq (ca. 1355 AD). Bhatner (now known as Hanumangarh) and Rawatsar on the highway from Delhi to Multan became major granaries for the sultanate, but subsequently, the canal suffered from mismanagement due to invasions and factional fights and remained disused for a long time despite some efforts by Emperors Akbar (1568) and Shah Jahan (1647). In 1826, the Marquis of Hastings got it repaired as a branch of the West Yamuna Canal, which helped to stabilize crop production in the area (Anonymous., 1884). In Punjab, a massive program called the Gray Inundation Canal of Ferozpur was undertaken between 1875 and 1883, which linked several abandoned courses of Sutlej (called the Naiwals) through Abohar, Bathinda, Malerkotla, Sangrur, etc. These were then linked by an inundation canal from Ferozpur and the Sirhind Canal from Ropar to distribute part of the Sutlej-Beas water into 13 canals for irrigating the Bangar land in between. Since all these Naiwals had natural outflow into the Ghaggar, the canals increased the discharge along the Ghaggar and also water supply in the Ganganagar area. The next major hydrographic change took place in 1927 when the Bikaner Canal (now Gang Canal) was constructed from the Sutlej's left bank in the erstwhile Bikaner State to cover a culturable area of ~263000 ha. Despite these development efforts, croplands in the desert were very few. In 1963, the Bhakra Dam was constructed on the Sutlej River, and the Bhakra Canal system was constructed to irrigate the eastern part of the present Hanumangarh district. This was followed by the development of the Rajasthan Canal (now the Indira Gandhi Nahar Pariyojana, IGNP), which brought the Himalayan water from Harike Barrage at the confluence of the Beas and the Sutlej to parts of Hanumangarh, Ganganagar, Bikaner, and Churu districts under its Stage I command (culturable command area, CCA: 541000 ha) and in parts of Bikaner, Jaisalmer, and Jodhpur districts under Stage II command (CCA: 1410000 ha). Work in Stage II area is still in progress (Anonymous., 2002). Presently, approximately 2.27 million ha area is under the CCA of different canal systems in the arid western part of Rajasthan for using 13716 million cubic meter of water per year. About 70% of the above allocations are in the IGNP.

Advent of Groundwater Irrigation

Groundwater became a major driver of change in agricultural land use mainly after the independence. The first major changes within western Rajasthan took place in the Shekhawati tract in the northeast and also in the Ned area in the deltaic tract of the Luni River. Farmers first opted for diesel pump sets to energize their wells. With time, as rural electrification progressed and the state Ground Water Department (GWD) moved in its wake to search for potential aquifers and sink tube wells for drinking water in far-flung villages, the farmers followed in their footsteps and started sinking their own tube wells for irrigation. In 1951, the net well-irrigated area in western Rajasthan was about 136000 ha only. By 1980, it increased to 620893 ha, by 2000 1.27 million ha (mha), by 2010 to 1.83 mha, and by 2016 to 2.37 mha. In 1980, about 48% of the total irrigated area was served by the wells and 52% by the canals. In the year 2000, the values changed to 56% and 44%, respectively; in 2010 to 60% and 40%, respectively; and in 2016 to 62% and 38%, respectively. Major expansion is now taking place in the western districts of Barmer, Jaisalmer, and Bikaner, as well as in the western part of Jodhpur and Churu districts. Not only the sandy plains but also the gravel plains with less than 15 cm of sediments have become the sites for irrigated cropping. Farmers first removed the coarse gravels from their fields through laborious hand-picking and then put about 30 cm thick dune sand and pond silt on the surface to create ambience for crop growth. In the early 1970s, when groundwater was tapped around Jodhpur city, a vast sandstone pediment near the city was transformed thus to grow mustard, chili, vegetables, etc., through irrigation, but the quantum of water applied was very high and often wasteful. The experiment paid rich dividends to the farmers, and the method began to be replicated in many new areas, but it depleted the aquifer. The rapid progress during the last few decades is related to Green Revolution. The overall groundwater exploitation in the region during 1991 was 48%, which rose to 120% by 2001 and 149% by 2011. About 87% of the exploitation is used for irrigation, and only 13% is available for drinking and other purposes.

Green Revolution

A major shift in the agricultural land use took place from the mid-1970s when the Green Revolution (GR) concept took roots in the desert of Rajasthan (Kar, 2014b). GR technology, developed since the 1960s, depended heavily on the availability of following four inputs: high-yielding crop varieties, irrigation water, chemical fertilizers, and chemical pesticides. Several drought-resistant, fast-growing, and high-yielding crop varieties were developed for the region, but many of these demanded more water, fertilizers, and pesticides than the traditional varieties. So the demand for irrigation water, inorganic fertilizers, and chemical pesticides also increased manifold. Rural electrification, groundwater exploration, and IGNP network helped farmers to convert large areas of rain-fed croplands and open rangelands into irrigated rabi croplands. Further boost was provided by amending the rules to enable cropping in certain categories of marginal land. Easy access to farm machinery (including tractors), hybrid seeds, fertilizers, pesticides, market, etc., was provided through soft loans, transport network, storage facilities, and value chains. Free advisory on appropriate agronomical practices during cropping seasons was also introduced.

The above investments paid good dividends as is reflected through increase in net sown area (NSA) and crop production. Summarizing the available revenue data from 1960–1961 to 2014–2015 into 5-yearly averages, we noticed that the mean NSA during 1960–1961 to 1964–1965 period was about 40.8% of the total area in western Rajasthan, which peaked to 49.6% in the period 1990–1991 to 1994–1995 but then declined through the next two 5-year periods to stand at 46.2% in 2000–2001 to 2004–2005. This was because the region experienced some large droughts during 1999, 2000, and 2002. Recovery since then led to expansion of NSA to about 54.9% by the period 2010–2011 to 2014–2015. Overall, the NSA increased over the 55-year period by 34.5%. Current fallow declined during the same period by 13.8%, and long fallow by 38.8%. By 20005–2006 to 2009–2010 period, the food crop production increased over that in 1960–1961 to 1964–1965 period by 76%, especially in case of the irrigated rabi crops. Production of spices registered an increase of 72% and industrial crops by about 98%. Yet there are many niche areas in the desert where rain-dependent dry farming is still practiced with an average land productivity of 150–250 kg grain per ha, plus 500–1000 kg fodder per ha (as compared to about 5000 kg grain per ha in irrigated rabi croplands, but with little fodder output).

As irrigated cropping involved land leveling, drainage, avoidance of competition from unwanted plants, uniform seeding, etc., the naturally grown trees and shrubs in the fields were uprooted first. These used to arrest the sand from blowing during summer, provide fodder for the livestock, supply emergency food items during lean months, and maintain soil nutrient balance to some extent. The traditional animal-driven wooden plow, which it replaced, used to churn up the sandy loam soil to about 30 cm depth. Tractors began to overturn the soil to a depth 1 m or more. As irrigated crops began to increase the farmers' income, tractors began to flatten more of sandy hummocks and dune slopes to expand the area under irrigation and even started operating on sand dunes. As their number swelled from about 14,500 in 1980 to more than 500,000 by 2018, more area came under deep plowing and hence more area under sand destabilization. The desert began to produce wheat, mustard, cumin, vegetables, spices, and medicinal plants in plenty. As demand for the products increased, so did the pressure on the finite land and water resources. In 1961–1962, about 8% area of the NSA in western Rajasthan was irrigated, of which about 36% was under well irrigation and the rest 64% under canal and tank irrigation. In 1991–1992, canals and tanks served 50% area of the total irrigated, and the wells served the rest 50%. Soon thereafter, the wells began to surpass the area served by canals, and at the beginning of this century, wells served 56% of the NSA. By 2016–2017, it served 62% of NSA, despite the large parts of the region being declared as overexploited "dark zone." Unlike in the canal command areas, farmers in well-irrigated areas have direct control on water supply, which has led to its almost uncontrolled use. Groundwater extraction increases tremendously during the periods of agricultural drought as farmers resort to lifesaving irrigation to kharif crops, which was not in vogue in the past. Presently, out of the total irrigated croplands, 35% occur on the sandy plains, 31% on sand dunes and interdune plains, and 23% on the older alluvial plains. The rain-fed croplands dominantly occur on sand dunes and interdune plains (49%), followed by sandy plains (32%) and older alluvial plains (8%; Moharana et al., 2013).

Challenges and Sustainability

Environmental Consequences of GR

The GR paradigm, despite being a remunerative venture for the farmers, has also brought some new challenges. The reduction in open rangeland areas from about 47% of the total revenue area in the 1960s to 32% now, and the destruction of natural vegetation in croplands, has drastically reduced the free fodder availability for small ruminants. Presently, farmers meet bulk of fodder requirement from their crop residues and from the boundary plantation of their fields. Shortfall is met through market purchases, which is becoming a costly proposal, especially during drought years. Also, crop production itself is facing new challenges from rising temperature, declining groundwater, and deteriorating land quality, making fodder availability more difficult and animal husbandry a risky venture for average farmers.

Continuous crop cultivation without keeping the land fallow is leading to gradual impoverishment of the soils. Among the macronutrients, phosphorus (P) content has reached very low level in the shallow to moderately deep sandy loam soils, especially in the rain-fed croplands, while among the micronutrients, zinc (Zn) and iron (Fe) have gone down drastically in both rain-fed and irrigated fields (Tsunekawa et al., 1997; Mahesh Kumar et al., 2019).

Too much pumping of groundwater has led to 61 groundwater development blocks out of 74 being labeled as "overexploited," leaving only seven as "critical" and six as "semicritical" blocks. This has prompted the state to restrict further groundwater exploration in several blocks, but groundwater-irrigated areas are increasing fast in the non-restricted blocks in the central and western parts. As groundwater level is falling over large areas, the water quality is deteriorating, which has started impacting the land quality. Deeper sourcing is also increasing the energy cost. Dry wells and unsustainable energy cost have compelled many farmers in the Shekhawati and the Godwar tracts to either revert back to rain-fed subsistence farming or to seek alternate livelihood. Currently, irrigation uses 4762.5 mcm of groundwater (87% of the total groundwater use), and drinking and other nonirrigation purposes 723.6 mcm (13%). Based on demand and supply projections, it has been estimated that by 2025, the state has to allocate 792 mcm of groundwater for drinking and other nonirrigation purposes in the region, but this will leave only 479.6 mcm in the aquifers for irrigation, a shortfall of 4282.8 mcm (Anonymous., 2014). The consequences will be unthinkable.

Excessive irrigation in the canal command areas has led to waterlogging and salinity in parts. When the farmers of the IGNP area were first introduced to canal irrigation five decades ago, they thought that crop yields would be proportional to the amount of water applied, and so practiced flood irrigation. The dominant crop rotations during the initial period were (a) cotton-wheat-cotton-fallow, (b) mung bean-mustard-fallow-mustard, and (c) groundnut-wheat-groundnut-fallow. With flood irrigation, the subsoil water level gradually came up to the root zone, which compelled farmers to shift from the wheat-mustard cropping system to the rice-wheat system, but soon wheat was also difficult to grow, and only rice could grow on high water table. As water began to appear on the surface, even rice was difficult to grow, and the land had to be abandoned (Kar, 2018). During 1981–1995, the average rate of rise in water table in the IGNP command areas was measured as 0.88 m per year, leading to a rise in the area under waterlogging and salinity buildup. Tank irrigation in the Pali-Jalor area of the Godwar tract has also salinized parts of their command areas. Another flash point is emerging in the Luni delta (Ned area) where the Narmada Canal has started irrigating the soils underlain at 2–10 m depth by highly saline groundwater. Although the original plan has been changed after our protest to allow only drip and sprinkler irrigation with limited period of supply, the soils are bound to get salinized soon after the groundwater level rises due to surface input.

Some controversial policy decisions also led to spread of waterlogging. In the 1960s, when Hanumangarh town faced waterlogging due to flood irrigation, a "Ghaggar Diversion Channel" was constructed to drain the excess water to some interdune areas between Rawatsar and Suratgarh, in the hope that the dune sand would absorb the excess water. Unfortunately, the engineers did not factor the likely impact of a thick gypsiferous and clay barrier at 1.0–1.5 m depth in the dumping area and that of a buried palaeochannel of the Drishadvati River in its vicinity. Soon after the channel was commissioned, the interdunes turned into vast lakes, and the farmers along the palaeochannel in Rawatsar-Suratgarh area noticed abnormal rise in groundwater level in their productive fields due to seepage from the new lakes. By 1990, the palaeochannel turned into a vast pool of toxic water, and cropping had to be totally abandoned, with devastating effects on the once-prosperous farmers (Kar et al., 2009).

The other major problem is deep plowing of the sandy soil and uprooting of natural vegetation with tractors, which has threatened a large area with sand mobilization and dust emission. This is despite the fact that the mean summer wind speed has gradually declined from the 1950s. As drying of wells is forcing farmers to revert back to rain-fed farming, the deep-plowed land bereft of natural plant cover is encouraging more of sand movement. The situation may worsen further as wind erosivity is likely to increase in next few decades due to global warming. The only silver lining is the increasing incidence of spring rains, which currently stabilizes the sandy landscape in the mid-May for about a week. If spring rains increase with time, and if plant cover is allowed to increase, we may expect some positive impacts (Kar, 2012, 2019).

Monitoring the Changes Through Remote Sensing and GIS

Visual interpretation of satellite sensor data has revealed that about 30% area of western Rajasthan is now slightly affected by land degradation, 41% is moderately affected, and 21% severe to very severely. Wind erosion has affected about 75% area, while other degradation processes like water erosion, salinization, etc., could be mapped in small areas (Kar et al., 2009). Results from visual remote sensing, however, tend to vary with the interpreter's perception, which is a major drawback for degradation mapping. In order to overcome the problem, especially for wind erosion mapping that covers the maximum area, we developed a digital method to estimate aeolian sand reactivation from the satellite sensor-derived radiance and the emissivity values of different land surfaces and related the output with indices of cultivation and grazing pressures at village level (Kar, 2011, 2019). The only limitation of this method for long-term monitoring is the non-availability of the sensor wavelength bands used in the calculation. Kundu et al. (2016, 2017) used the long-term data on rain use efficiency of plants, as well as vegetation health index (VHI) and residual trends of vegetation index, to assess vegetation degradation in the region. For the present study, we used a method of regional-scale time-series mapping of VHI to understand the likely implications of changes in irrigated croplands on sand reactivation.

We first mapped the major irrigated rabi croplands in Thar Desert from a cloud-free MODIS Terra satellite image of 25 February 2019 at 500 m pixel, and then overlaid on it the six traditional agroecological zones (AEZs) within the desert that we mapped from a merged SRTM DEM-Landsat ETM+ digital image, based on our knowledge of natural resource endowment and the administrative boundaries (Fig. 12.7). Considering the satellite-derived VHI as a surrogate for crop condition, we then attempted a time series analysis of the weekly VHI values for February for the last four decades (1982–2018), because the irrigated rabi crops in the desert attain maximum greenness during February, when signature from most other major plant cover declines. The VHI is calculated globally by NOAA (National Oceanic and Atmospheric Administration, USA) from the AVHRR (Advanced Very High-Resolution Radiometer) sensors aboard a series of polar-orbiting satellites, in which the surface radiance and temperature data in different wavelengths are converted into Normalized Difference Vegetation Index (NDVI) and Brightness Temperature (BT) values, and then further processed into a Vegetation Condition Index (VCI), which is a proxy for moisture condition for plant vigor, and a Temperature Condition Index (TCI), which is a proxy for thermal condition. Combining VCI and TCI through algorithm, one derives VHI. We sourced from NOAA the

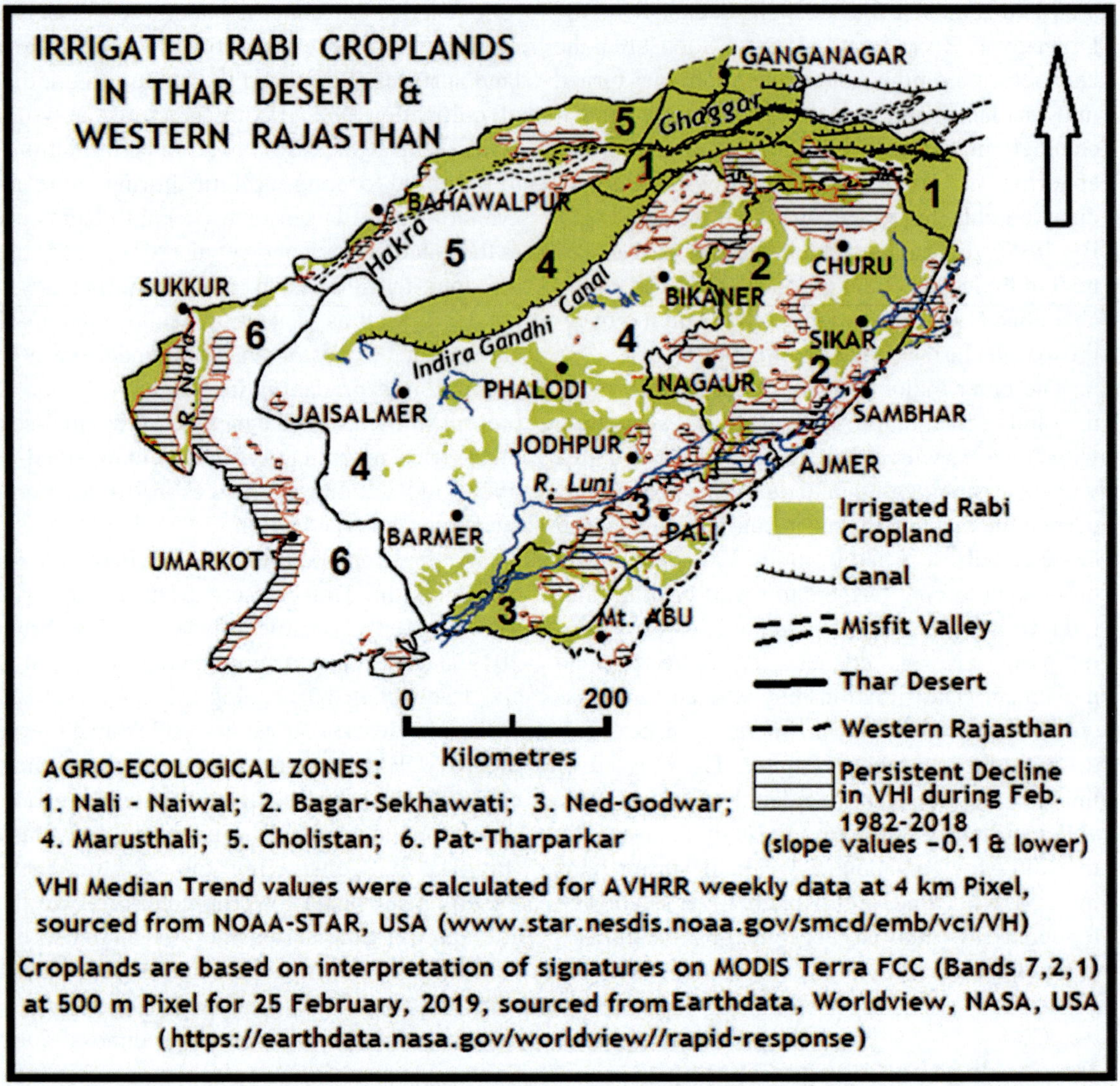

Fig. 12.7. Irrigated rabi croplands in Thar Desert and western Rajasthan, with areas showing persistent decline in vegetation health index during February (1982-2018). Details of map variables and sources are given in the map

weekly global VHI data for the month of February (from 1982 to 2018) at 4 km pixel resolution and amalgamated the four-weekly data into Maximum Value Composites for February for each pixel location every year. We then stacked the yearly VHI images to perform a median trend analysis (Theil-Sen protocol). This statistical method calculates the slope between every pairwise combination of each pixel in the stacked yearly data to find the median value through several iterations and keeps only those values that persist to show a trend for more than 29% of the length of the series. Thus, it provides a much robust output on the trend.

Based on the slope values of the median trend, we determined that 40% area of the desert had registered over the last four decades a slightly positive trend in VHI (trend slope of 0.01 to 0.74), 10% a moderately positive trend (0.75 to 1.00), and 33% strongly positive trend (1.01 or more). At the same time, 16% area registered a persistent negative trend (13% slightly negative, -0.01 to 0.74, and 3% moderate to strongly negative, >0.74). Some of this decline took place in the irrigated croplands of the Bagar-Shekhawati and the Ned-Godwar zones (Fig. 12.7). In Pakistan part, the most persistent decline took place in the Pat-Tharparkar zone, while another flash point is emerging in the Cholistan, leading to higher sand reactivation and land salinization.

To monitor the sand reactivation pattern from satellite sensors, we developed an Aeolian Sand Reactivation Index, which we fortified with the pattern of broadband emissivity from the different land surfaces (ASRI_bbe). Using the 8-day summary of surface reflectance values at 500 m pixel resolution over the desert from mid-March to mid-June (2000–2015), as available from the MODIS sensor, we calculated the ASRI_bbe for each year's summer months (Kar, 2019). We then evaluated the mean sand reactivation pattern with the pattern of annual rainfall, summer wind speed, and satellite-derived Aerosol Index in ultraviolet wavelength (UV_AI). Fig. 12.8 summarizes the pattern over Thar Desert. ASRI_bbe could not be calculated for the pre-2000 years due to nonavailability of required sensor data, but we find a gradual decline in wind speed over time (despite the occasional peaks). UV_AI, which increases with dust storm and is sensed by the satellites at 7 km above the surface, is found to peak after a poor monsoon year. It is also found to vary with the recent ASRI_bbe values. What is surprising is that the patterns of UV_AI and ASRI_bbe do not mimic the pattern of wind speed. We argue that this is due to the widespread tractor plowing and baring of sandy surface. We feel that expansion of irrigated farming will further encourage sand destabilization, which will exacerbate the situation. Apart from regulating groundwater irrigation, there is a strong need for water and plant cover management in canal command areas in Marusthali.

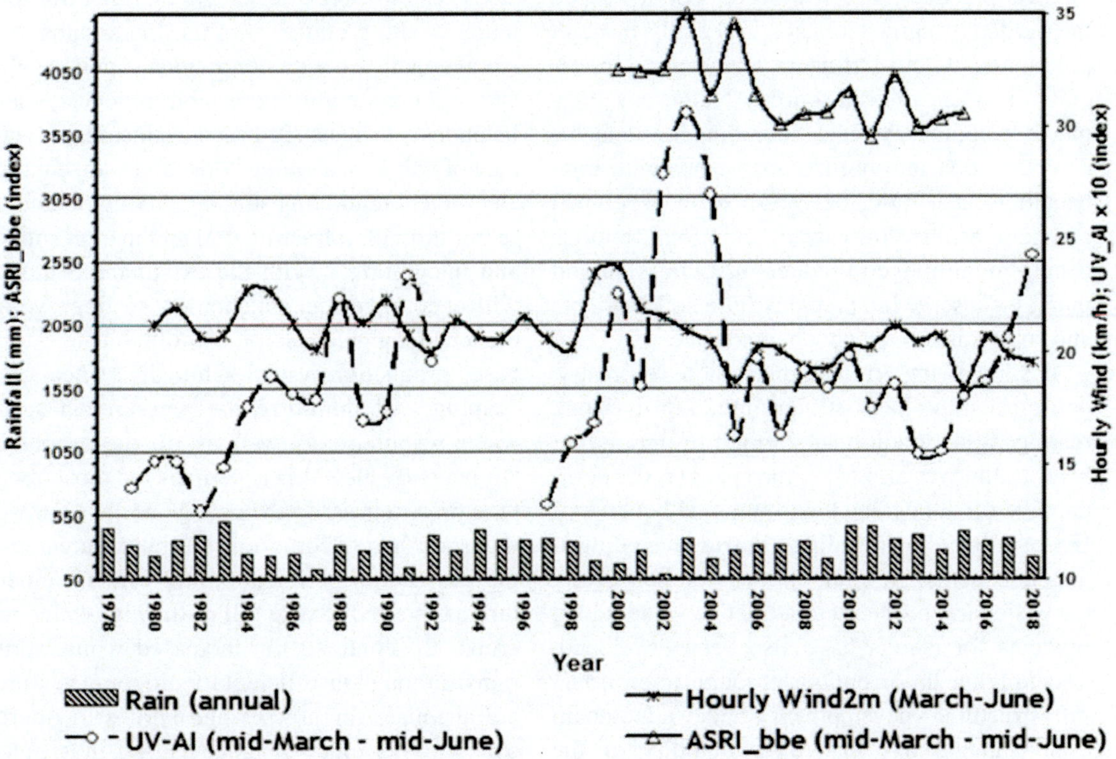

Fig. 12.8. Mean rainfall (annual total, mm), hourly surface wind speed (March-June, km/h), UV Aerosol Index (mid-March to mid-July) and ASRI_bbe index (mid-March to mid-July) for Thar Desert (1978-2018). Basic data was accessed from the NASA Giovanni site

The Road Ahead

The emerging problems can still be tackled with some research-mediated policy interventions. Wasteful use of irrigation water has so far been a major negative outcome of GR. Since the demands for farm outputs will remain high and the need for feeding the growing population will also continue, groundwater exploitation will continue. What is needed under such a scenario is to increase the water use efficiency of crops and to encourage drip and sprinkler irrigation through subsidized provisioning. Fortunately, farmers have started to make course correction. Sprinkler and drip systems have spread over large parts of western Rajasthan. In the canal-dominated Nali-Naiwal zone, farmers have stopped flood irrigation and are reclaiming their land.

Agricultural research has come out with several improved crop varieties that require less water and can withstand the impacts of erratic rainfall, drought, heat, etc. Several viable agronomical practices have been developed to reduce the burden of agrochemicals, especially through production of bio-fertilizers and organic pesticides using cheap local sources. Numerous alternative cropping systems, including options for diversification into horticultural crops with high benefit to cost ratio, have also been developed and demonstrated for ease of adoption. Methods of minimum tillage to reduce dust emission and thereby conserve the precious little soil nutrients and soil moisture are now being tested.

IFS, involving optional mixes of agroforestry, agri-horticulture, agri-silvipasture, and livestock rearing, was a traditional strength of agriculture here. Rathore et al. (2019) have shown the comparative economic and environmental benefits of IFS over the GR, including climate change mitigation benefits. Several options for converting the basic farm produces to attractive value-added products for market have also been developed. Encouraging those options through appropriate infrastructural development and value-chain establishment may ensure profitability of the farm produces and sustainability of the systems. It is in this context that the agricultural scientists are now advocating the adoption of an "Ever-green Revolution" concept (Swaminathan, 2006; Paroda, 2018). Although the adoption rates of new technologies are low in the desert, the emerging constraints are forcing the farmers to gradually adopt the suggested technologies. Improved varieties of horticultural crops like pomegranate, jujube (Ber), date palm, etc., have recently started to replace the traditional crop cultivation in many areas of Marusthali, especially in parts of Barmer and Jaisalmer districts. Farmers have even started to lease a part of their land for solar panel installation, which allows the land parcels to regain part of their lost vitality.

A major problem in our desert since the introduction of GR is an utter neglect of the livestock sector, despite the sector holding much higher potential than irrigated cropping. As we discussed earlier, livestock-based farming is still a better option economically, especially in the west. Unlike crop cultivation under GR, which requires grain storage facilities and market access for postproduction growth, development in livestock sector is more dependent on faster transportation of the produce from the hinterlands; cold storage facilities with continuous supply of electricity for the highly perishable products; establishment of livestock-based industries to take care of dairy, meat, bone, hide, and skin, etc.; and an efficient marketing strategy. Strengthening the sector through infrastructural and market support and integrating it with the existing IFS models will encourage the stakeholders to improve the degraded rangelands for a profitable livestock-based production system, while diversification of farming will minimize the pressure on scarce water resources, as well as on the croplands. Climatically also, this is perhaps the most opportune time to usher in the change as the southwest monsoon has become more erratic, threatening the kharif crop production, and winter temperature has started to take toll of the high-value rabi crops. By contrast, the increased winter/spring rains, although insufficient for cropping ventures, are adequate for grasses and shrubs to flourish over the vast open rangelands and thus potentially beneficial for the livestock sector. The inhabitants here had reached such a climatic situation during the mature and the late Harappan periods and had wisely diversified to more efficient crop-and-animal-based farming systems to survive sustainably over the next few millennia. With enough scientific results proving the model's success under the prevailing climate, there is no reason why it should not succeed now.

Conclusions

Our short review of agricultural land use in the Thar Desert shows that despite being constrained by aridity, the natural resources that the region has inherited through the landscape evolutionary processes possess the potentials to provide strength and stability to its agriculture-based livelihood, provided the land users do not exceed the critical thresholds of pressure on the land and do not excite the inherent vulnerability of the landscape. The history of land use strengthens our view that the inhabitants have always successfully negotiated the environmental constraints through traditional wisdom and upcoming skills, which can be emulated now to surmount the new challenges.

References

Anonymous. (1874). Notes on the lost river of the Indian Desert. *The Calcutta Review, 59*, 1–29.

Anonymous. (1884). *Gazetteer of the Hisar District* (p. 75). Govt. of Punjab, Lahore.

Anonymous. (2002). *History of Indira Gandhi Nahar* (p. 405). Indira Gandhi Nahar Board, Govt. of Rajasthan.

Anonymous. (2011). *CAZRI Vision 2030* (p. 24). Central Arid Zone Research Institute.

Anonymous. (2014). *Dynamic Ground Water Resources of India (as on 31st March 2011)* (p. 283). Central Ground Water Board, Faridabad.

Ashwal, L. D., Solanki, A. M., Pandit, M. K., Corfu, F., Hendriks, B. W. H., Burke, K., & Torsvik, T. H. (2013). Geochronology and geochemistry of Neoproterozoic Mt. Abu granitoids, NW India: Regional correlation and implications for Rodina paleogeography. *Precambrian Research, 236*, 265–281.

Bates, J., Petrie, C. A., & Singh, R. N. (2018). Cereals, calories and change: Exploring approaches to quantification in Indus archaeobotany. *Archaeological and Anthropological Sciences, 10*, 1703–1716.

Bharadwaj, O. P. (1961). The arid zone of India and Pakistan. In L. D. Stamp (Ed.), *A History of Land Use in Arid Regions* (pp. 143–174). UNESCO.

Bhati, T. K., & Joshi, N. L. (2007). Farming systems for the sustainable agriculture in Indian arid zone. In K. P. R. Vittal, R. L. Srivastava, N. L. Joshi, A. Kar, V. P. Tewari, & S. Kathju (Eds.), *Dryland Ecosystem: Indian Perspective* (pp. 35–52). Central Arid Zone Research Institute and Arid Forest Research Institute.

Bhati, T. K., Rathore, V. S., Singh, J. P., Beniwal, R. K., Nathawat, N. S., Birbal, D., Shamsudheen, M., Ram, B., Rohilla, P. P., Rao, S. S., Meghwal, P. R., Kumar, P., Sharma, A. K., & Painuli, D. K. (2009). Alternative farming systems for hot arid regions. In A. Kar, B. K. Garg, M. P. Singh, & S. Kathju (Eds.), *Trends in Arid Zone Research* (pp. 354–382). Central Arid Zone Research Institute.

Clift, P., Carter, A., Giosan, L., Durcan, J., Duller, G. A. T., Macklin, M. G., Alizai, A., Tabrez, A. R., Danish, M., VanLaningham, S., & Fuller, D. Q. (2012). U-Pb zircon dating evidence for a Pleistocene Sarasvati River and capture of the Yamuna River. *Geology, 40*, 211–214.

Dhir. (2003). Thar Desert in retrospect and prospect. *Proceedings, Indian National Science Academy, 69A*, 167–184.

Erskine, K. D. (1909). *Rajputana Gazetteers, Volume III-A: The Western Rajputana States Residency and Bikaner Agency* (p. 399). The Pioneer Press.

Francklin, W. 1805. Military Memoirs of Mr. George Thomas. John Stockdale, (reprint), 392.

Fuller, D. Q. (2011). Finding plant domestication in the Indian Subcontinent. *Current Anthropology, 52 Supplement 4*, S347–S362.

Ghose, B., Kar, A., & Husain, Z. (1979). The lost courses of the Saraswati River in the Great Indian Desert - New evidence from Landsat imagery. *The Geographical Journal, 145*, 446–451.

Goudie, A., Allchin, B., & Hegde, K. T. M. (1973). The former extensions of the Great Indian Sand Desert. *The Geographical Journal, 139*, 243–257.

Green, A. S., & Petrie, C. (2018). Landscapes of urbanization and de-urbanization: A large-scale approach to investigating the Indus Civilization's settlement distributions in northwest India. *Journal of Field Archaeology, 43*, 284–299.

Haynes, E. S. (1999). Land use, natural resources and the Rajput state, 1780-1980. In R. Hooja & R. Joshi (Eds.), *Desert, Drought and Development* (pp. 53–119). Rawat Publications.

Jodha, N. S. (1985). Population growth and the decline of common property resources in Rajasthan, India. *Population and Development Review, 11*, 247–264.

Kar, A. (2011). Geoinformatics in spatial and temporal analysis of wind erosion in Thar Desert. In S. Anbazhagan, S. K. Subramanian, & X. Yang (Eds.), *Geoinformatics in Applied Geomorphology* (pp. 39–62). CRC Press.

Kar, A. (2012). GCM-derived future climate of arid western India and implications for land degradation. *Annals of Arid Zone, 51*, 147–169.

Kar, A. (2014a). The Thar or the Great Indian Sand Desert. In V. S. Kale (Ed.), *Landscapes and Landforms of India*. World Geomorphological Landscape Series. Springer, pp. 79–90.

Kar, A. (2014b). Agricultural land use in arid western Rajasthan: Resource exploitation and emerging issues. *Agropedology, 24*, 179–196.

Kar, A. (2018). Desertification: Causes and effects. In D. Bartlett & R. P. Singh (Eds.), *Exploring Natural Hazards: A Case Study Approach* (pp. 159–205). CRC Press, Taylor & Francis Group.

Kar, A. (2019). Measuring land degradation: A quantitative approach for better understanding of the problem in Thar Desert. In *Proceedings, 13th International Conference on Development of Drylands, Jodhpur* (pp. 149–158). International Dryland Development Commission, Alexandria, and Arid Zone Research Association of India.

Kar, A. (2020). Identification and mapping of the Saraswati River system: A review and new findings. In S. K. Acharyya, K. Ghosh, & A. Kar (Eds.), *Saraswati: The River par Excellence* (pp. 29–98). The Asiatic Society.

Kar, A., & Ghose, B. (1984). The Drishadvati River system of India: An assessment and new findings. *The Geographical Journal, 150*, 221–229.

Kar, A., & Kumar, A. (2020). Evolution of arid landscape in India and likely impact of future climate change. *Episodes, 43*, 511–523.

Kar, A. and Shukla, J. P. (2000). Remote sensing of ground water along the Saraswati palaeochannels in western part of Jaisalmer district, Thar Desert. *MAEER's MIT Pune Journal, Special Issue (1999-2000)*, 28–36.

Kar, A., Moharana, P. C., Raina, P., Kumar, M., Soni, M. L., Santra, P., Ajai, A., & A.S. and Dhinwa, P.S. (2009). Desertification and its control measures. In A. Kar, B. K. Garg, M. P. Singh, & S. Kathju (Eds.), *Trends in Arid Zone Research* (pp. 1–47). Central Arid Zone Research Institute.

Kar, A., Singhvi, A. K., Juyal, N., & Rajaguru, S. N. (2004). Late Quaternary aeolian sedimentation history of the Thar Desert. In H. S. Sharma, S. Singh, & S. De (Eds.), *Geomorphology and Environment* (pp. 105–122). ACB Publications.

Kar, A., Singhvi, A. K., Rajaguru, S. N., Juyal, N., Thomas, J. V., Banerjee, D., & Dhir, R. P. (2001). Reconstruction of the late Quaternary environment of the lower Luni plains, Thar Desert, India. *Journal of Quaternary Science, 16*, 61–68.

Kenoyer, J. M. (1995). Interaction systems, specialise crafts and culture change: The Indus Valley tradition and the Indo-Gangetic tradition in South Asia. In A. Wezler & M. Witzel (Eds.), *Indian Philology and South Asian Studies* (Vol. 1, pp. 213–257). Walter de Gruyter.

Kumar, M., Kar, A., Raina, P., Singh, S. K., Moharana, P. C., & Chauhan, J. C. (2019). Spatial variability of available soil nutrients in the Sekhawati region of Thar Desert, India. *Journal of the Indian Society of Soil Science, 67*(1), 21–33.

Kundu, A., Dwivedi, S., & Dutta, D. (2016). Monitoring the vegetation health over India during contrasting monsoon years using satellite remote sensing indices. *Arab Journal of Geosciences, 9*, 144. https://doi.org/10.1007/s12517-015-2185-9

Kundu, A., Patel, N. R., Saha, S. K., & Dutta, D. (2017). Desertification in western Rajasthan (India): An assessment using remote sensing derived rain-use efficiency and residual trend methods. *Natural Hazards, 86*, 297–317.

Madella, M., & Fuller, D. Q. (2006). Palaeoecology and the Harappan Civilisation of South Asia: A reconsideration. *Quaternary Science Review, 25*, 1283–1301.

Moharana, P. C., Gaur, M. K., Choudhary, C., Chauhan, J. S., & Rajpurohit, R. S. (2013). A system of geomorphological mapping for western Rajasthan with relevance for agricultural land use. *Annals of Arid Zone, 52*, 163–180.

Mughal, M. R. (1982). Recent archaeological research in the Cholistan Desert. In G. L. Possehl (Ed.), *Harappan Civilization* (pp. 84–96). Oxford and IBH.

Neogi, S., French, C. A. I., Durcan, J. A., Singh, R. N., & Petrie, C. A. (2019). Geoarchaeological insights into the location of Indus settlements on the plains of northwest India. *Quaternary Research, 1-19*. https://doi.org/10.1017/qua.2019.70

Paroda, R. S. (2018). Reorienting Indian Agriculture: Challenges and Opportunities. *CABI Publishing*, 336.

Pawar, V., Parmar, N., & Sharan, A. (2013). Harappan civilization: Emerging picture in Hanumangarh district, Rajasthan. Heritage: Journal of Multidisciplinary Studies. *Archaeology, 1*, 475–485.

Petrie, C., Singh, R. N., Bates, J., Dixit, Y., French, C. A. I., Hodell, D. A., Jones, P. J., Lancelotti, C., Lynam, F., Neogi, S., Pandey, A. K., Parikh, D., Pawar, V., Redhouse, D. I., & Singh, D. P. (2017). Adaptation to variable environments, resilience to climate change. *Current Anthropology, 58*, 1–30.

Petrie, C. A., & Bates, J. (2017). 'Multi-cropping', intercropping and adaptation to variable environments in the Indus civilisation. *Journal of World Prehistory, 30*, 81–130.

Petrie, C. A., Parikh, D., Green, A. S., & Bates, J. (2018). Looking beneath the veneer. Thoughts about environmental and cultural diversity in the Indian civilization. In D. Frenez, G. M. Jamison, R. W. Law, M. Vidale, & R. H. Meadow (Eds.), *Walking with the Unicorn – Social Organization and Material Culture in Ancient South Asia* (pp. 453–474). Archaeopress Publishing Ltd.

Prabhakar, V. N. (2013). *Harappans and their Mesopotamian contacts* (p. 26). IIC Occasional Publication 48, India International Centre.

Rao, S., Chan, Y., Bugler-Lacap, D. C., Bhatnagar, A., Bhatnagar, M., & Pointing, S. B. (2015). Microbial diversity in soil, sand dune and rock substrates of Thar Monsoon Desert, India. *Indian Journal of Microbiology*. https://doi.org/10.1007/s12088-015-0549-1

Rathore, V. S., Tanwar, S. P. S., Kumar, P., & Yadav, O. P. (2019). Integrated farming system: Key to sustainability in arid and semi-arid regions. *Indian Journal of Agricultural Sciences, 89*, 181–192.

Rennell, J. (1788). *Memoir of a Map of Hindoostan; or the Mogul Empire* (2nd ed., p. 472). M. Brown.

Roy, M. (2009). Agriculture in the Vedic period. *Indian Journal of Historical Science, 44*, 497–520.

Sarkar, A., Deshpande-Mukherjee, A., Bera, M. K., Das, B., Juyal, N., Morthekai, P., Deshpande, R. D., Shinde, V. S., & Rao, L. S. (2016). Oxygen isotope in archaeological bioapatites from India: Implications to climate change and decline of Bronze Age Harappan civilization. *Scientific Reports, 6*, 26555. https://doi.org/10.1038/srep26555.

Schuldenrein, J., Wright, R., & Khan, M. A. (2007). Harappan geoarchaeology reconsidered: Holocene landscapes and environments of the Greater Indus Plain. In E. C. Stone (Ed.), *Settlement and Society: Essays dedicated to Robert McCormick Adam* (pp. 83–116). Cotsen Institute.

Sharma, D. (1966). *Rajasthan through the Ages. Volume 1: From the Earliest Time to 1316 AD*. Rajasthan State Archives. 914 pages.

Singh, A., & Sinha, R. (2019). Fluvial response to climate change inferred from sediment cores from the Ghaggar-Hakra palaeochannel in NW Indo-Gangetic plains. *Palaeogeography, Palaeoclimatology, Paleoecology, 532*, 109247. https://doi.org/10.1016/j.palaeo.2019.109247

Singh, A., Thomsen, K. J., Sinha, R., Buylaert, J.-P., Carter, A., Mark, D. F., Mason, P. J., Densmore, A. L., Murray, A. S., Jain, M., Paul, D., & Gupta, S. (2017). Intuitive influence of Himalayan river morphodynamics on Indus Civilisation urban settlements. *Nature Communications, 8*, 1617. https://doi.org/10.1038/s41467-017-01643-9, 14.

Singhvi, A. K., & Kar, A. (2004). The aeolian sedimentation record of the Thar Desert. *Proceedings Indian Academy of Sciences (Earth & Planetary Science), 113*, 371–401.

Sinharoy, S., Malhotra, G., & Mohanty, M. (1998). *Geology of Rajasthan* (p. 278). Geological Society of India.

Swaminathan, M. S. (2006). An evergreen revolution. *Crop Science, 46*, 2293–2303.

Tod, J. (1832). *Annals and Antiquities of Rajasthan, or the Central and Western Rajpoot States of India* (Vol. 2). Smith & Elder.

Tsunekawa, A., Kar, A., Yanai, J., Tanaka, U., & Miyazaki, T. (1997). Influence of continuous cultivation on the soil properties affecting crop productivity in the Thar Desert. *India. Journal of Arid Environments, 36*(6), 367–384.

Wadhawan, S. K. (2018). Palaeogene – Neogene tectonics and continental aggradational basins in north-western India: Implications for geological evolution of Thar Desert. In S. Bajpai, S. C. Tripathi, & V. Prasad (Eds.), *The Indian Paleogene* (pp. 151–166). Springer.

13 Human Response Toward COVID-19 Pandemic in Terms of Adaptation, Displacement, and Climate Change in India

Shambhavi Krishna and Shailendra Rai

Background

The latest challenge in front of the human beings is the global pandemic called coronavirus disease 2019 (COVID-19). The disease started in Wuhan City, China. The countries largely affected by this disease are the USA, European countries, South Asian, African countries, Western Pacific, and Eastern Mediterranean countries having millions of cases of coronavirus and more than 1.5 million deaths till now. The children; women especially pregnant women; elderly people especially those who are suffering through other diseases like heart and respiratory diseases, diabetes, and cancer (Cohen et al., 2017); and people with low immunity are first to get affected. This is a kind of communicable disease which could only be stopped by following the rule of social distancing (Chatterjee et al., 2020). Since this disease creates a chain through its host, it is important to break the chain for which lockdown was introduced all over the globe. COVID-19 affected the human health, but at the same time, since the whole world got into the lockdown situation which caused a great reduction, air and water pollution, displacement of migratory workers, and more tremors of earthquake are felt in the Delhi region (Singh, 2020); thus, it made an enormous impact in changing the lifestyle and working habits of the human beings.

S. Krishna (✉)
K Banerjee Centre of Atmospheric and Ocean Studies, University of Allahabad, Prayagraj, India

S. Rai
K Banerjee Centre of Atmospheric and Ocean Studies, University of Allahabad, Prayagraj, India

M. N. Saha Centre of Space Studies, University of Allahabad,
Prayagraj, India

COVID-19 Pandemic in India

Figure 13.1 shows the current situation of coronavirus disease in India state-wise, and the states of India which have become the major hotspots of coronavirus disease are, namely, Maharashtra, Gujarat, Rajasthan, Madhya Pradesh, Kerala, Karnataka, Tamil Nadu, West Bengal, Andhra Pradesh, and Telangana. In order to prevent the situation to get even worse due to continuous spread of coronavirus disease, the Government of India enforced the complete lockdown in the country. It started with a 21-day lockdown which has continued till the month of June as the cases of coronavirus are continuously rising. The government then started unlocking the situation on June 1, 2020 on certain relaxation criteria. During the lockdown, the government made three zonal classifications depending upon the number of corona patient found in those area. These zones were red zone with highest number of patients with double growth rate of corona patient. Then comes the orange zone with medium number of

A. R. Siddiqui, A. Sahay (eds.), *Climate Change, Disaster and Adaptations*, Sustainable

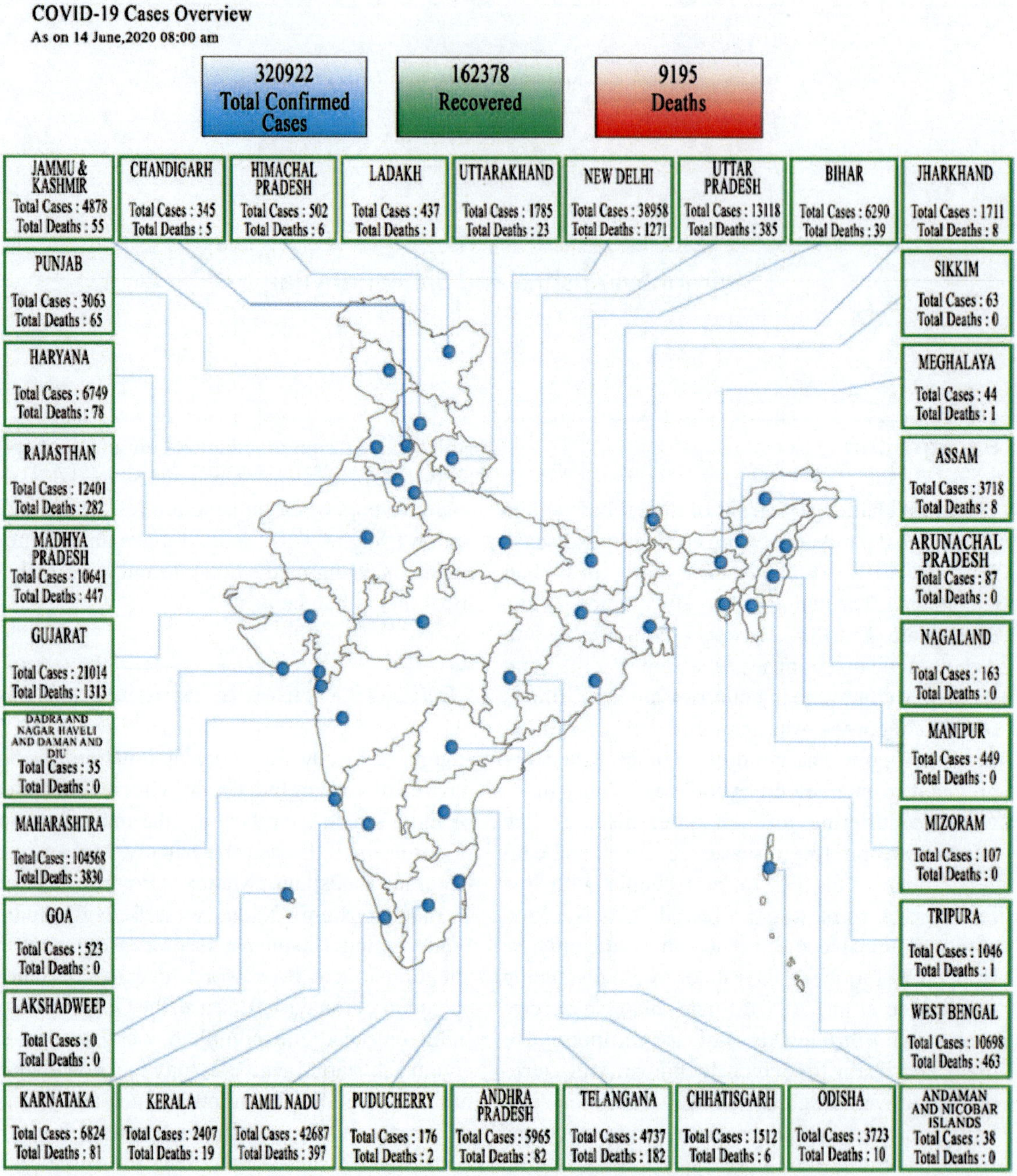

Fig. 13.1 The current status of COVID pandemic in India till June 14, 2020. (Source MoHFW)

patients and lastly the green zone being a safe zone.

Due to the closing of all the industries and factories all over the country, many people lost their jobs especially the daily waged workers which caused their migration by road to their home. Since many humanitarians raised voice for the migratory workers, the Government of India started the "Shramik Special" Trains in early May all across the country so that each and every such people reach their hometowns. But this step led to the increase in the cases of coronavirus disease through the exodus of migratory workers to their hometowns due to which rate of disease spread increased as well as the green zones soon got turned into orange and red zone at a faster rate. The lockdown caused the improvement in the quality of the environment which was damaged due to pollution and reduced the harmful effects of climate change within a very short period of time such as reduction in air pollution in New Delhi (Mahato et al., 2020) and drop in the Earth's ambient seismic activity (Imperial College London, 2020, July 23). The work-from-home concept has gained a lot of acceptance, and the people have started giving it importance in the schools, colleges, universities, and offices.

Data and Methodology

The migration data is collected from the Ministry of Home Affairs, the Government of India, the Census of India 2011 website, and the Economic Survey of India report of 2017. The data of coronavirus disease patients and the number of deaths is collected from the website of Ministry of Health and Family Welfare (MoHFW). On the other hand, the hourly data of various pollutants like PM2.5, PM10, NO_2, CO_2, SO_2, and ozone of all the 36 stations of New Delhi is collected from the Central Pollution Control Board (CPCB), and the average of data of each hour is taken to get a single value for a day, and then the average of data from every station is taken in order to get single value of a parameter for a day of New Delhi region. The seismological data related to the earthquake occurrence in New Delhi and the region surrounding the New Delhi is collected from the National Center of Seismology, the Ministry of Earth Sciences, and the Government of India. The analysis technique applied is the time-series analysis with the help of chronological data collected from various sources mentioned above in the cumulative manner, and the analysis of pollutants has been done by taking averages of the daily and hourly data collected from all the 36 stations of CPCB. Some of the other data such as data related to earthquake have been represented in the form of table as well.

Result and Discussion

The lockdown in India caused the closing of industries, factories, and construction works leading to loss of jobs and livelihood to daily waged workers in the country. According to the data of Census of India 2011, there are about more than 300 million migrants in the country of which nine million are interstate migrants as per the Economic Survey of India 2016–2017 based on the data of passengers traveling through railways. The states in India from where the maximum migrants belong to are Uttar Pradesh, Bihar, Madhya Pradesh, Rajasthan, West Bengal, and Jharkhand (Singh et al., 2011), and the states where these migrants travel for search of jobs include Delhi, Maharashtra, Gujarat, Kerala, Karnataka, Tamil Nadu, and Andhra Pradesh (Bhagat & Keshri, 2020). As per the data of coronavirus disease cases in India shown in Fig. 13.1, the states receiving the maximum interstate migrants are also the hotspots of coronavirus disease. As the government imposed the lockdown, the workers began to migrate to their homes through roads in absence of transportation, and it is one of the factors responsible for the spread of coronavirus disease to new locations especially to states from where the large number of migrants belong to. The start of Shramik Special Trains which carried workers to their respective homes led to increase of coronavirus patients in different states of India.

From Fig. 13.2a, in Uttar Pradesh, during the first lockdown phase from March 25 to April 14, 2020, the total cases recorded in 21 days were 625 with 8 deaths, whereas in the second lockdown phase from April 15 to May 3, 2020, the total cases recorded in 19 days were 1985 with 35 deaths. In the third lockdown phase from May 4 up to May 17, 2020, cases decreased to 1819 with 69 deaths in just 14 days, which increased to 3611 with 105 deaths during the fourth lockdown phase from May 18 to May 31. And from June 1 to June 14, the total cases in Uttar Pradesh reached to the mark of 5043 in only 14 days with 168 deaths.

From Fig. 13.2b, in Bihar, during the first lockdown phase from March 25 to April 14,

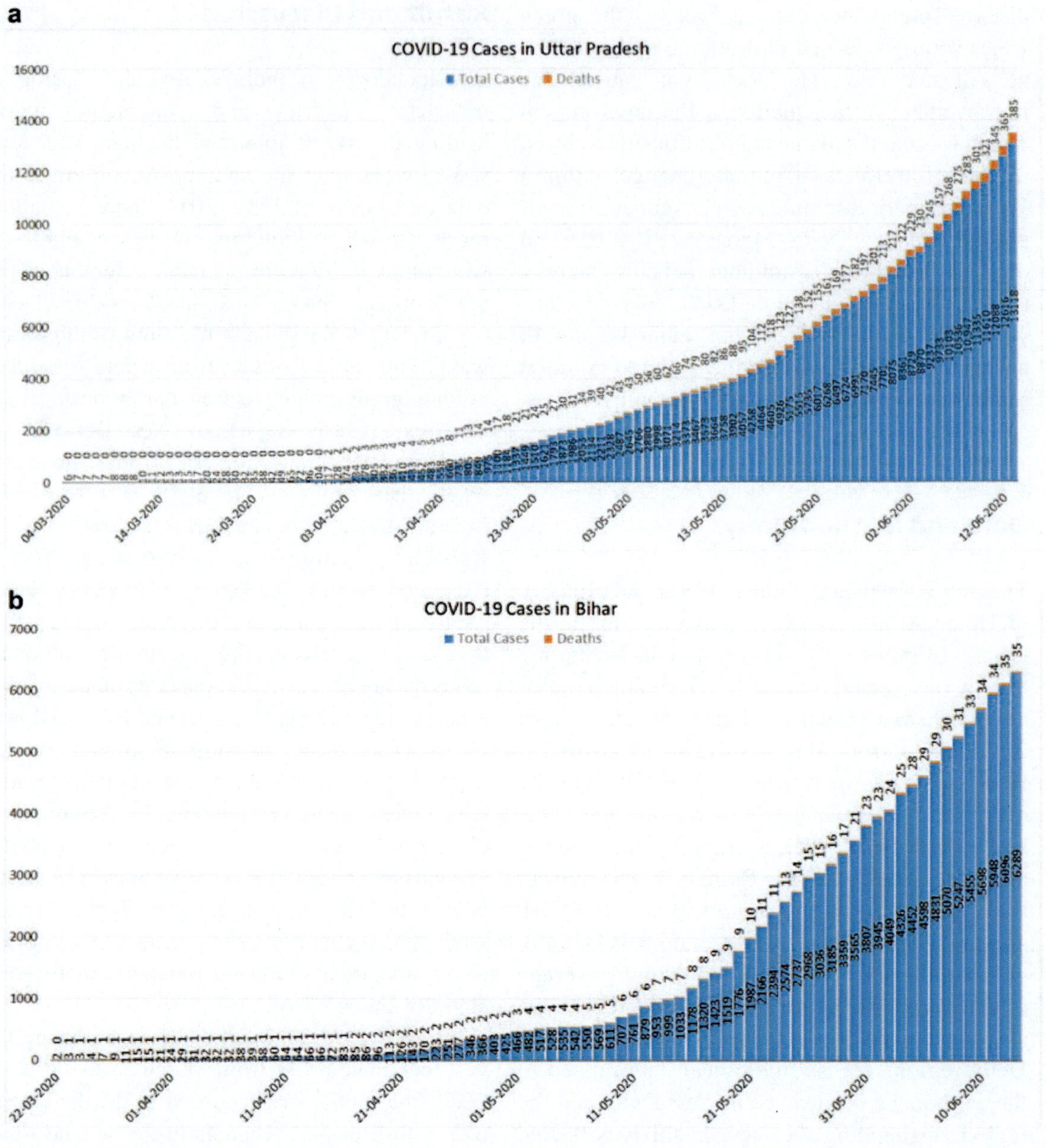

Fig. 13.2 The time-series presentation of coronavirus disease cases and deaths in the states of (**a**) Uttar Pradesh, (**b**) Bihar, (**c**) Madhya Pradesh, (**d**) Rajasthan, (**e**) West Bengal, and (**f**) Jharkhand till June 14, 2020. (Source MoHFW)

2020, the total cases recorded in 21 days were 63 with zero deaths, whereas in the second lockdown phase from April 15 to May 3, 2020, the total cases in 19 days were 451 and 3 deaths. In the next lockdown phase from May 4 up to May 17, 2020, the cases in 14 days were 803 with 4 deaths, which increased to 2487 with 15 deaths within 14 days during the fourth lockdown phase from May 18 to May 31, and in June 1 to June 14, 2020, the total cases in only 14 days were 2482 with 12 deaths.

From Fig. 13.2c, in Madhya Pradesh, during the first lockdown phase from March 25 to April 14, 2020, the total cases recorded in 21 days were 734 with 53 deaths, whereas in the second lockdown phase from April 15 to May 3, 2020, the

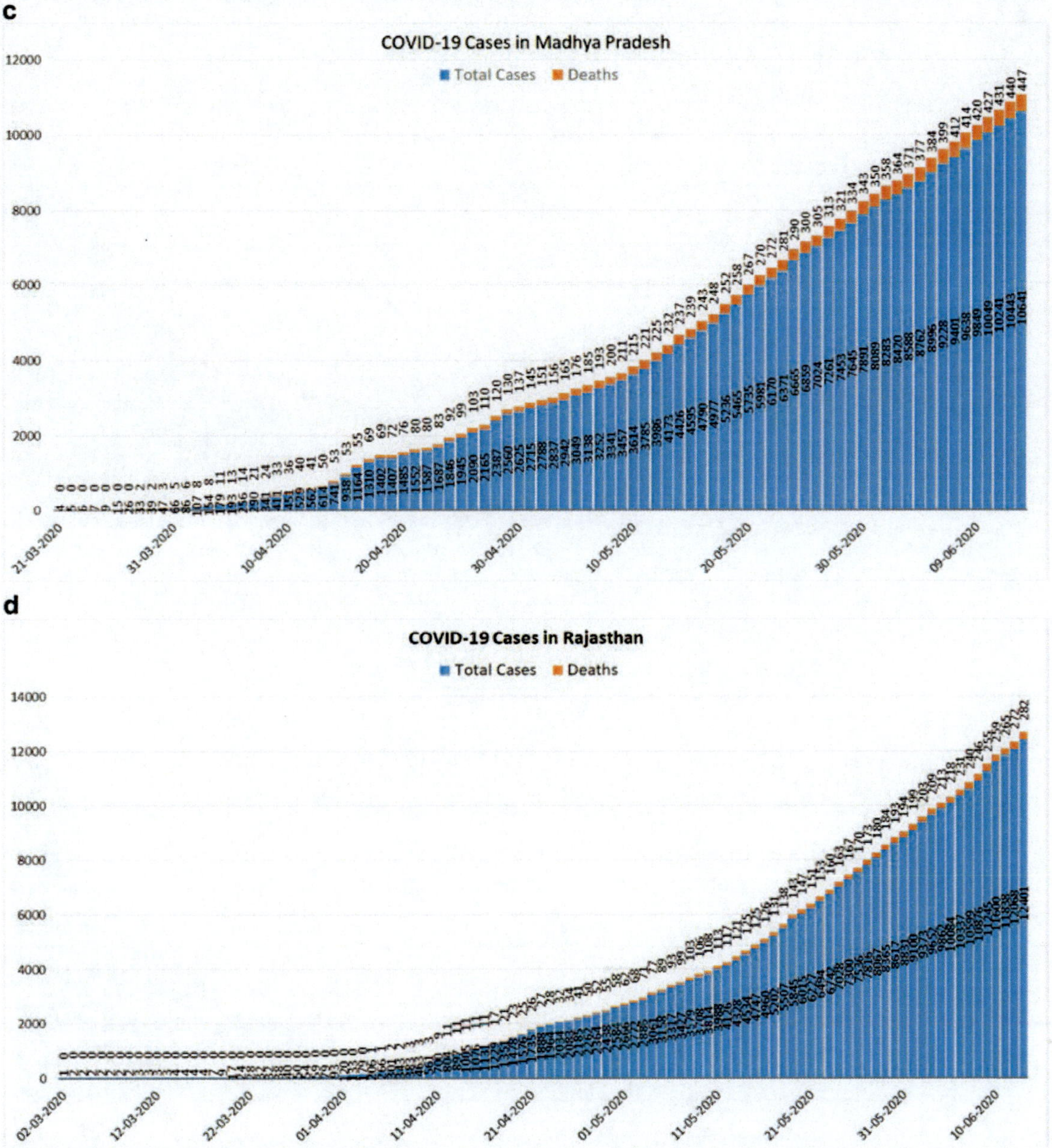

Fig. 13.2 (continued)

total cases recorded in 19 days were 2096 with 103 deaths. In the third lockdown phase from May 4 up to May 17, 2020, the cases in 14 days were 2140 with 92 deaths, which further increased to 3112 with 102 deaths within 14 days during the fourth lockdown phase from May 18 to May 31, and in June 2020, the total cases recorded counted 2552 in 14 days and 97 deaths.

From Fig. 13.2d, in Rajasthan, in the first lockdown phase from March 25 to April 14, 2020, the total cases recorded in 21 days were 973 with 11 deaths, whereas in the second lockdown phase from April 15 to May 3, 2020, the total cases recorded in 19 days were 1881 with 60 deaths. In the third lockdown phase from May 4 up to May 17, 2020, the cases in 14 days were 2316 with 60 deaths. During the fourth lockdown

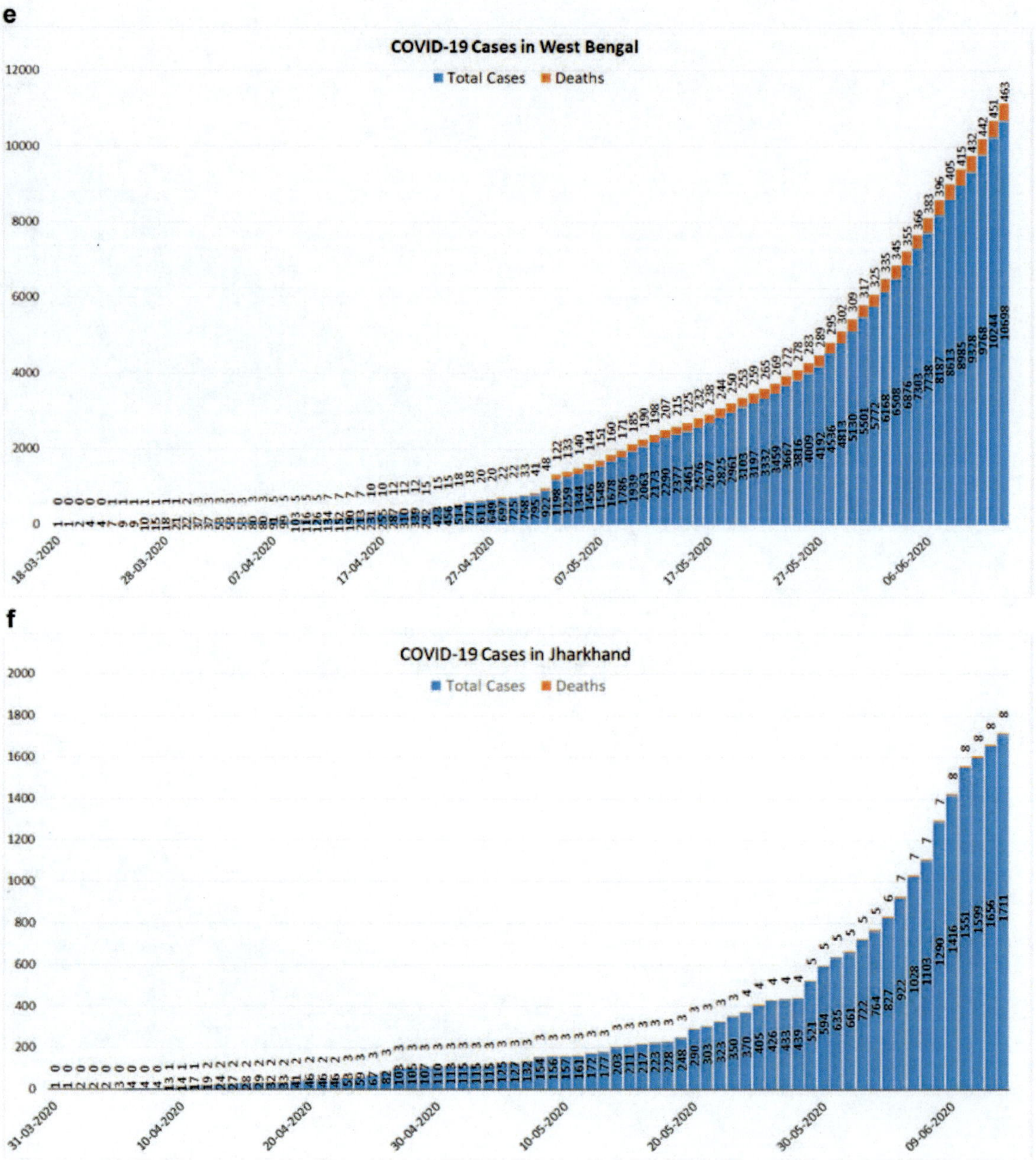

Fig. 13.2 (continued)

phase from May 18, to May 31, the total cases recorded were 3629 with 63 deaths within 14 days, and in June 1 to June 14, 2020, the total cases were 3570 in only 14 days with 88 deaths.

From Fig. 13.2e, in West Bengal, in the first lockdown phase from March 25 to April 14, 2020, the total cases recorded in 21 days were 181 with 6 deaths, whereas in the second lockdown phase from April 15 to May 3, 2020, the total cases recorded in 19 days were 1008 with 115 deaths which is a very high growth. In the third lockdown phase from May 4 up to May 17, 2020, the cases in 14 days were 1479 with 116 deaths. During the fourth lockdown phase from May 18 to May 31, the total cases recorded were 2824 with 79 deaths within 14 days, and in June

1 to June 14, 2020, the total cases were 5197 in only 14 days with 146 deaths.

Lastly, from Fig. 13.2f, in Jharkhand, in the first lockdown phase from March 25 to April 14, 2020, the total cases recorded in 21 days were 27 with 2 deaths, whereas in the second lockdown phase from April 15 to May 3, 2020, the total cases recorded in 19 days were 88 with 1 death. In the third lockdown phase from May 4 up to May 17, 2020, the cases in 14 days were 108 with no deaths. During the fourth lockdown phase from May 18 to May 31, the total cases recorded were 412 with 2 deaths within 14 days, and in June 1 to June 14, 2020, the total cases were 1076 in only 14 days with 3 deaths.

From this study, it is evident that Madhya Pradesh and Rajasthan have more coronavirus disease cases and deaths in the first phase of lockdown because of sharing boundaries with states such as Maharashtra, so the daily waged workers who were traveling by road by walking reached early in these states in comparison to other states. The analysis shows that cases recorded in April were less compared to the cases in May as on May 1, 2020, the Indian government run the Shramik Special Train and other transportation facility in the country. But since these workers, especially the daily waged workers, were traveling from hotspot regions of coronavirus disease, many of them became a carrier and let the spread of disease to the states from where maximum population of migrants belong to. Therefore, Uttar Pradesh, Bihar, Madhya Pradesh, Rajasthan, West Bengal, and Jharkhand reported a breakout of corona patients in the month of May and first two weeks of June. In this way, these daily waged migrants became the most vulnerable section of the population during COVID-19 pandemic as well as the carrier of coronavirus, responsible for spreading the disease to new areas.

Though COVID-19 pandemic proved to be fatal to the humans, the lockdown during this time served as a much-needed step in reducing the pollution in India. With the start of the year 2020, the capital city of India, which is Delhi, became the center of air pollution. The main reason behind this was the pollution caused due to vehicle exhaust, industries and factories exhaust release, and increase in dust due to the construction activities combined with fog in winters that led to the formation of smog and decreased the air quality of Delhi.

Most of the PM2.5 pollution mainly comes from combustion. In Fig. 13.3a, in January, the value of PM2.5 was recorded to be more than 350 μg/m^3, but it suddenly drooped to a value ranging from 50 μg/m^3 to 100 μg/m^3 in the months of March to June mainly because during these months, due to lockdown, there has been a decrease in the activity of combustion at the local level. Similarly, Fig. 13.3b shows a decrease in the trend of pollutant PM10, whose main contributions are the road dust and construction dust, from 485 μg/m^3, the highest value recorded in the month of January, to the range of 100–200 μg/m^3 in the lockdown phase from March to June that led to the stop of construction works. And since there was a decrease in vehicle, the road dust also reduced, but we see that there is a gradual increase in the values after May 3 because of the relaxation provided in the lockdown and the starting of the construction work as well as the movement of vehicles on the road. The next Fig. 13.3c is of the pollutant carbon monoxide (CO), which is also a very critical pollutant and is the main constituent of vehicle exhaust. The highest value of 3.2 mg/m^3 was recorded in January which decreased suddenly to 0.5 mg/m^3 after the complete lockdown was imposed from March 25, 2020, but we see that it started to slowly increase in the month of May because people started using vehicles as per the relaxation provided by the government in May 2020. Figure 13.3d shows that there has been no difference in the values of sulfur dioxide (SO_2) before and after the lockdown. The main reason behind this is that Delhi is using BS-6 fuel since many years to minimize the pollution from sulfur dioxide which mainly comes from combustion of coal and diesel. The next figure is Fig. 13.3e which shows the impact of lockdown on the emission of nitrogen dioxide. The main factors responsible for emission of NO_2 are road traffic and fossil fuel combustion. It can be seen that on March 21, 2020, more than 50 μg/m^3 value was observed which reduced to less than 20 μg/m^3 with the start of lockdown, but with the

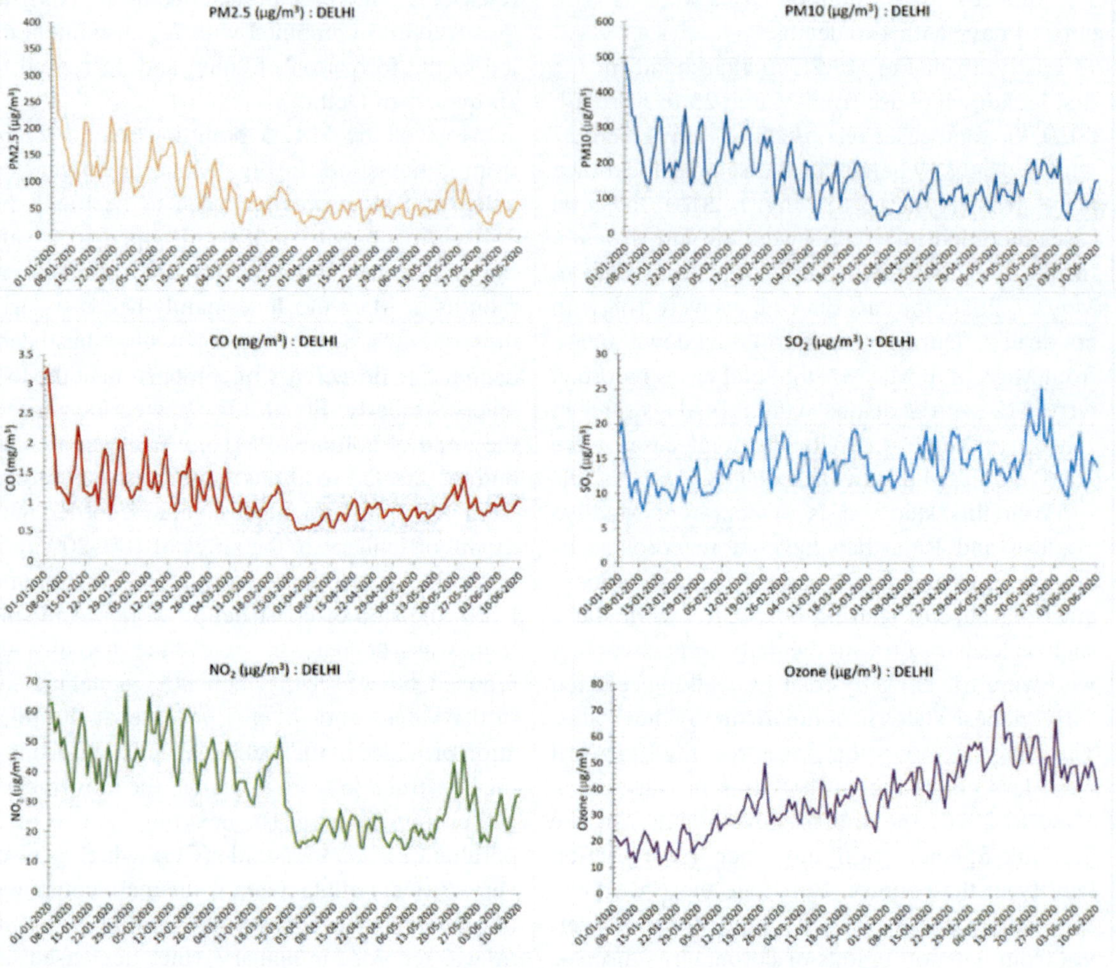

Fig. 13.3 The graphical presentation of air pollutants. (**a**) PM2.5, (**b**) PM10, (**c**) CO, (**d**) SO_2, (**e**) NO_2, (**f**). Ozone. The graph represents the daily average of hourly data of these pollutants from CPCB

start of the fourth lockdown phase, there has been a gradual increase in the value of NO_2 because of increase in traffic and combustion activities. Lastly, Fig. 13.3f shows the effect of lockdown on the values of ozone. Ozone is basically a secondary pollutant, and NO_x is responsible for destroying ozone, but NO_2 forms ozone. Since before lockdown there was more NO_x in the air due to more transportation, low value of ozone is seen in the analysis. Where due to absence of transportation during lockdown there is no NO_x, high values of ozone is available (Table 13.1) (R. P. Singh, 2020).

It was observed that during the lockdown, more frequent earthquakes originated in and around the region of the capital of India, New Delhi. The following table provides the details related to the earthquakes that occurred during the lockdown period. With the use of a Richter scale, it is observed that there has been a total of 16 earthquakes of low magnitude ranging from 1.8 to 4.3 during the lockdown compared to the earthquakes that have occurred in Delhi previously. This is because there has been a significant drop in the ambient noise and vibration on Earth due to the lockdown. Therefore, earthquakes with much small frequency can be detected having more accuracy and specificity. Thus, it could provide us the opportunity to do more study about the Earth crust and effects of ocean waves.

Table 13.1 List of earthquakes that occurred during lockdown from April to June with date and area of occurrence

Date of occurrence of earthquake	Regions that experienced earthquake	Magnitude of earthquake experienced
07 April 2020	Delhi	4.1
12 April 2020	Delhi	3.5
13 April 2020	Delhi	2.7
16 April 2020	Delhi	2
22 April 2020	Delhi	4.3
03 May 2020	Delhi	3
05 May 2020	Faridabad	2.3
10 May 2020	Faridabad	3.4
15 May 2020	Faridabad	2.2
28 May 2020	Faridabad	2.5
29 May 2020	Rohtak	2.9
31 May 2020	Faridabad	4.3
01 June 2020	Rohtak	1.8, 3
03 June 2020	Delhi, Gurgaon Border	3.2
04 June 2020	Rohtak	2.1
08 June 2020	Rohtak	2.1

Source: National Center of Seismology, Ministry of Earth Sciences, Government of India

Conclusion

The above study reveals how the humans have responded during the time of COVID-19 pandemic toward the exodus of migratory workers and reduction in impact of climate change. The decision of the government to run "Shramik Special Trains" during COVID-19 pandemic was a good effort, but due to absence of proper management and lack of implementation measures especially in terms of social distancing inside the train and getting migratory workers quarantined after reaching the destination point, it served as one of the factors in the spread of coronavirus cases to new locations because large migratory workers population belong to the areas that are under red hotspot zones of coronavirus cases, and many of them returning from there acted as a carrier of the disease to the destination where they finally reached. This can be concluded from the above analysis that there has been following increase in the average cases of coronavirus disease in the abovementioned states. That is, in Uttar Pradesh, on a daily average, 30 cases per day were observed during the first lockdown phase, 104 cases in the second phase, 130 cases in the third phase, and 258 cases in the fourth phase, and 389 cases per day were noticed up to June 14, 2020. Similarly, in Bihar Pradesh, on a daily average, 3 cases per day were observed during the first lockdown phase, 24 cases in the second phase, 58 cases in the third phase, and 178 cases in the fourth phase, and 191 cases per day were noticed up to June 14, 2020. In the state of Madhya Pradesh, on a daily average, 35 cases per day were observed during the first lockdown phase, 110 cases in the second phase, 153 cases in the third phase, and 223 cases in the fourth phase, and 196 cases per day were noticed up to June 14, 2020. In Rajasthan, on a daily average, 46 cases per day were observed during the first lockdown phase, 99 cases in the second phase, 165 cases in the third phase, and 259 cases in fourth phase, and 278 cases per day were noticed up to June 14, 2020. In West Bengal, on a daily average, 9 cases per day were observed during the first lockdown phase, 54 cases in the second phase, 106 cases in the third phase, and 202 cases in the fourth phase, and 400 cases per day were found till June 14, 2020. In the state of Jharkhand, on a daily average, 1 case per day were observed during the first lockdown phase, 5 cases in the second phase, 8 cases in the third phase, and 30 cases in the fourth phase, and 83 cases per day were noticed up to June 14, 2020. Since the Shramik Special Trains began in May, this has caused an increase in the coronavirus disease cases in the states from where maximum population of migratory workers belong to after they started reaching there, and it is also observed that daily average cases of coronavirus disease are found to be in more number in months of May and June as compared to in April. Hence, it is proved that exodus of migratory workers is one of the reasons behind the spread of coronavirus disease in India.

Another issue highlighted in this chapter is the impact of coronavirus disease on the pollution status of India during lockdown. The above analysis shows that there has been about 70% decrease in the pollution in the capital city of India, New Delhi, due to decrease in effects of the major critical pollutants as mentioned above as the sources of these pollutants were reduced or cut off during lockdown phase due to COVID pandemic. A series of earthquakes in Delhi being recorded during the lockdown is mainly due to decrease in the ambient seismic noise and vibration. It also enlightens the fact of hitting of major earthquake in the future or prohibits occurrence of major earthquake in the future. However, no scientific basis has been found till now to prove this point. But Delhi and nearby areas are close to the Himalayas and have presence of three fault lines making it a zone of high seismic activity, so there is need to take this matter seriously which is possible as in the time of lockdown, the accuracy of detection of earthquake has increased, providing us opportunity to study the earth's crust and ocean waves.

The issues described above can be sorted by making new adaptation practices in our day-to-day life. The coronavirus disease spread through the "Shramik Special Trains" could have been controlled by executing the proper implementation of the plans of the government such as providing alcohol-based sanitizers, masks, and face shields, proper thermal screening at the arrival and departure stations of a particular train, following the rule of social distancing inside the train, appointing a medical team in each and every train, and assuring that each and every passenger of the Shramik Special Trains get quarantined after reaching its destination and before stepping out in public. The downfall in the air pollution in New Delhi can be continued in the future by putting control on the human activity and subtracting the source of pollutants. This can be the new chapter in the adaptation strategies in the working of humans which can save our environment in the near future. Therefore, it is worthwhile for air pollution models to test the current scenario where only background level is present. The new adaptation that can be implemented and will be a good idea for the government is the "work-from-home concept" more frequently and so widely so that air pollution can be minimized. Though lockdown has cut air pollution, still it has an indirect effect on controlling COVID-19. Higher air pollution levels could have caused more respiratory problems for people prone to such illness and also elderly people and could have intensified the situation among the patients. Since Delhi and the surrounding areas are getting more sensitive toward the occurrence of earthquakes, there needs to be adaptation in the building and maintenance of the infrastructure, and there is an urgent need to invest in anti-earthquake preventive measures that can be done by promoting the retrofitting of old buildings by the government. The government should lay emphasis on investing in advancing our early warning system related to the seismic activity. Thus, it is the time for the strong human response toward adapting according to the current situation in terms of displacement and climate change scenario in the era of coronavirus disease.

References

Bhagat, R. B., & Keshri, K. (2020). Internal migration in India. In M. Bell, A. Bernard, C.-E. Elin, & Y. Zhu (Eds.), *Internal migration in the Countries of Asia: A cross-national comparison*. Springer International Publishing.

Chatterjee, K., Chatterjee, K., Kumar, A., & Shankar, S. (2020). Healthcare impact of COVID-19 epidemic in India: A stochastic mathematical model. *Medical Journal Armed Forces India*.

Cohen, A. J., Brauer, M., et al. (2017). Estimates and 25-year trends of the global burden of disease attributable to ambient air pollution: an analysis of data from the Global Burden of Diseases Study 2015. *Lancet, 17*, 30505–30506.

Imperial College London. (2020). COVID-19 lockdown caused 50 percent global reduction in human-linked Earth vibrations. *ScienceDaily*. Retrieved 24 July 2020 from www.sciencedaily.com/releases/2020/07/200723143728.htm.

Mahato, S., Pal, S., & Ghosh, K. (2020). Effect of lockdown amid COVID-19 pandemic on air quality of the megacity Delhi, India. *Science of The Total Environment, 730*, 139086. https://doi.org/10.1016/j.scitotenv.2020.139086

Singh, R. P., 2020, Pandemic lockdown sensitizes New Delhi to earthquake risk, Temblor. https://doi.org/10.32858/temblor.087

Singh, V. K., Kumar, A., Singh, R. D., & Yadava, K. N. S. (2011). Changing pattern of internal migration in India: Some evidence from census data. *International Journal of Current Research, 3*(4), 289–295.

14 Human Adaptation During Covid-19 Pandemic: The Role of Perceived Stress and Resilience

Vipul Kumar, Neena Kohli, and Pankaj Tripathi

Introduction

Coronavirus, a novel virus, spread rapidly in India and worldwide. It afflicts the individual's life and has become a life's crisis. Herzlich (1973) observed that when people face illness, particularly a long-term one, their self-perceptions are blurred. Illness signifies many times as a crisis in life. Similarly in the Covid-19 outbreak, an individual's perception is blurred about the consequences. Coronavirus spread from person to person, which came out as a terrible pandemic in the year 2019, and most of the countries of the world were affected by this virus along with India (WHO, 2020). The onset of the epidemic started in China and spread worldwide. India is a developing country in which social and economic as well as political life are affected by this Covid-19.

What Is Pandemic?

Pandemics are widespread outbreaks of infectious diseases that may increase morbidity and mortality worldwide and affect the psychological, economic, social, and political disruption. Morse (1995) and Jones et al. (2008) reported that in the past centuries, an increasing probability of pandemics depended on global travel and integration, urbanization, changes in land use, and a large amount of exploitation and destruction of the natural environment.

The Dictionary of Epidemiology definition of pandemic is accepted worldwide, and it defines a pandemic as "an epidemic occurring worldwide, or over a very wide area, crossing international boundaries and usually affecting a large number of people" (Harris, 2000). Firstly, there is a need to understand epidemics which are defined as "the occurrence in a community or region of cases of an illness … clearly in excess of normal expectancy" (Porta, 2014). Besides, a pandemic is conceptualized as "an epidemic occurring over a very wide area, crossing international boundaries, and usually affecting a large number of people" (Porta, 2014). In a pandemic, a major component is a geographical scale rather than the severity of illness. Covid-19 transmitted from person to person through widespread human infection worldwide. This pandemic has covered a large geographical area and affected the global economy.

However, coronavirus is not the first epidemic or pandemic to affect India as there have been a number of health emergencies experienced in India in the past, and Table 14.1 lists out some of the major epidemics of last two centuries.

V. Kumar (✉) · N. Kohli · P. Tripathi
Department of Psychology, University of Allahabad, Prayagraj, Uttar Pradesh, India

A. R. Siddiqui, A. Sahay (eds.), *Climate Change, Disaster and Adaptations*, Sustainable

Table 14.1 Summary of major epidemics in India

Name	Years	Number of cases	Mortality
First cholera pandemic	1817–1824	4050 million	18 million
Sixth cholera pandemic	1899–1924	8 million	2.1 million
Seventh cholera pandemic	1964–1975	675,188	20,256
Plague epidemic	1896	20 million	440,000
Spanish flu	1918	2050 million	1020 million
Smallpox	1974	61,482	15,000
Plague	1994	693	56
Flu H1N1	2009–2015	50,000	2700
Nipah virus	2018	753 (exposed)	17

Sources: Chaturvedi et al. (2020)

Stages of a Pandemic

Pandemic affects a large number of people and increases morbidity and mortality. It causes more loss of life than an epidemic across various countries on a large geographical scale. As we know, the World Health Organization (WHO, 2020) declared Covid-19 as a pandemic when the disease was severe and transmitted from person to person and spreading quickly at a global level. In a pandemic, the death of people depends on (a) how many persons are infected (numbers of an infected person), (b) severity of disease (its virulence), (c) the number of vulnerable groups or individuals in society, and (d) their prevention efforts and how effective they are. WHO (2009) suggested stages of the pandemic alert system as given below:

Stage 1: A virus is in an animal, and there is no transmission to humans.

Stage 2: An animal virus transmits from animal to human.

Stage 3: In this phase, cases of the virus are scattered in small clusters to humans with limited outbreaks. Also, the virus is confirmed to be spreading from human to human.

Stage 4: Sustained outbreaks at the community level.

Stage 5: Widespread human-to-human transmission in more than one country of one of the WHO regions.

Stage 6: Community-level outbreaks in at least one more country and is different from Stage 5 (Fig. 14.1).

Covid-19 Pandemic: Origin, History, Spread, Characteristics

In November 2019, a novel coronavirus emerged which was named severe acute respiratory syndrome coronavirus 2 (SARS-CoV-2). The first case was diagnosed in Wuhan, China. SARS-CoV-2, a newly named coronavirus disease 2019 (COVID-19), is identified as the cause of illness. In the beginning, people thought that it is transmitted from animal (or birds) source to human, but after a significant increase in the number of infected people, evidence showed that it is transmitted from person to person via airborne droplets (WHO, 2020). Covid-19 is threatening the whole of mankind because its rate of infection and transmission patterns increases rapidly.

Effects of Covid-19

Pandemic affects the individuals as well as society at the cognitive, emotional, social, and economic levels. Covid-19 has affected a large number of people who are facing illness or are being killed by this infectious disease. It affected the day-to-day life because infected persons are mostly asymptomatic and showed same conditions as a normal viral infection such as fever, cold, cough, body malaise, and breathing problems in the initial state (Haleem et al., 2020a, 2020b). But it leads to pneumonia and increases the mortality. To stop the spread of the deadly virus, WHO (2020) suggested several strategies such as adopting hygienic habits (e.g., regularly washing the hands, avoidance of face-to-face interaction, etc.), social distancing, wearing of masks, and

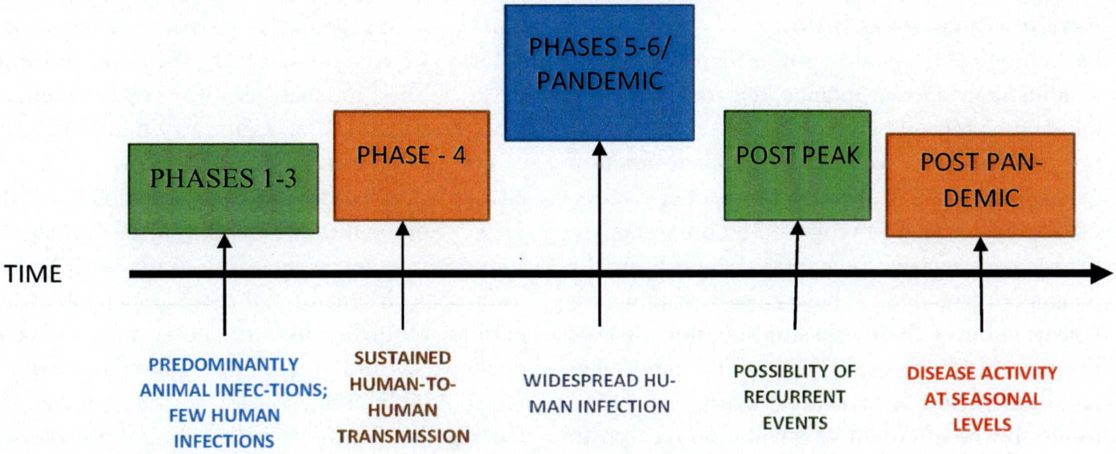

Fig. 14.1 Pandemic influenza phases. (Adapted from WHO (2009))

so on. Challenges of coronavirus infection entered into the lives of individuals with many uncertain consequences in health care, economic, and social aspects (Haleem et al., 2020a, 2020b).

Covid-19 Pandemic as a Stressful Event

Perceived stress is defined as feelings and thoughts of individuals about certainty and uncertainty in their life and how much they appraise changes in their life, problems, and one's ability to deal with problems or difficulties (Phillips, 2013). The uncertainty of symptoms and recovery from infection of Covid-19 leads to negative physical and psychological health (Li et al., 2020). Covid-19 breakout is correlated with psychological distress and symptoms of mental illness or negative mental health (Bao et al., 2020). As a consequence, there is a need to adapt to the challenges to reduce the problems at a physical and psychological level. Biesecker and Erby (2008) defined adaptation as the dynamic and multidimensional process, which is the outcome of the process of appraisal of a health threat. However, adjustment often refers to a desirable state or endpoint (Brennan, 2001).

Personality as Resilient to Adapt the Covid-19 Pandemic

Resilience is conceptualized as a positive adaptation, or the ability to manage or recover mental health, even though facing adverse conditions (Wald et al., 2006). Covid-19 consequences are uncertain and create a negative impact on the individual's life, and it needs to bounce back. Luthar et al. (2000) defined that resilience is a dynamic process allowing for positive adaptation in a context of significant adversity. The framework of disaster resilience states that "*resilience provides an opportunity to confront the social-ecological foundations of risk and development; yet it has been vaguely conceptualized, without offering a concrete approach to operationalization* (Keating et al., 2016). Keating et al. (2016) suggested that resilience is the process of meaningful change in the situation. Covid-19 situations are challenging for individuals in society, so they need to adapt to the situation on a psychological and sociocultural level.

Outcomes of Covid-19 Pandemic

The chapter focuses on issues of livelihood, whether they have perceived threat or risk of coronavirus, and how they manage the crisis and involvement in psychological and sociocultural adaptation.

Psychological Adaptation

Psychological adaptation is considered as similar to adjustment and acceptance. It is measured with the help of related but distinct constructs including psychological well-being, physical functioning, anxiety, and depression (Biesecker & Erby, 2008). Taylor (1983) proposed a comprehensive theory of cognitive adaptation suggesting that people are motivated to have a positive subjective construction of their stressful situation. A positive cognitive appraisal is essential to have a better adjustment in a stressful situation. According to the theory of cognitive adaptation (search for meaning, mastery, and self-enhancement), people try to adapt to the threatening situation (Taylor, 1983). People search for the meaning of their stressful situation and indulge in the process of sensemaking.

Carmer (1998) pointed out in the theory of defense mechanism development that traumatic experiences are faced by individual processes at a conscious and unconscious level, which promotes them to underpin the adaptation process. Cramer proposes that the adaptation process incorporates psychological defense mechanisms and coping strategies as the appropriate means of satisfying individual needs for adaptation to reality. In the scenario of Covid-19, the defense mechanism has a significant role in changing the perception of reality, and it may protect individuals from excessive anxiety, irrational fear, and worry caused by the perception of external events or internal destroying mental states. In this aspect, psychological adaptation is understood as overt or covert actions performed to deal with psychological distress.

Sociocultural Adaptation

Because Covid-19 spreads through human interactions, social distancing rules ranging from voluntary social distancing to forced lockdowns have been widely applied throughout the world to stop the spread of virus. However, such an approach leads to a sudden decrease in human-to-human interactions, a process on which our society is totally dependent. Our sociocultural way of living and thinking such as cultural values (Hofstede, 1980; Schwartz & Bilsky, 1987; Triandis, 1995), self-construals (Markus & Kitayama, 1991), thinking styles (Ji et al., 2001; Nisbett et al., 2001; Spencer-Rodgers et al., 2010), regulatory focus (Higgins et al., 2008; Kurman & Hui, 2011), and so on, have all been upended by the spread of the virus. Sociocultural adaptation is framed as social competence, learning skills, and behavior across circumstantial categories such as situations requiring assertiveness, formal social occasions, and meeting strangers (Argyle, 1969; Argyle et al., 1981; Trower et al., 1978). During the Covid-19 outbreak, people faced a lot of problems at the individual level as well as social level, and the situation demanded individualistic skills rather than holistic skills (group skills). Since cultural values reflect the desired end states that are worth pursuing (Hofstede, 1980; Schwartz & Bilsky, 1987; Triandis, 1995), they are likely to influence members' attentiveness to and prioritization of stressors in the appraisal processes. For example, in a country that values individualism (vs. collectivism), people tend to form an independent (vs. interdependent) self-construal (Markus & Kitayama, 1991) and prefer to use ideal self (vs. ought self) to guide their behaviors (Higgins et al., 2008; Kurman & Hui, 2011).

Theoretical Approaches

Theoretical approaches suggested that perceived stress involves evaluation of the situation (Cohen et al., 1983), which provide the meaning of the situation (psychological adaptation) and improve their health and well-being in terms of psychological and social functioning (Taylor, 1983). Social psychologists used the theory of reasoned action (TRA) as a behavior prediction theory for understanding and predicting the determinants of health behavior (Ajzen & Fishbein, 1980; Fishbein & Ajzen, 1975). Covid-19 demands changes in lifestyle and behavior to improve immunity for better psychological and social functioning. TRA approach focuses on how people use to behave certain patterns because they evaluate situations and behaviors along with using a rational decision-making process in planning appropriate actions (Redding et al., 2000).

As stated by Redding et al. (2000), individuals evaluate the situation and involve in meaningful appropriate behavior through the rational decision to successful adaptation. Bandura's Social Cognitive Theory (SCT), also known as Social Learning Theory, explained further than individual factors in health behavior change to integrate environmental and social factors. SCT has been extensively used to predict health behavior to prevention, health promotion, and modification of an unhealthy lifestyle to protect them for many different risk behaviors. Péry-Woodley (1990) and Baranowski et al. (1997) reveal that SCT explains the thought of people and its effect on their behavior. The approach of SCT is formed by reciprocal determinism—the dynamic interaction between individual, environment, and behavior—where changes in one of these factors may influence the other two. Bandura's STC focuses on personal factors related to cognitive human capacities as the determinants of behavior; it focuses on personal characteristics (demographics, personality, cognitive factors like attitude, thoughts, beliefs, and knowledge), emotional arousing/coping (ability to deal effectively with emotional stimuli with specific techniques, strategies, and activities), behavioral capacity (individual's knowledge and skills), self-efficacy (individual's belief about own capacity to perform a behavior in the various situation), expectations (beliefs about the outcome of a behavior), expectancies (individual's capacity to predict the outcome of a performing behavior), self-regulation (ability to manage/control behavior), observational/experiential learning (acquired behavior through observation and experiences), and reinforcement (consequences of behavior that may increase/decrease the occurrence of behavior) (Péry-Woodley, 1990; Baranowski et al., 1997). Therefore, SCT proposed that the interaction between that individual, environment, and behavior and individual's cognitive and behavioral skills for coping with the situation changes the patterns of behavior during the Covid-19 pandemic.

Covid-19, a novel virus, attacks not just at the individual level, but it affects the whole society on economic, social, and health issues. This research is aimed to (a) see the differences in perceived stress, resilience, psychological adaptation, and sociocultural adaptation among residing areas; (b) study the differences in perceived stress, resilience, psychological adaptation, and sociocultural adaptation among major landforms; (c) study the relationship between perceived stress, resilience, psychological adaptation, and sociocultural adaptation; and (d) find out the predictors of psychological adaptation and sociocultural adaptation.

Method

Research Design

An online survey was designed to study the relationship between perceived stress, resilience, psychological adaptation, and sociocultural adaptation. Covid-19 pandemic situation attracts the researcher to explore the physical, psychological, and social health due to distress, fear, and worry.

The sample consisted of 250 participants from UP and Bihar. The age ranges from 18 to 35 years. Data was collected through an online survey designed with the help of Google Forms application. The questionnaire tapped information regarding the demographics, such as age, gender, and geographical location (residing areas and landforms), including perceived stress scale (PSS-4), resilience scale, psychological adaptation scale, sociocultural adaptation scale, and a general health questionnaire (GHQ-12).

The details of the demographics are given below:

Variables	Descriptives
Age	M = 25.89 (SD = 4.76)
Gender	Male = 146; female = 104
Residing area	City = 163; metro = 35; village = 52
Landforms	Hill/mountain = 20; plane = 187; plateau = 18; sea coast = 25

Measures

1. *Perceived stress scale (PSS-4)*: The perceived stress scale (PSS) consisted of four items, developed by Cohen et al. (1994), which is the most widely used psychological instrument for measuring the perception of stress. It is a measure of the degree to which situations in one's life are appraised as stressful. Cronbach's alpha of this scale was 0.656.
2. *Brief resilience scale (BRS)*: It consisted of six items related to assessing the ability to bounce back, developed by Smith et al. (2008). Cronbach's alpha of this scale was 0.747.
3. *Psychological adaptation scale*: Adaptation of the participants was measured using the 20-item psychological adaptation scale (PAS) (Biesecker et al., 2013). The PAS is used to identify effective coping (coping efficacy), self-esteem, social integration, and spiritual well-being. It focuses on the cognitive, behavioral, and emotional way to deal with risk and major problems; it covers psychological adaptation associated with health risk. It is previously used for assessing the process of adapting to a significant health threat. Cronbach's alpha of this scale was 0.955.
4. *Sociocultural adaptation scale*: It consisted of 21 items which are widely used to measure the competence, learning new skills, and behavior of the individual, developed by Wilson (2013). Cronbach's alpha of this scale was 0.931.

Procedure Firstly, the Google Forms link was generated by the researcher with the instruction and sent to 550 contacts through WhatsApp, Facebook, and messages. But 130 participants did not properly complete the form, 170 participants did not give a response, and 250 participants responded completely.

Results

Data was analyzed through SPSS 22.0. Descriptive analysis was done to calculate the mean and standard deviation. Correlational analysis was done to understand the relationship between variables. Thereafter, hierarchical regression was carried out to find the predictors of psychological and sociocultural adaptation (Tables 14.2, 14.3, 14.4, and 14.5).

For identifying the dominant predictors of psychological adaptation, the Enter Method Model-wise regression analysis was used, and this resulted in five models: Model 1: age and gender; Model 2: age, gender, and perceived stress; and Model 3: age, gender, perceived stress, and resilience. In psychological adaptation, Models 1 and 2 were not predicted significantly. When, Model 3 was introduced, resilience was emerged as the significant predictor of psychological adaptation.

Similarly, for identifying the dominant predictor of sociocultural adaptation, the Enter Method Model-wise regression analysis was used, and this resulted in five models: Model 1: age and gender; Model 2: age, gender, and perceived stress; and Model 3: age, gender, perceived stress, and resilience. In sociocultural adaptation, Model 1 was not found significant, but Model 2 & 3 were predicted significantly. In Model 2, perceived stress was emerged as the significant predictor of sociocultural adaptation, and in Model 3, perceived stress and resilience, both were emerged as significant predictors of sociocultural adaptation.

Discussion

The present research endeavor aimed to see the differences in perceived stress, resilience, psychological adaptation, and sociocultural adaptation among residing areas. Findings suggested that based on residing area, there were no significant differences in personal factors like perceived stress, resilience, and psychological adaptation. Coronavirus, labeled as a deadly virus, induces similar emotions in individuals, so they evaluate the situation in the same way. But on sociocultural adaptation, there were significant differences among people who live in the city, metro, and village. People living in metro city showed higher sociocultural adaptation than the city and village. People from the city were greater on

Table 14.2 One-way ANOVA for showing the differences among residing areas

Variables	City ($N = 163$) Mean (SD)	Metro ($N = 35$) Mean (SD)	Village ($N = 52$) Mean (SD)	F-value
Perceived stress	2.45(0.73)	2.51(0.81)	2.32 (0.70)	0.78
Resilience	3.17(0.65)	3.20(0.87)	3.16(0.75)	0.32
Psychological adaptation	3.62(0.90)	3.87(0.87)	3.50(1.05)	1.32
Cultural adaptation	2.79(0.85)	3.16(0.70)	2.71(0.86)	3.58[a]

[a]Significant at the 0.05 level (two-tailed)

The result showed that there were significant differences in cultural adaptation among different residing areas, but no differences were found on perceived stress, resilience, and psychological adaptation

Table 14.3 One-way ANOVA for showing the differences among major landforms

Variables	Hill/mountain ($N = 20$) Mean (SD)	Plane ($N = 187$) Mean (SD)	Plateau ($N = 18$) Mean (SD)	Sea coast ($N = 25$) Mean (SD)	F-value
Perceived stress	2.68(0.73)	2.40(0.73)	2.45(0.68)	2.41(0.80)	0.88
Resilience	3.35(0.58)	3.18(0.72)	2.97(0.53)	3.12(0.58)	1.06
Psychological adaptation	3.85(0.98)	3.58(0.94)	3.79(0.74)	3.71(1.01)	0.74
Cultural adaptation	2.92(0.85)	2.84(0.83)	2.88(0.86)	2.60(0.865)	0.73

Results revealed that there were no significant differences in perceived stress, resilience, psychological adaptation, and cultural adaptation across major landforms

sociocultural adaptation than village people. Because they can easily learn the thing required in Covid-19 pandemic, they can access the information and better use the available resources. In contrast, village people have a lack of information and accessibility of resources, so they are unable to adapt to the situation successfully.

The second objective was to study the differences in perceived stress, resilience, psychological adaptation, and sociocultural adaptation among major landforms. Findings elaborated that there were no significant differences on perceived stress, resilience, psychological adaptation, and sociocultural adaptation among major landforms, i.e., hill/mountain, plane, plateau, and sea coast. Many reasons can be counted for these similar responses on perceived stress, resilience, psychological adaptation, and sociocultural adaptation. The information regarding Covid-19 was threatening and people have been forced to change their habits, to promote preventive behaviors/measures, and to maintain their immunity (health and well-being). Landforms distribution does not have a significant role in changing the perception of the threat about Covid-19 and their resilience and psychological and sociocultural adaptation.

The third objective was to study the relationship between perceived stress, resilience, psychological adaptation, sociocultural adaptation, psychological functioning, and social function. The findings suggested that perceived stress was significantly associated with sociocultural adaptation and psychological functioning, and it means that the evaluation of actual threat increases the engagement in social behavior to adapt to the situation. Further, resilience was significantly associated with psychological and sociocultural adaptation, which indicates resilience may lead to psychological and sociocultural adaptation.

The psychological adaptation was significantly correlated with sociocultural adaptation and psychological functioning, but it was not significant with social functioning. It indicates that psychological adaptation increases sociocultural adaptation and also leads to psychological functioning but not related to social functioning. However, sociocultural adaptation was also significantly related to psychological functioning, but it was not correlated with social functioning. Due to Covid-19 pandemic, people need more psychological adaptation as well as a social adaptation to achieve greater psychological functioning, but due to maintaining distances, it fails to achieve social functioning. Previous researches indicated that fear of uncertainty or disease may

Table 14.4 Descriptive and correlations among variables

	Mean (SD)	Perceived stress	Resilience	Psychological adaptation
Perceived stress	2.43 (0.74)	1		
Resilience	3.17 (0.68)	0.124	1	
Psychological adaptation	3.63 (0.93)	0.084	0.378[a]	1
Sociocultural adaptation	2.83 (0.83)	0.237[a]	0.0.337[a]	0.397[a]

[a]Correlation is significant at the 0.01 level (two-tailed)
Results showed that perceived stress was positively correlated with sociocultural adaptation, whereas resilience was significantly correlated with psychological and sociocultural adaptation

Table 14.5 Predictors of psychologicaladaptation and sociocultural adaptation (standardized beta)

	Psychological adaptation			Sociocultural adaptation		
Predictors	Model 1	Model 2	Model 3	Model 1	Model 2	Model 3
Age	0.057	0.068	0.060	−0.003	0.026	0.019
Gender	0.009	0.016	0.018	−0.036	−0.019	−0.017
Perceived stress		0.094	0.047		0.239[a]	0.200[a]
Resilience			0.372[a]			0.312[a]
R^2	0.003	0.012	0.148	0.001	0.057	0.153
Adjusted R^2	−0.005	0.00	0.134	−0.007	0.046	0.139
F-value	0.415	0.999	10.63[a]	0.163	4.98[a]	11.06[a]

[a]Correlation is significant at the 0.01 level (two-tailed)

predict the physical/mental health outcomes and negative societal behavior which includes psychological distress, mental health disorders, poor physical health, and health risk behaviors (Shigemura et al., 2020). Findings revealed that fear about coronavirus is a significant mediator of the relationship between perceived risk and mental health problems.

The fourth objective was to find out the predictors of psychological adaptation and sociocultural adaptation. The results indicated that age and gender do not predict the psychological and sociocultural adaptation, whereas resilience emerged as the predictor of psychological adaptation. According to the characteristic approach, resilient personality may have a flexible approach to overcome the adverse condition (Connor & Davidson, 2003; Ong et al., 2006). But if they have no flexible approach to deal with stress, they fail to cope with stress or adversity due to Covid-19 pandemic. Due to pandemic situations, stress arises in individuals which initiate the body to manage the functioning, to recover from obstructions, and to improve abilities to adapt to the threatening situations (Zautra et al., 2010). Luthar et al. (2000) revealed that successfully adapting to the stress and coping with adverse situations positively and effectively are the outcome of resilience. As per the situation of coronavirus infectious disease, people deal with a lot of psychological threats as well as social dysfunctions. Researches highlighted that resilience significantly mediated the relationship between various psychological factors and subjective well-being and psychological health in the context of COVID-19 (Yildirim & Arslan, 2020), revealing that resilience may buffer the negative effects of stress on psychological health (Ong et al., 2006). Further results showed that perceived stress and resilience were significant predictors of sociocultural adaptation. Resilience is a personal factor affecting the psychological functioning in adverse situations and has a buffering role in pandemic-related stress and well-being (Maunder et al., 2008). Researchers suggested that resilience is positively related to stressful situations and life outcomes (Hu et al., 2015), but its impacts on psychosocial functioning and psychological and sociocultural adaptation during coronavirus outbreaks remain understudied with a few exceptions. There is a need to understand why the sociocultural adaptation is essential for surviving because sociocultural adaptation promotes new skills and strategies to improve their well-being.

As per the theory of cognitive adaptation, individuals attempt to manage threats in their life through positive illusions, which contribute to protecting their psychological health (Taylor & Brown, 1988, 1994). From a theoretical perspective, adaptation is a process which is initiated by an external stimulus or life events, and it causes physiological and social responses. Physiological adaptation needs the changes at the physical level to respond to the threatening situation (Helson, 1948, 1964) and demands the changes at the psychological level to respond to the life events (Frederick & Loewenstein, 1999; Wilson & Gilbert, 2008).

Covid-19 induces fear in individuals which demands the psychological and social adaptation through physical distancing, social distancing, and individual's skills. Sociocultural adaptation conceptualized as "the anxiety that results from losing all our familiar signs and symbols of social intercourse" (Oberg, 1960, p. 177). Searle and Ward (1990) suggested that the adjustment with inexperienced signs and symbols of social interaction may be characterized as either psychological or sociocultural. Due to Covid-19 pandemic situation, it demands changes in people's behavior in different aspects, which never happen in their life. This also demands on large scale changes in behavior which requires skills and competence.

Overall results indicated that people's lives are endangered due to Covid-19 pandemic, and its uncertain consequences of health may impact the adaptation, psychological, and social functioning. The work status as work from home emerged, which signifies a new work culture. It requires a lot of new skills and abilities to survive because job requirements are increasing these days to achieve the goal. The current study revealed that resilient people adapt to situations psychologically, and by learning and developing competence in them may improve their sociocultural adaptation. Finally, successful adaptation leads to psychological functioning but not social functioning because Covid-19 decreases people's interaction in society to follow physical or social distancing.

Conclusions

Pandemic affects everyone in society physically, psychologically, and socially. They desire to adapt to the situation successfully to achieve greater psychological and social functioning. It demands a greater level of changes in daily life activities and routines, which dramatically leads to various psychological problems such as worry, disturbance in sleep routine, distress and emotional demands, or social problems like ability to deal with people, learn techniques to work from home, and other social demands through social media or Internet. These changes in Indian culture lead to an imbalance in demand and availability of resources, which may impact the people's ability to adapt to the situation and its consequences. However, a person's own ability to evaluate the situation and to bounce back may have a role in psychological and sociocultural adaptation, whereas successful adaptation improves their psychological functioning.

Limitations and Implications

Limited personal factors are investigated such as perceived stress and resilience. The findings highlight that an individual's ability is more important than demographic variables. People need to improve them physically, psychologically, and socially to achieve greater well-being. This chapter contributes to literature, society, and educators. Further, it gives enlightenment to investigate the role of some social factors in adaptation and psychosocial functioning.

References

Ajzen, I., & Fishbein, M. (1980). Understanding attitudes and predicting social behavior.

Argyle, W. J. (1969). European nationalism and African tribalism. In P. H. Gulliver (Ed.), *Tradition and transition in East Africa* (pp. 41–57). Routledge.

Argyle, M., Furnham, A., & Graham, J. A. (1981). *Social situations*. Cambridge University Press.

Bao, Y., Sun, Y., Meng, S., Shi, J., & Lu, L. (2020). 2019-nCoV epidemic: Address mental health care to empower society. *The Lancet, 395*(10224), e37–e38.

Baranowski, T., Bar-Or, O., Blair, S., Corbin, C., Dowda, M., Freedson, P., Pate, R., Plowman, S., Sallis, J., Saunders, R., Seefeldt, V., Siedentop, D., Simons-Morton, B., Spain, C., Tappe, M., & Ward, D. (1997). Guidelines for school and community programs to promote lifelong physical activity among young people. *Morbidity and Mortality Weekly Report, 50*(RR-6), 1–36.

Biesecker, B. B., & Erby, L. (2008). Adaptation to living with a genetic condition or risk: A mini-review. *Clinical Genetics, 74*(5), 401–407.

Biesecker, B. B., Erby, L. H., Woolford, S., Adcock, J. Y., Cohen, J. S., Lamb, A., Lewis, K. V., Truitt, M., Turriff, A., & Reeve, B. B. (2013). Development and validation of the Psychological Adaptation Scale (PAS): Use in six studies of adaptation to a health condition or risk. *Patient Education and Counseling, 93*(2), 248–254.

Brennan, J. (2001). Adjustment to cancer—coping or personal transition?. *Psycho-Oncology: Journal of the Psychological, Social and Behavioral Dimensions of Cancer, 10*(1), 1–18.

Carmer, P. (1998). Coping and defense mechanisms: What's the difference? *Journal of Personality, 66*(6), 919–946. https://doi.org/10.1111/1467-6494.00037

Chaturvedi, P., Ramalingam, N., & Singh, A. (2020). Is COVID-19 man-made? *Cancer Research, Statistics, and Treatment, 3*(2), 284.

Cohen, S., Kamarck, T., & Mermelstein, R. (1983). A global measure of perceived stress. *Journal of Health and Social Behavior, 24*, 385–396.

Cohen, S., Kamarck, T., & Mermelstein, R. (1994). Perceived stress scale. *Measuring Stress: A Quide for Health and Social Scientists, 10*, 1–2.

Connor, K. M., & Davidson, J. R. T. (2003). Development of a new resilience scale: The Connor-Davidson Resilience Scale (CD-RISC). *Depression and Anxiety, 18*, 76–82.

Fishbein, M., & Ajzen, L. (1975). *Belief, attitude, intention and behavior: An introduction to theory and research* (pp. 181–202). Addison-Wesley.

Frederick, S., & Loewenstein, G. (1999). 16 Hedonic adaptation. Well-being: The foundations of hedonic psychology, 302–329 Russell Sage Foundation.

Haleem, A., Javaid, M., & Vaishya, R. (2020a). Effects of COVID 19 pandemic in daily life. *Current Medicine Research and Practice, 10*(2), 78–79.

Haleem, A., Javaid, M., Vaishya, R., & Vaish, A. (2020b). Effects of COVID-19 pandemic in the field of orthopaedics. *Journal of Clinical Orthopaedics & Trauma, 11*(3), 498–499.

Harris, S. S. (2000). *A dictionary of epidemiology* (4th ed.). Oxford University Press.

Helson, H. (1948). Adaptation-level as a basis for a quantitative theory of frames of reference. *Psychological Review, 55*(6), 297.

Helson, H. (1964). *Adaptation-level theory: An experimental and systematic approach to behavior*. Harper and Row.

Herzlich, C. (1973). *Health and illness: A social psychological analysis* (Vol. 5). European Association of Experimental Social Psychology by Academic Press.

Higgins, E., Pierro, A., & Kruglanski, A. (2008). Chapter 8. Re-thinking culture and personality. https://doi.org/10.1016/B978-0-12-373694-9.00008-8.

Hofstede, G. (1980). *Culture's consequences: International differences in work-related values*. Sage.

Hu, T., Zhang, D., & Wang, J. (2015). A meta-analysis of the trait resilience and mental health. *Personality and Individual Differences, 76*, 18–27.

Ji, L. J., Nisbett, R. E., & Su, Y. (2001). Culture, change, and prediction. *Psychological Science, 12*(6), 450–456.

Jones, K. E., Patel, N. G., Levy, M. A., Storeygard, A., Balk, D., Gittleman, J. L., & Daszak, P. (2008). Global trends in emerging infectious diseases. *Nature, 451*(7181), 990–993.

Keating, A., Campbell, K., Mechler, R., Magnuszewski, P., Mochizuki, J., Liu, W., Szoenyi, M., & McQuistan, C. (2016). Disaster resilience: What it is and how it can engender a meaningful change in development policy. *Development Policy Review, 35*(1), 65–91.

Kurman, J., & Hui, C. (2011). Promotion, prevention or both: Regulatory focus and culture revisited. *Online Readings in Psychology and Culture, 5*(3), 1–16.

Li, S., Wang, Y., Xue, J., Zhao, N., & Zhu, T. (2020). The impact of COVID-19 epidemic declaration on psychological consequences: A study on active Weibo users. *International Journal of Environmental Research and Public Health, 17*(6), 2032.

Luthar, S. S., Cicchetti, D., & Becker, B. (2000). The construct of resilience: A critical evaluation and guidelines for future work. *Child Development, 71*(3), 543–562.

Markus, H. R., & Kitayama, S. (1991). Culture and the self: Implications for cognition, emotion, and motivation. *Psychological Review, 98*(2), 224.

Maunder, R. G., Leszcz, M., Savage, D., Adam, M. A., Peladeau, N., Romano, D., Rose, M., & Schulman, R. B. (2008). Applying the lessons of SARS to pandemic influenza. *Canadian Journal of Public Health, 99*(6), 486–488.

Morse, M. (1995). *U.S. Patent No. 5,380,968*. U.S. Patent and Trademark Office.

Nisbett, R. E., Peng, K., Choi, I., & Norenzayan, A. (2001). Culture and systems of thought: Holistic versus analytic cognition. *Psychological Review, 108*(2), 291.

Oberg, K. (1960). Cultural shock: Adjustment to new cultural environments. *Practical Anthropology, 4*, 177–182.

Ong, A. D., Bergeman, C. S., Bisconti, T. L., & Wallace, K. A. (2006). Psychological resilience, positive emotions, and successful adaptation to stress in later life. *Journal of Personality and Social Psychology, 91*(4), 730.

Péry-Woodley, M. P. (1990). Contrasting discourses: Contrastive analysis and a discourse approach to writing. *Language Teaching, 23*(3), 143–151.

Phillips, A. C. (2013). Perceived stress. *Encyclopedia of Behavioral Medicine*. 1453–1454.

Porta, M. (Ed.). (2014). *A dictionary of epidemiology*. Oxford University Press.

Redding, C., Rossi, J., Rossi, S., Velicer, W., & Prochaska, J. (2000). Health behavior models. *The International Electronic Journal of Health Education, 3*(Special Issue), 180–193.

Schwartz, S. H., & Bilsky, W. (1987). Toward a universal psychological structure of human values. *Journal of Personality and Social Psychology, 53*(3), 550.

Searle, W., & Ward, C. (1990). The prediction of psychological and sociocultural adjustment during cross-cultural transitions. *International Journal of Intercultural Relations, 14*(4), 449–464.

Shigemura, J., Ursano, R. J., Morganstein, J. C., Kurosawa, M., & Benedek, D. M. (2020). Public responses to the novel 2019 coronavirus (2019-nCoV) in Japan: Mental health consequences and target populations. *Psychiatry and Clinical Neurosciences, 74*(4), 281.

Smith, B., Dalen, J., Wiggins, K., Tooley, E., Christopher, P., & Bernard, J. (2008). The brief resilience scale: Assessing the ability to bounce back. *International Journal of Behavioral Medicine, 15*(3), 194–200. https://doi.org/10.1080/10705500802222972

Spencer-Rodgers, J., Williams, M. J., & Peng, K. (2010). Cultural differences in expectations of change and tolerance for contradiction: A decade of empirical research. *Personality and Social Psychology Review, 14*(3), 296–312.

Taylor, S. E. (1983). Adjustment to threatening events: A theory of cognitive adaptation. *American Psychologist, 38*(11), 1161.

Taylor, S. E., & Brown, J. D. (1988). Illusion and well-being: A social psychological perspective on mental health. *Psychological Bulletin, 103*(2), 193.

Taylor, S. E., & Brown, J. D. (1994). Positive illusions and well-being revisited: Separating fact from fiction. *Psychological Bulletin, 116*(1), 21–27.

Triandis, H. C. (1995). A theoretical framework for the study of diversity. In M. M. Chemers, S. Oskamp, & M. A. Costanzo (Eds.), *Diversity in organizations: New perspectives for a changing workplace* (pp. 11–36). Sage Publications, Inc.

Trower, P., Bryant, B., & Argyle, M. (1978). *Social skills and mental health*. University of Pittsburgh Press.

Wald, J., Taylor, S., Asmundson, G. J. G., Jang, K. L., & Stapleton, J. (2006). *Literature review of concepts: Psychological resiliency* (Contract Report W7711–057959/A). Toronto: Defence Research and Development Canada-Toronto.

Wilson, J. (2013). *Exploring the past, present and future of cultural competency research: The revision and expansion of the sociocultural adaptation construct* (Unpublished doctoral dissertation). Victoria University of Wellington.

Wilson, T. D., & Gilbert, D. T. (2008). Explaining away: A model of affective adaptation. *Perspectives on Psychological Science, 3*(5), 370–386.

World Health Organization. WHO Director-General's opening remarks at the media briefing on COVID-19, 11 March 2020. Available at https://www.who.int/dg/speeches/detail/who-director-general-s-opening-remarks-at-the-media-briefing-on-covid-19%2D%2D-11-march-2020. Accessed on 18 March 2020.

World Health Organization, Special Programme for Research, Training in Tropical Diseases, World Health Organization. Department of Control of Neglected Tropical Diseases, World Health Organization. Epidemic, & Pandemic Alert. (2009). *Dengue: Guidelines for diagnosis, treatment, prevention and control*. World Health Organization.

Yildirim, M., & Arslan, G. (2020). Exploring the associations between resilience, dispositional hope, subjective well-being, and psychological health among adults during early stage of COVID-19. *Current Psychology, 39*, 1–11.

Zautra, A. J., Arewasikporn, A., & Davis, M. C. (2010). Resilience: Promoting well-being through recovery, sustainability, and growth. *Research in Human Development, 7*(3), 221–238.

Evolutionary Forces and Human Migration: How Sapiens Conquered the World

15

Syed Ibrahim Rizvi

Introduction

Life evolved on this earth around 3.5 billion years back. The events which might have preceded this most important event are still not fully understood. Perhaps our minds are not tuned to fathom the time span which was involved in the almost infinite reactions which took place between the primordial molecules before a self-replicating structure emerged. This self-replicating molecule could have been the first nucleic acid providing the basis for creation of initial life forms. Single-celled organisms evolved into multicellular organisms. Mutations in the DNA led to formation of new species, and as time passed, a huge kaleidoscope of life forms emerged inhabiting the earth.

Through a process of natural selection, evolutionary forces continued to exert their effect, and new traits emerged in organisms which made them better suited to sustain the prevailing environment (Darwin, 1859). The same process created human ancestors from primates, the record of which are found in Africa. A 200,000-year-old fossil record at Omo Kibish in Ethiopia provides the first documentation of modern *Homo sapiens*.

S. I. Rizvi (✉)
Department of Biochemistry, University of Allahabad, Allahabad, India

Earliest Human Migration

Migration is a natural human behavior inherently associated with the origins of humanity. Humans and their closest ancestors have been known to be migrating to new regions since the history of the genus Homo almost two million years back. Such a historical factual event suggests that migration was in a way related to the process of natural selection, linking success in migration with successful reproduction. Around 110,000 to 60,000 years ago, there was cold and arid condition recurring every 20,000 years in the Northern Hemisphere. The wobble in the earth's axis became a reason for climatic shifts. These climatic conditions made the green corridors between Africa and Eurasia possible, thus making the migration of Homo sapiens to uncharted territory possible.

According to the paleontological record, humans started to migrate out of Africa between 60,000 and 70,000 years ago. The primary reason for this huge milestone in humankind's march on this earth is uncertain, but it is speculated that it was somehow dictated by major changes in climatic conditions that were happening around that time; a sudden cold interlude in the earth's climate. This was subsequent to the onset of one of the dark periods of the last ice age. Harsh conditions made it inhospitable for African dwellers. There is genetic evidence to show that there was a sharp decline in population size of Homo

A. R. Siddiqui, A. Sahay (eds.), *Climate Change, Disaster and Adaptations*, Sustainable

sapiens around this time. In fact, the human population came down to less than 10,000. That humankind survived this period is one of the biggest achievements of life on this earth.

Qafzeh and Skhul, the caves in Israel, provide proof of the first archaeological evidence of human migration beyond Africa. These archeological sites, discovered in the 1930s, contain the remains of 11 modern humans. Most of these humans appear to have been buried following predetermined rituals.

With the improvement in the climate sometime around 70,000 years ago, humans prospered in population from a near-extinction level. The population expanded, and some novelty seeking groups of individual explorers started to venture beyond Africa. The first humans to reach the Eurasian landmass traveled across the Bab-al-Mandab Strait (the strait that separates present-day Yemen from Djibouti). These early migrators colonized rapidly along the coast to India and reached Southeast Asia and Australia around 50,000 years ago. This is the first documented migration of humans and was a definite step toward the human colonization of the globe.

Archeological artifacts found at Jwalapuram, a 74,000-year-old site situated in southern India, match those found in Africa during the same period. According to Michael Petraglia, the celebrated anthropologist of the University of Cambridge, who led the archeological expedition, there is no record to confirm the presence of humans at Jwalapuram. It suggests the earliest known settlement of modern humans outside the African continent. There are however some notable exceptions, for example, the dead enders at Israel's Qafzeh and Skhul prehistoric sites (Field et al., 2007).

Around 50,000 years ago, a second wave of African settlers made another journey into migrating to the Middle East and southern Central Asia. It was from these positions that further colonization of the northern regions of Europe and Asia were undertaken by new migratory settlers (Fig. 15.1).

Some 20,000 years back, a group of Asian settlers headed toward the Arctic during the period of the end of ice age. It was during this period that the ice sheets covered the far north and the condensation of the earth's moisture caused the drop in sea levels by more than 300 feet. This significant climatic change created a bridge that connected the Asia to the American continent. The crossing of this land barrier was the last hurdle which humans achieved in their quest for migration to farthest corners of this earth. As early as 15,000 years ago, humans had colonized the land beyond south of the ice, and within the next 1000 years, they had reached all the way to the tip of South America. Interestingly, a very recent study based on radiocarbon dating of some early archaeological sites have provided evidence pointing out that the dates for human occupation of the American continent could be between the past 20,000 and 30,000 years (Gruhn, 2020).

Biological Reasons to Explain Human Migration

Fatty Acids in Nutrition

According to archaeologist Christopher Henshilwood et al. (2004), about 80,000 years ago, modern humans entered into a "dynamic period" which was characterized with an array of innovation. There is evidence for such an event from the South African cave sites at Blombos, Klasies River, Diepkloof, and Sibudu. A strong corroborative evidence is available with regard to the large-scale use of seashells for ornamental purposes, perhaps the first known jewelry. At Klasies River, burned vegetation traces suggest that the earliest hunter-gatherers had the idea that by clearing land, they could stimulate the growth of edible plants.

The extensive presence of seashells suggests that seafood was a major source of nutrition for the humans during this critical period. This nutritional availability at a crucial juncture in human history provided the necessary fatty acids that humans needed to sustain the energy needs of their large brains: "This was the evolutionary driving force for humans to become more cognitively aware, faster-wired, faster-brained, and smarter," says University of Cape Town

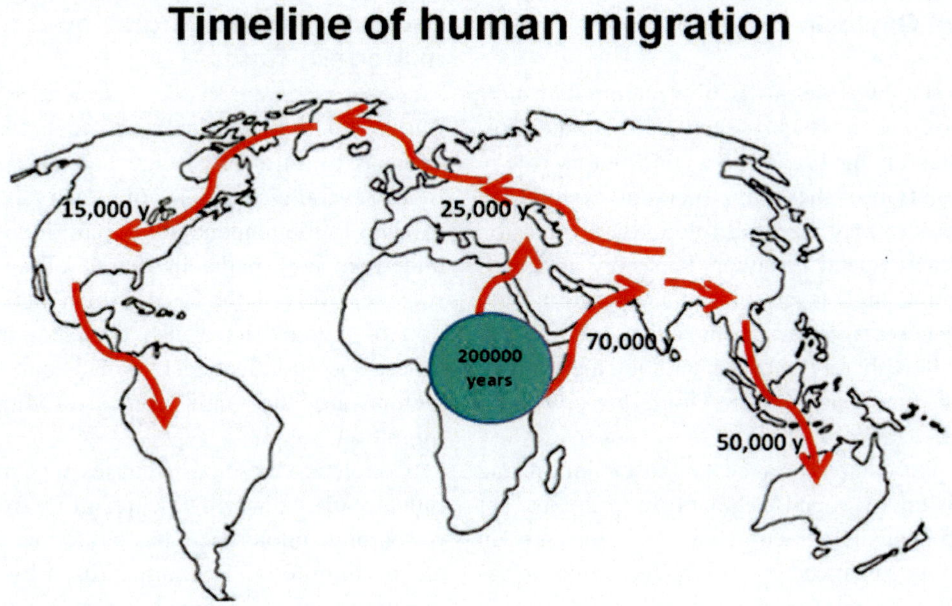

Fig. 15.1 Timeline of human migration

archaeologist John Parkington (Crawford et al., 1999). This period was also linked to the onset of speech.

Evolution of Dopamine Receptor

With the evolution of cognitive abilities subsequent to expansion of brain, other biochemical events also happened which played decisive roles in encouraging the tendency of humans to migrate. Scientists have found a strong association between the dopamine receptor 7R+a gene (DRD4) and the distance to which a population has moved from its original location (Matthews & Butler, 2011). Dopamine is a chemical which is produced by the brain. It plays a role in how we feel pleasure. It plays a big part of our unique human ability to think and plan. Dopamine is known to help humans to strive, focus, and find things interesting. The DRD4 7R+ receptor helps the brain to recognize the availability of dopamine and activate related sensations.

Interestingly, individuals with an active and functional DRD4 7R+ receptor were more inclined to experience pleasure from new activities which translated into exploring new habitats and places. The strong link between migratory tendency and the DRD4 7R+ receptor provides a very interesting insight into the biochemical impulsive factors which made early humans leave their habitat and migrate.

Traverse Arch in Foot

Another biological factor which may have played a role in facilitating human migration was the evolution of the traverse arch in the human foot. Humans evolved to walk and run effectively on the ground using two feet. The arched foot is a characteristic unique to humans compared from other primates. This unique feature promoted bipedalism. This salient feature of human evolution not only made locomotion easier but also freed the two hands which became important tools in communication. The traverse arch in the foot provides it with the stiffness required to act as a lever that transmits the forces generated by muscles of the leg as they push against the ground (Lichtwark & Kelly, 2020). The arch also retains flexibility to function like a compressed spring to store and then release mechanical energy.

Role of Oxytocin

There is a hypothesis that the mammalian hormone oxytocin, which is secreted from the pituitary gland in the brain, had a contributory role in the evolution of the human nervous system and continues to play a central role in the expression of intimate social behavior. It is now acknowledged that high levels of social cognition, and complex social interactions and social bonds would have barely evolved without the physiological functions of the hormone oxytocin (Carter, 2014).

Of particular relevance to the evolution and expression of social behavior in primates are selective social interactions, which in turn rely on sensitivity, enhanced cognition, and communication. Human migratory behavior relied heavily on creation of social bonds. A higher level of oxytocin thus provided the necessary catalyst for social interactions leading to higher survival strategies.

Mutation Affecting Pain Threshold

Neanderthals were tough and resilient, and these physical attributes made them endure harsh weather conditions and protect themselves from dangerous animals. Despite higher physical strength, the Neanderthals were more sensitive to pain. Due to lower pain threshold, their capacity to endure sustained physical activity was limited. Recent findings have established that genetic mutations in DNA some 60,000 years back made humans with higher pain threshold. Genetic mutations happened in a gene called *SCN9A*, and this gene codes for a protein that dictates how signals of painful stimuli are sent to nerves in the spinal cord and brain (Hugo Zeberg et al., 2020). This important mutation in Homo sapiens made possible that humans could endure more hard work for a longer duration compared to more physically strong Neanderthals. Perhaps the higher pain threshold also contributed to humans taking migratory paths to far-off lands.

Factors Affecting Migration in Modern Times

The first human migration was fueled by factors dictated by climate change and availability of food. Several biological evolutionary events contributed to the tendency of certain individuals to undertake new paths in search of novelty. As societies evolved, the forces which dictate migration of humans have also changed. In present times, sociopolitical, economic, and ecological factors are the main forces driving human migration.

Coupled to these basic causes, a spurt in communal violence worldwide, resulting from ethnic or religious intolerance, has led to increased levels of migration. Economic disparity between developing and developed countries often results in the movement of skilled labor from the former to the latter. Scarcity of food and water resources also trigger people to migrate to countries where these resources are more easily available.

Examples of Human Migration in the Modern Times

Human migration has never been uniform, nor there has been a single reason to explain this phenomenon. Several factors have been responsible for human migration. Migration has been caused by war, enslavement, and persecution.

Jews moved out of their lands after repeated exile and the destruction of Jerusalem in 70 AD, creating a huge and widespread diaspora. Historical evidence suggests that 12 million Africans were enslaved and forced to move to the Americas during the slave trade between the 1500 and the 1860 AD. After the World War II in 1945, hundreds of thousands of holocaust victims were displaced to Europe. After the Vietnam War, over 125,000 people from Vietnam migrated to the United States causing a humanitarian crisis of great magnitude.

In the mid of twentieth century, the world saw one of the biggest human migrations during the partition of India and Pakistan. Sadly, this migration was both political and religious. The pangs of this partition are still visible in the societies which were affected by large-scale migration.

Driven by famine, natural disasters, and human rights abuses, migration continues in the twenty-first century. In 2013, migration took place from North Africa and the Middle East into Europe, seeking to escape poverty and political instability in their homelands. Hundreds of thousands of Rohingya inhabitants have been coerced to migrate from Myanmar to Bangladesh.

The homecoming of millions of migrants from Indian metropolitan cities to their villages after the government imposed lockdown in 2020 in the wake of COVID-19 pandemic is an example of reverse human migration. The displacement and the human suffering are prime examples of perils of the modern society.

Conclusion

Intolerance of faith, thought, and speech, growing economic disparities between countries, as well as climate change due to environmental insult would be the key determinants that will drive human migration in the twenty-first century. A 2018 World Bank report observed that around 143 million people may become "climate migrants," displaced from their homes by water scarcity, floods, and droughts. Whatever the reasons be, human migration will continue as long as there are humans and as long as there are opportunities and places to go.

References

Carter, C. S. (2014). Oxytocin pathways and the evolution of human behavior. *Annual Review of Psychology, 65*, 17–39.

Crawford, M. A., Bloom, M., Broadhurst, C. L., Schmidt, W. F., Cunnane, S. C., Galli, C., Gehbremeskel, K., Linseisen, F., Lloyd-Smith, J., & Parkington, J. (1999). Evidence for the unique function of docosahexaenoic acid during the evolution of the modern hominid brain. *Lipids, 34*, S39–S47.

Darwin, C. (1859). *On the origin of species by means of natural selection*. John Murray.

Field, J. S., Petraglia, M. D., & Lahr, M. M. (2007). The southern dispersal hypothesis and the south Asian archaeological record: Examination of dispersal routes through GIS analysis. *Journal of Anthropological Archeology, 26*, 88–108.

Gruhn, R. (2020). Evidence grows for early peopling of the Americas. *Nature*. https://doi.org/10.1038/d41586-020-02137-3

Henshilwood, C., d'Errico, F., Vanhaeren, M., van Niekerk, K., & Jacobs, Z. (2004). Middle stone age shell beads from South Africa. *Science, 304*, 404.

Hugo Zeberg, H., Dannemann, M., Sahlholm, K., Robinson, H. P. C., Kelso, J., & Paabo, S. (2020). A Neanderthal Sodium Channel Increases Pain Sensitivity in Present-Day Humans. *Current Biology, 30*, 3465–3469.

Lichtwark, G. A., & Kelly, L. A. (2020). Ahead of the curve in the evolution of human feet. *Nature, 579*, 31–32.

Matthews, L. J., & Butler, P. M. (2011). Novelty-seeking DRD4 polymorphisms are associated with human migration distance out-of-Africa controlling for neutral population gene structure. *American Journal of Physical Anthropology, 145*, 382–389.

Check for updates

16 Adjustment of the Coastal Communities in Response to Climate Variability and Sea Level Rise in the Sundarban, West Bengal, India

Ashis Kumar Paul and Anurupa Paul

Introduction

Coastal communities of the Sundarban and adjacent areas of the Hugli estuary complex are suffering from the effects of climate variability (e.g., intensity of the SST and increased frequency of severe cyclones), sea level rise vulnerability, shoreline retreat and related displacement of settlements, active land subsidence in and around the southwestern coasts, and the occurrences of tidal floods and associated inundations (Fig. 16.1). About 76 lakhs people are now inhabited in different island sectors of the Sundarban reclaimed areas in West Bengal, India. The height of the delta plain surface is ranging from 1.7 m elevation to 6.0 m elevation above sea level over which the saltwater inundation frequently takes place during the period of astronomic tides and particularly in the events of cyclonic storms. Several studies (Paul & Bandyopadhyay, 1987, Paul, 2002, 2016; Paul, 2018; Sahana et al., 2019, Jana, 2020, Karpytchev et al., 2018; Warrick & Oerlemans, 1990; Nicholls et al., 1995) revealed that only the amount of 26.5 °C sea surface temperature (SST) in the Bay of Bengal can provide the favorable condition for generating low pressure center over the sea surface, but the estimated SST of the northern Bay of Bengal shows 32 °C to 34 °C temperature range during the first week and second week of the month of May 2020. Such high values of the SST are related to the effects of global warming in the tropical seas and provide vulnerable condition for generating and intensification of the tropical cyclone storms at present. The high SST values are inducive to the mechanism of thermal expansion of seawater in the local area to raise the sea level. Relative sea level vulnerabilities of the northern Bay of Bengal fringe coastal belts are also influenced by deltaic subsidence, sediment-filled estuaries and tidal rivers, and also the fluctuations of atmospheric pressure over the sea surface of the region. The studies on the shifting of cyclonic tracks and their landfall points along the coastline also represent dynamic behaviors in the northern Bay of Bengal particularly from the Mahanadi Delta (Odisha, India) to Irrawaddy Delta (Mayanmar) since 1970 to 2020.

It is also observed in the Sundarban coast during and after the events of a few cyclone landfalls (Aila Cyclone, 2009, Bulbul Cyclone, 2019; & Umpoon Cyclone, 2020) that the reclaimed parts of the delta plain of the Sundarban are severely affected by salt water inundations or flooding

A. K. Paul (✉)
Department of Geography, Vidyasagar University, Midnapore, West Bengal, India

A. Paul
Department of Remote Sensing & GIS, Vidyasagar University, Midnapore, West Bengal, India

A. R. Siddiqui, A. Sahay (eds.), *Climate Change, Disaster and Adaptations*, Sustainable

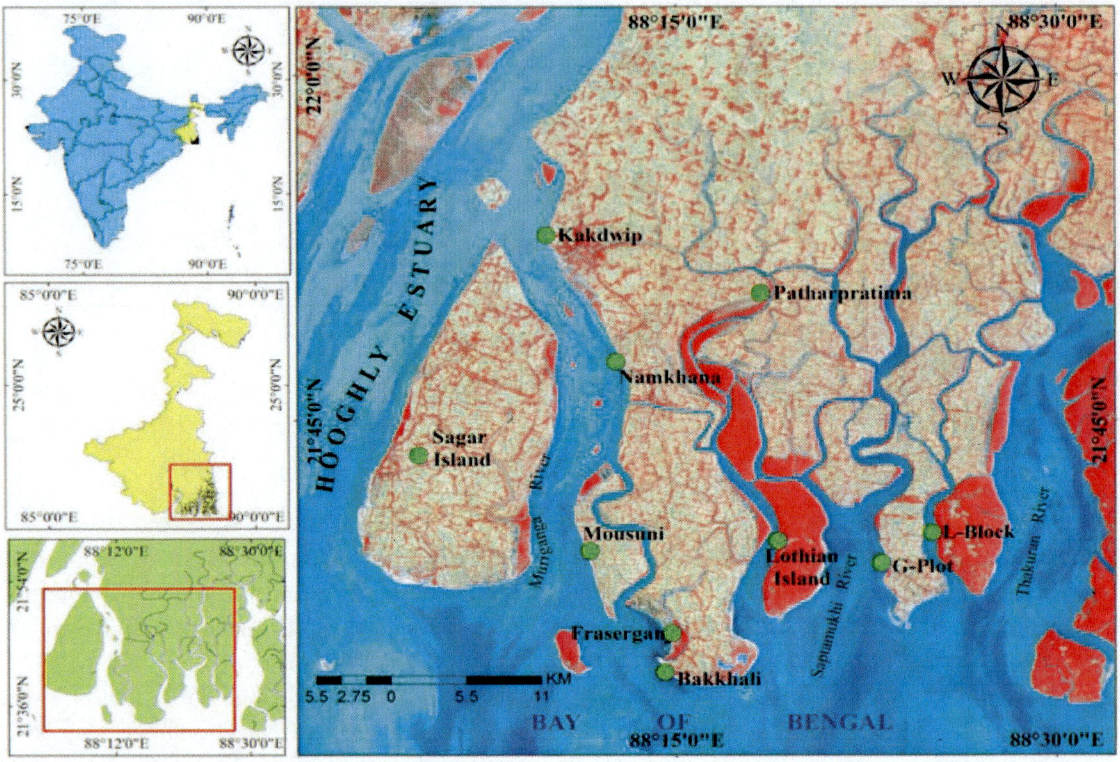

Fig. 16.1 The study area of the western Sundarban

(1.50 m to 2.20 m depths). The coastal villages of the seafront areas, estuary fringes, and bank margins of tidal rivers are demolished by flooding, sediment movements, and shoreline retreats. If there was no physical protection of long earthen embankments (> 3500 km) along the reclaimed margins of the Sundarban, the entire areas (5555 km^2 of the reclaimed tract) might be inundated even in the astronomic tides under present situation. The younger tidal flats of mangrove-dominated delta plain surface with inter-distributary flood basins were reclaimed historically (1793–1941) for permanent settlements of the local people in the region by the British (East India Company) policies in pre-independent India. Younger alluviums were cultivated by rice paddy crops, and village settlements were extended up to the sea faces and also into the remote interior islands of undivided Bengal (24 parganas, Khulna, Bakarganj divisions). Such unfavorable tracts of the deltaic lowlands usually flooded by tides twice daily and densely forested by mangroves were reclaimed by protective covers of earthen embankments and construction of a number of sluice gates across the tidal creeks for drainage and sediment control management in the polderized environment.

Gradual clearings of the mangrove buffers and other wetland vegetation tracts of active tidal spill grounds with overutilization of younger fertile alluviums by producing rice paddy crops over the periods of more than 200 years have exposed the protective environment to the vulnerabilities of rising sea, advancing sea, climate variability, land subsidences, coastal erosion, bank erosion, and coastal flooding at present. Current survey by the present authors depicts that most of the village settlements fringed with the seashores and bank margins of estuaries and tidal rivers are demolished in the cyclone landfalls and engulfed by the advancing sea in the previous decades (1970–2020). A large number of coastal communities have been migrated inlands for their rehabilitation after the occurrences of repeated cyclone landfall and the role played by pushing effects of marine environment along the vulnerable sites of the deltaic shorelines.

Materials and Methods

The present work is conducted in the West Bengal portion of Indian Sundarban adjacent to the Hugli-Matla estuary complex, considering the existing literatures, application of remote sensing techniques, use of the IMD cyclone vulnerability data, NOAA sources of the SST data on Bay of Bengal, settlement records of the Sundarban coastal tract, and analysis of the information sheets prepared in uninterrupted field survey activities over the previous decades about the responses of the inhabited people, mangrove buffers, and hydro-geomorphological dynamics of the deltaic coast to the tropical cyclones and rising sea level trend of the region (Fig. 16.2).

Review of existing literatures, satellite images (1990, 2010, and 2020), DEM, field survey (1990–2020), contour map, nature of geomorphology, shoreline change assessment, settlement map with villages, SST map of the Bay of Bengal, cyclonic tracks between 1970 and 2020 provided information to carry out the present study.

Result and Discussion

Climate Variability in the Northern Bay of Bengal Coasts (Indian, Sundarban Coasts)

The long-term weather records of the grid data of the Sundarban coastal belts depicted the average increased rate of temperature and rainfall amount (1982–2020). The rate of evaporation during the month of May is highest, and as long as the duration of summer months (March–May) prevails over the region, the impregnation of salts occurs into the topsoil due to encroachment of tide water into the surface depressions by astronomic tides (spring and neap tide phases) and high evapo-

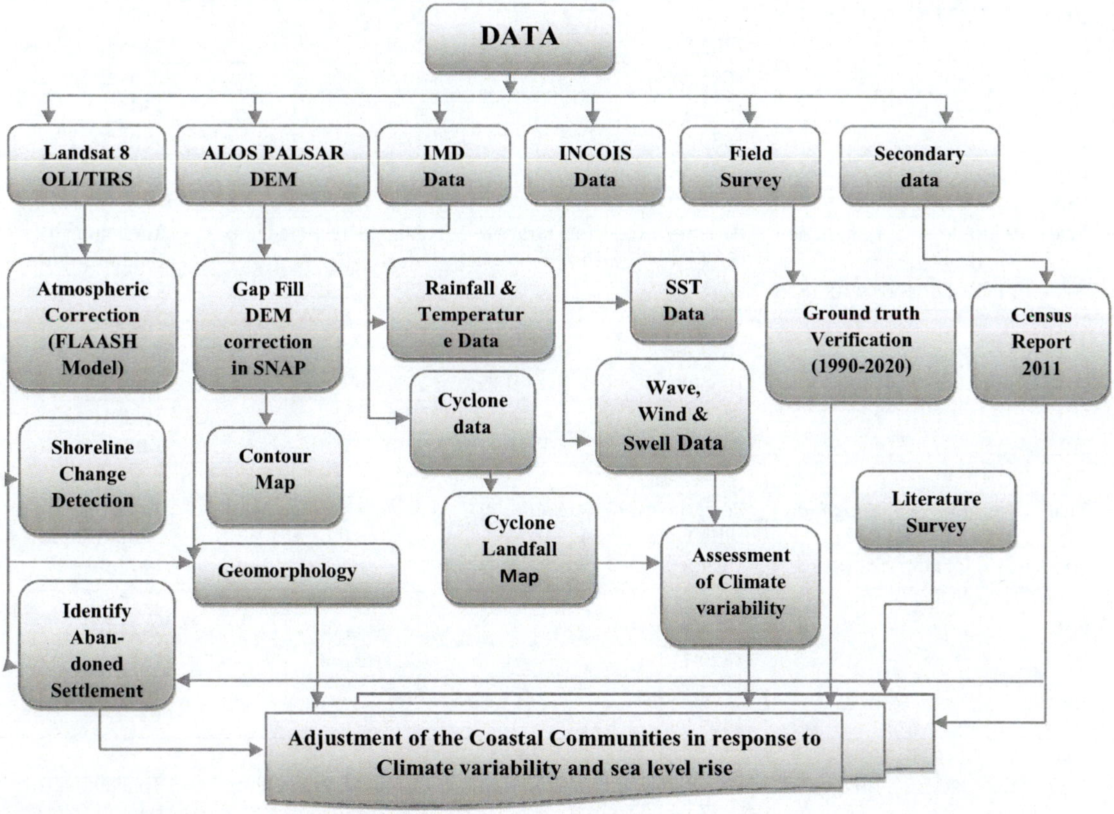

Fig. 16.2 Methodology of the Study area

ration rate of the pre-monsoon phase, respectively (Table.16.1). A large tract of salt pans or salt flats has been developed and expanded over the southwestern part of the Sundarban (Paul et al., 2017). Though the summer monsoon rainfall is reduced due to reduction of mangrove buffers and resulted to limitation of evapotranspiration rate in the region and subsequently decrease in the recycled component of precipitation, the occurrences of storm rainfall in the tropical cyclones sometimes provide high amount of precipitation in the Sundarban coasts (e.g., 1988, 1995, 2007, 2009, 2019, and 2020). Changes in precipitation are also observed (monsoon precipitation) in the coastal belts by the researcher (Supantha Paul et al. 2016) in one of the works on Indian summer monsoonal rainfall. Above all the global warming, induced climate change has increased the variability of temperature and precipitation amount over the land surface areas and also in the sea surface areas of the Bay of Bengal.

The SST data of northern Bay of Bengal since 2002 and up to 2020 reveals the maximum rise of 30.50 °C during the month of May (2009 to 2016). The SST of the month of April is also showing as high as 29 °C to 29.5 °C in the regional extent between 78.0469 E, 5.625 N and 95.625 E, 25.3125 N. However, the mean low SST of the regional extent is also showing as high above 26.5 °C, which allows to develop low pressure centers over the sea (Fig. 16.3).

Table 16.1 Assessment of the climate variability condition of Canning, Kakdwip, Sagar Island, and Namkhana Abad in Sundarban (2000–2018)

Canning					
Year	Latitude	Longitude	Potential evapotranspiration (PET) total (mm)	Annual average temperature (degree C)	Total annual rainfall (in mm)
2000	22°18’37”N	88°39’15”N	1618.5	26.5	2120.5
2010			1702.8	26.9	1135.6
2013			1696.3	26.87	2164.2
2014			1812.1	27.37	1344.7
2015			1880.4	27.64	1894
2017			1862.3	27.57	1668
Kakdwip					
Year	Latitude	Longitude	Potential evapotranspiration (PET) total (mm)	Annual average temp (degree C)	Total annual rainfall (mm)
2000	21°52’41”N	88°10’48”N	1844.2	27.5	1300.5
2010			1869.7	27.6	1280
2013			1724.8	27	1476.5
2014			1950.02	27.9	1490
2015			1869.7	27.6	1670
2017			1922.6	27.8	1480.9
Sagar Island					
Year	Latitude	Longitude	Potential evapotranspiration (PET) total (mm)	Annual average temp (degree C)	Total annual rainfall (mm)
2000	21°43’41”N	88°06’30”N	1724.8	27	1276.9
2010			1864.4	27.58333333	1276.97
2013			1704.7	26.91666667	1421.05
2014			1829.1	27.41666667	1471.21
2015			1844.1	27.5	1664.46
2017			1869.6	27.6	1472.73
Namkhana Abad					
Year	Latitude	Longitude	Potential evapotranspiration (PET) total (mm)	Annual average temp (degree C)	Total annual rainfall (mm)
2018	21°46’14”N	88°13’47”N	1792.3	27.29	1750

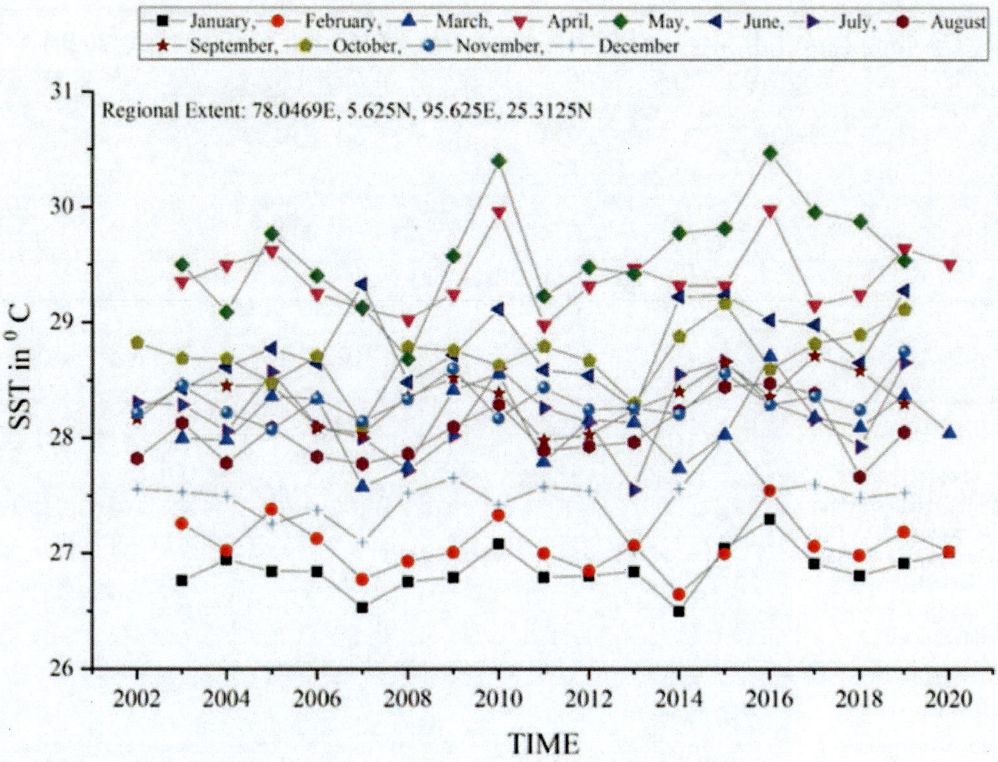

Fig. 16.3 The variation of SST over the years 2002–2020 in the northern part of the Bay of Bengal (Source: Diagram prepared based on the NOAA data)

The frequency and magnitude of cyclones have been increased (e.g., 1970, 1988, 1991, 2007, 2009, 2019, and 2020) in the region of West Bengal and Bangladesh coasts during the previous decades (50 years period). Such landfalls of notorious cyclones along the Bay of Bengal fringed coasts in between "Bhola Island" of Bangladesh and "Sagar Island" of West Bengal have reduced over 350 km^2 forest lands of the Sundarban from the sea face by high rate of coastal erosion. The shoreline change analysis with the temporal remote sensing data also reveals that settlement areas of island fringes on the sea face and on the estuary mouths are devastated and engulfed into the sea. Remnants of settlement structures and submerged forests are visible in the inshore waters exposed in the extreme low-tide phase (Highest Astronomical Tide, or HAT, phase low water limit) during September 15 to October 15 of each year (Fig. 16.4).

Deltaic Subsidence and Relative Sea Level Rise

Delta always tends to subside due to auto compaction method of sediments deposited in the region by Ganga-Brahmaputra-Meghna systems. The subsidence rate was predicted by a group of researchers (Karpytchev et al., 2018) for the case of Ganga-Brahmaputra delta considering the diversity of lithospheric thickness and mantle viscosity of the region. From the outcome of this research, it was predicted that the areas of southwestern Sundarban and the adjacent Hugli estuarine coast will be subsided at the rate of 22 mm per year. A number of quantitative estimates of future sea level has been developed in the previous century with estimate ranging from as low as 0.1 m to as high as 3.7 m by the year 2100 (Warrick & Oerlemans, 1990). The reasons for acceleration of the sea level rise are explained from the global perspectives by a few workers (Nicholls et al., 1995). The subsidence of Ganga

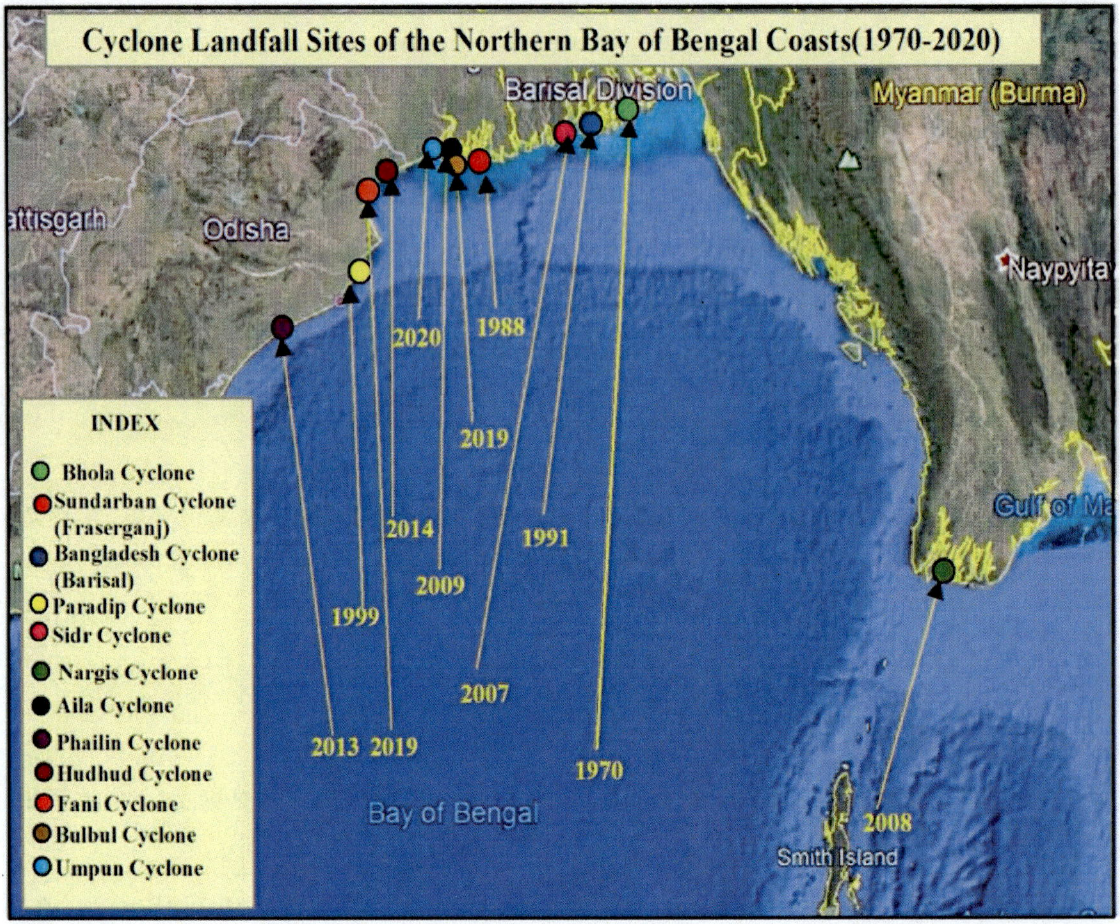

Fig. 16.4 The landfall of devastating cyclones along the coastlines of the Sundarban (1970–2020)

Delta in West Bengal and Bangladesh could be accelerated by relative thickness variations of sediments over the structural framework of the Bengal Basin (Paul, 2000) and their dynamisms. This process can accelerate the relative sea level changes with dynamic sedimentary depositional environment over the deltaic tract in the region. Milliman et al. (1987) estimated the subsidence rate of Ganga Delta in portion of Bangladesh by lithological method as there would be 0.70 m by 2050, and after considering with the global sea level rise of 0.40 m of this period, the relative sea level rise would be 1 m 10 cm (1.10 m) in the region. Intergovernmental Panel for Climate Change (IPCC) also predicted in its last working report (Bahr et al., 2018) regarding the effects and accelerated rates of sea level rise for the case of coastal Sundarban and Ganga-Brahmaputra delta.

Coastal Erosion Rates

The highest rate of coastal erosion is recorded in the region of tropical coasts by shoreline change analysis with temporal remote sensing data. Research predicts that there will be loss of nearly 50% of sandy beaches by the end of the century (Brooks et al., 2020). The shore transects of over two million stations at a spacing of 250 m along the shore around the world's entire coastline from 1984 to 2015 provide a global scale prediction of the position of future shorelines. The maximum shoreline changes depicted from the study area are 45.9 m and 128 m by 2050 and 2100, respectively, due to sea level rise and storminess of the sea in the tropics. Studies in the region of Sundarban coast (Paul, 2000 and Paul et al. 2017) also indicate that the islands of the sea face and estuary banks are affected by rapid erosion after each annual high-energy events (southwest monsoon waves and currents, tidal waves, and swell waves) and episodic cyclone landfall associated with storm surge activities along the seashores of the northern Bay of Bengal. The average annual shoreline retreat is recorded as 16 m in between 1971 and 2000 and similarity estimated as 7 m to 10 m per year in between 2000 and 2019 in the deltaic alluvium coasts (Tables. 16.2 and 16.3). The wider and thicker sandy sea beaches backed by sand dunes of natural buffers with wave energy dissipation mechanisms existed in the early part of 1980s in and around the shoreline of Sagar Island, Mousuni Island, Fraserganj, Bakkhali, Fedric Island, and Henry's Island under the Hugli-Saptamukhi estuarine complex (Paul, 2000, Paul et al. 2016).

However, the underlying mud banks of earlier sandy sea beaches fringed with sand dunes are now exposed due to removal of topsoil by sea level rise, tidal current sources, high wave energies, and related longshore currents parallel to the shoreline in the region. There are no significant sand dunes as well as sandy sea beaches along the sea face areas of the western Sundarban at present particularly within the Hugli-Saptamukhi estuarine complex. Sand size sediments of the earlier sea beaches and degraded sand dunes are now mostly moved in the nearby areas in the forms of sandbars, sandspits, overwash sand fan lobes, inlet filling, and ebb delta accumulations of sands (Paul 2019), and finally, the remaining amount are transported into the offshores and drifted into the estuaries by cross shore currents.

As the beach sands are removed from the top surface and underlying mud banks with tree trunks or mangrove stumps are lowered in surface elevation, the breaking waves during rising and falling tide have etched the mud banks with rilling, pitting, and furrowing mechanism on the seaward face. In general, the older layer was exposed to erosion for an interval of time before deposition of the younger layer and indicated a break in sedimentary geologic record along the shoreface of the Sundarban as an erosional unconformity (Figs. 16.5a and b). The adjustment of the shoreface against the wave and current energies resulted from the landward advancing sea, and vertically rising sea will reduce the foreshore areas or intertidal areas in between the backshores and near shores along the coastal belt with the development of poor resilience capacity against the climate sensitivity. The deeply rooted mangroves (i.e., Excoecaria sp. and Phoenix sp.) over the compact mud banks are now unable to hold the older sediments on the sea face due to horizontal retreat of the shoreline and vertical lowering of the local surface by rapid erosion. Exposed roots of the mangroves along the shoreline over the erosive mud banks represent the critical conditions of the modern Sundarban (Figs. 16.6 and 16.7).

The Coastal Communities of the Sundarban Tract

The Sundarban coastal tracts were not favorable for human habitations before 1810 due to the location of dense mangrove forests and presence of intertidal environments of active saltwater spill grounds in the delta plain islands. The waterlogged forests and swampy tracts of the lower Ganga-Brahmaputra delta were surveyed and demarcated by Dampier and Hodges in 1830 after the imposition of a new rule of landed property for Bengal under the terms of the "Permanent Settlement Act" in 1793 by Cornwallis, the British Governor General of Bengal for the East India Company. Since then, land reclamation and settlement in the Sundarban progressed systematically year after year by putting up marginal embankments along the tidal creeks for stoppage of saltwater intrusion and bringing the reclaimed lands under cultivation as the wetlands proved to be generally fertile in the deltaic alluviums (Fig. 16.8).

However, with the development of "Sagar Island Society" in the western Sundarban in and around the island tracts of downstream section of

Table 16.2 Diversity in coastal erosion in the island of the Sundarban

Sl. no	Name of the island	Area existed and eroded at different period				Record changes during 1971–1991	Present area during 1995–2000
		1971	1975	1986	1991		
1	Sagar Island	233.178 km^2	232.434 km^2	232.125 km^2	223.94 km^2	(–) 9.338 km^2	211.08 km^2
2	Ghoramara Island	8.141 km^2	7.334 km^2	6.794 km^2	5.191 km^2	(–) 2.944 km^2	4.50 km^2
3	Lohachara Island	1.943 km^2	0.536 km^2	Eroded off	Eroded off	Entire island eroded	No trace of the island extension of subtidal sand bodies entire island eroded
4	Nayachar Island (Agnimari char)	17.999 km^2	28.492 km^2	39.791 km^2	40.344 km^2	(+) 22.343 km^2	
5	Suparibhanga Island (Bedford Island)	4.239 km^2	2.925 km^2	1.432 km^2	0.489 km^2	(–) 3.75 km^2	

Table 16.3 The rate of annual shoreline retreat from 1995 to 2020

Sl. no	Location	Total extent of erosion (in m)	Annual rate of erosion (in m/year) 1995–2020	Predicted extent of erosion (2020–2030) in m/year
1	Gobardhanpur	201.678	10.0839	302.517
2	Fraserganj	165.974	8.2987	248.961
3	Bakkhali	161.168	8.0584	241.752
4	Fedric Island	167.624	8.3812	251.436
5	Henry Island	186.142	9.3071	279.213
6	Bishalaxmipur	191.458	9.5729	287.187
7	Beguakhali	203.254	10.1627	304.881
8	Mousuni	143.944	7.1972	215.916
9	Ghoramara	162.351	8.11755	243.5265

Fig. 16.5 (**a**). Exposure of mud bank with erosional unconformity on the sea face (Boastkhali and Sagar Island) and (**b**). removal of topsoils and exposure of mangrove roots on the erosive mud bank (Henry's Island)

the river Hugli, the reclamation works were initiated in a few parts of Sagar Island in the year 1810. Gradually, the rapid land use changes in the lower Bengal over the past centuries resulted in steady depletion of wetlands of brackish water and saltwater environments. The last reclamation tract of "Mousuni Island" (1941 to 1942) is also located in the Hugli-Saptamukhi estuary complex.

The settlements were extended along the banks of tidal rivers and sea faces of the island tracts for easy communication by river transport and also for cultivation of younger deltaic alluviums to grow the paddy crops in the Tropical Moist Climate. The settlement infrastructures gradually developed in the forms of protective embankment structures, water control sluice gates across the embankments, villages, and village roads, historical "storm refuges" for protection of drinking waters and providing the higher grounds for temporary shelters of villagers, plots of agricultural lands after sufficient time of rainwash process, construction of village ponds or freshwater tanks, and mud-built houses along the margins of embankments in the low-lying areas of the deltaic coast. The notorious cyclones of

Fig. 16.6 Shoreline change detection of the seaward side of Sundarban, India (2000–2010)

previous centuries (i.e., 1862, 1864, 1942, 1978 etc.) also devastated the village settlements in the past, but settlers were experienced to adjust with the storms and saltwater inundations of the region by their struggles and dependency on various livelihood opportunities (Table.16.4).

The coastal communities of village settlements located at and around the estuary margins and sea-facing areas of Sagar island, Ghoramara-Lohachara Islands, Mousuni Island, Patibunia Island, Freserganj-Bakkhali Islands, Tat Island, and Namkhana Abad have been suffering since 1970s due to increased frequency and magnitude of tropical cyclones, advancing sea, high rate of shoreline retreats, and saltwater flooding. A number of village settlements have been abandoned in the region, and many villages are partially damaged along the island fringe areas by repeated embankment failures, shoreline shifting processes, saltwater encroachments, and cyclone landfalls in the previous decades. The damaged structures of houses, village tanks, and embankments are visible in the lower parts of "intertidal region" at the astronomic low tides from which the high-tide shoreline has been shifted a few hundred meters inland at present.

The settlements along the coastal belt of Sagar Block, Namkhana Abad, Patharpratima Block, and other adjacent areas of the western Sundarban are mostly affected. The number of household is reduced in the villages partially eluviated by erosion and totally or mostly eluviated in the sites of abandoned villages along the sea face. The large-scale abandonment of village settlements and significant reduction of households in the coastal villages are positively related with the advancing sea, shoreline retreat, and repeated cyclone landfalls in the coastal belt fringed with the northern Bay of Bengal.

Response of the Coastal Communities to the Climate Variability

The shoreline buffers have been weakened all along the coastal fringes of the reclaimed Sundarban due to erosion, sea level rise, and anthropogenic activities. As the qualities of the buffers of sandy sea beaches, shore fringe elevated sand dunes (12–14 m), backshore mangrove wetlands drained with tidal creeks are sufficiently degraded, the sea dykes or the protective embankments behind the natural buffers have been exposed to wave attack immediately along the sea face and estuary banks in the present Sundarban. In many cases, the three parallel embankments are located along the shorelines in degraded conditions in front of the repaired embankments due to repeated storm damage activities and advancing sea with rapid rate of erosion. Villages of the reclaimed Sundarban were located far behind the natural buffers (at least 2 or 3 km inland) of the sea-facing areas even before 1970s in the adjacent areas of the Hugli-Saptamukhi estuary complex but presently abandoned and partially damaged due to eating up of the land margins by erosion and inundations.

Field survey in the islands of Ghoramara, Sagar, Namkhana Abad, and Patharpratima depicted that the increased number of displaced

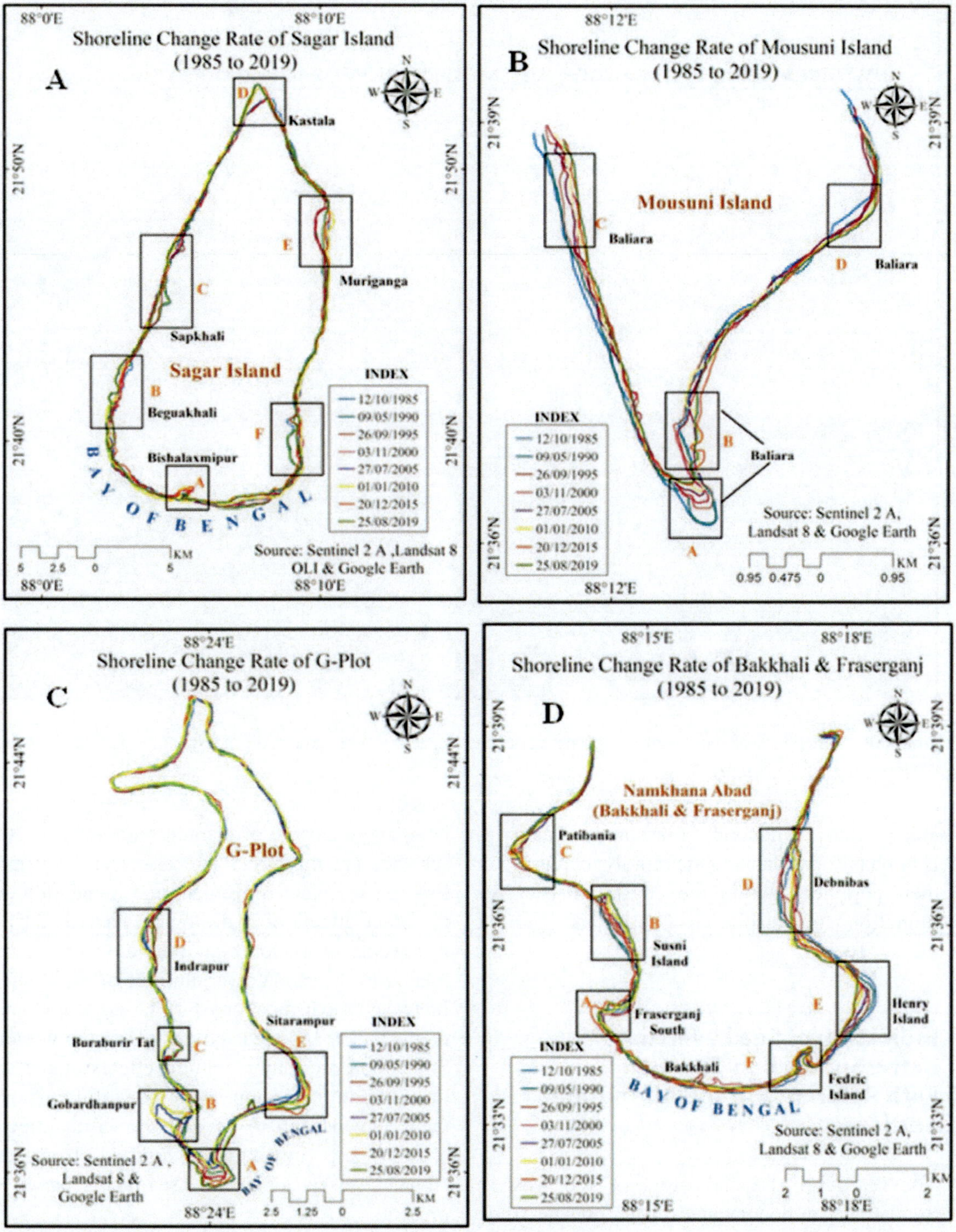

Fig. 16.7 Shoreline change detection of the seaward side to estimate annual rate of erosion in A. Sagar Island, B. Mousuni Island, C. G-plot and D. Bakkhali and Fraserganj, Sundarban, India (1985–2019)

people are totally not rehabilitated in the adjacent areas of open land but mostly migrated inland and resettled in different parts of the smaller urban fringes of the coastal districts. Some urban

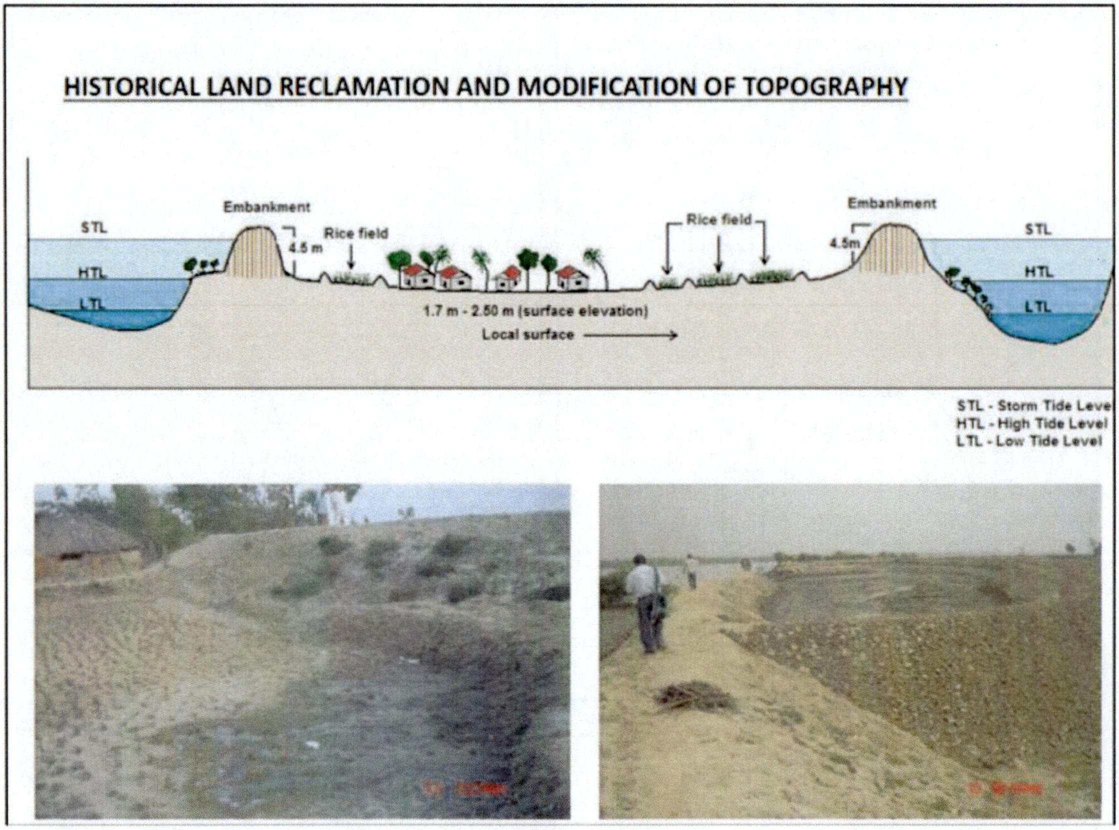

Fig. 16.8 Settlements of reclaimed Sundarban under the impacts of sea level rise process along the banks of tidal rivers

areas like Diamond Harbour, Kakdwip, Namkhana, and Canning are rapidly increasing in their periphery within the coastal district for migration of displaced people at present (Table. 16.5).

Indication of Sea Level Rise in the Sundarban Coasts with Reference to the Abandoned Settlements

The sea level rise indicators of the western Sundarban are documented in the present study with consideration of the significance of abandoned settlements along the coastal fringe landscapes. Removal of beach sands by longshore transport, active overwash process across the shore parallel sand dunes, lowering of the beach dune landscapes, emerging sandspits and sandbars, and exposure of planted mud banks on the sea face are the direct effects of sea level rise. Erosion sediment movement and inundation are the other effects of hydrological hazards during the events of cyclone landfalls along the coast related to the climate variabilities of the tropics. Increased repairment coast of damaged embankments and abandoned settlements of the western Sundarban are major cultural effects of the sea level rise in the region. More powerful embankments are needed to prevent the storm effects; erosion and inundations resulted from the advancing sea along the sea-facing areas at present.

Thus, the coasts of the powerful embankments are increasing after year along the shoreline with increased sea level rise vulnerabilities in the Sundarban. One estimate shows that four hundred crores rupees were to needed to repair the weaker embankments in

Table 16.4 Village settlements in abandoned and in partially damaged conditions in the western Sundarban

Sl. no.	Abandoned villages	Village settlements partially damaged	Year	Number of household
1	Khasimara		1988	0
2	Lohachara		1988	0
3	Baishnabpara		2000	0
4		Ghoramara	2020	1050(5000)
5		Mandirtala	2020	1323(6000)
6		Sapkhali	2020	1370(6598)
7	Boatkhali	Boatkhali	2011	0
8		Shibpur	2020	1930(9209)
9	Bishalakshipur		1988	0
10		Beguakhali	2020	1024(5434)
11	Dublat	Dublat	2020	1400(6098)
12	Rashpur		2000	0
13		Muriganga	2020	590(2570)
14		Baliara	2020	1650(8192)
15		Bagdanga	2020	831(3810)
16	Lakshmipur		1995–2020	232(926)
17		Amarabati	2011	1400(5385)
18	Gobardhanpur		2020	43(970)
19		Indranarayanpur	2020	750(4075)
20		Sitarampur	2020	1030(4228)
21		Buraburir Tat	2020	750(3335)

Table 16.5 The displacement of settlements from the island of the Sundarban in different times

Sl no.	Amount of land loss due to storm-induced erosion	Year of cyclonic storm event during which the land lost	No of displaced resident people or families	No. of people or families rehabilitated
1	Lohachara Island (Hugli estuary) area 11.20 km^2	1981, 1982, and 1985 (entire island eroded off)	350 families	200 families
2	Khasimara Block (part of Ghoramara Island in Hugli estuary) area 8.7 km^2	1965, 1969, 1970, and 1981 (entire area eroded off)	120 families	70 families
3	Ghoramara Island (Hugli estuary) area of erosion 3.5 km^2	1970, 1978, 1981, 1982, 1985, 1988, 1989, 1993, 1995, and 1997 (vast area eroded and the island is reduced in size)	207 families	122 families
4	Suparibhanga Island, Bedford (Hugli estuary) area 4.2 km	1985, 1988, 1989, 1993, and 1995 (entire island eroded off)	10 families	4 families
5	Bisalakshipur Block (Sagar Island) area 2.47 km^2	1932, 1934, 1935, 1936, 1937, 1940, 1942 and 1943, 1948, and 1950 (entire area eroded off)	40 families	8 families
6	Shibpur-Dublat Boatkhali (Sagar Island) area 1.5 km^2	1988, 1989, 1993, 1995, and 1997 (vast area eroded off and still getting eroded at significant rate)	198 families	133 families on a public road
7	Sagar Island (Muriganga Channel Bank) Muriganga, Kachuberia, Shikarpur, Ramkrishnapur Blocks) area eroded 5.45 km^2	1970, 1978, 1981, 1982, 1985, 1988, 1989, 1993, 1995, and 1997 (partially eroded)	400 families	170 families

Data source: Settlement Records,1925 (Govt. of West Bengal), Human perception survey about the problems of the displaced resident people; Report of the District Development Board (Zilla parisad S-24 PGS DT) Change detection study of islands (Vinod Kumar et al. 1994) IMD Pune (1979), Tracks of storms 1877–1970 the Sundarban in 1984; similar estimation was also made by the irrigation department as one thousand crores rupees were needed to repair the weaker embankments in the Sundarban reclaimed tracts in 2000, and again after the event of "Aila" cyclone in the Sundarban in 2009, the total estimated costs to repair the damaged embankments were six thousand crores rupees by the Irrigation Department and Sundarban Development Board. Thus, gradually, the costs of embankment repairment were increased (1984–2000–2010) in the region with increased threats of sea level rise, and it is assessed that there was a positive relationship between the increased protection costs and the increased rate of sea level rise in the low-lying deltaic coasts (Figs. 16.9 and 16.10).

The coastal fringe settlements were mostly or partially abandoned due to the progressive change in shoreline retreat; concentration of waves, tides, and current energies; episodic erosion; inundations; actives of overwash encroachments; and sea level rise process. The erosive land margins are now lying under1.5 m to 2.0 m in elevation after the removal of topsoil and washed by tides twice daily with gradual landward shift of high tides lines (HTL) in the low-lying alluvium coasts. The remaining settlements of the coastal fringe environment affected by repeated inundations are storm damage activities and are liable to tidal floods. Large number of settlements will be displaced or abandoned in the decades to come due to the advancement of sea in the Sundarban-reclaimed tracts (Paul, 2019). Thus, the abandonment of settlements along the seashore is also another indicator of sea level rise in the deltaic coast (Fig. 16.11).

Fig. 16.9 The increased sea level rise vulnerability has increased protection coasts of embankments in the Sundarban

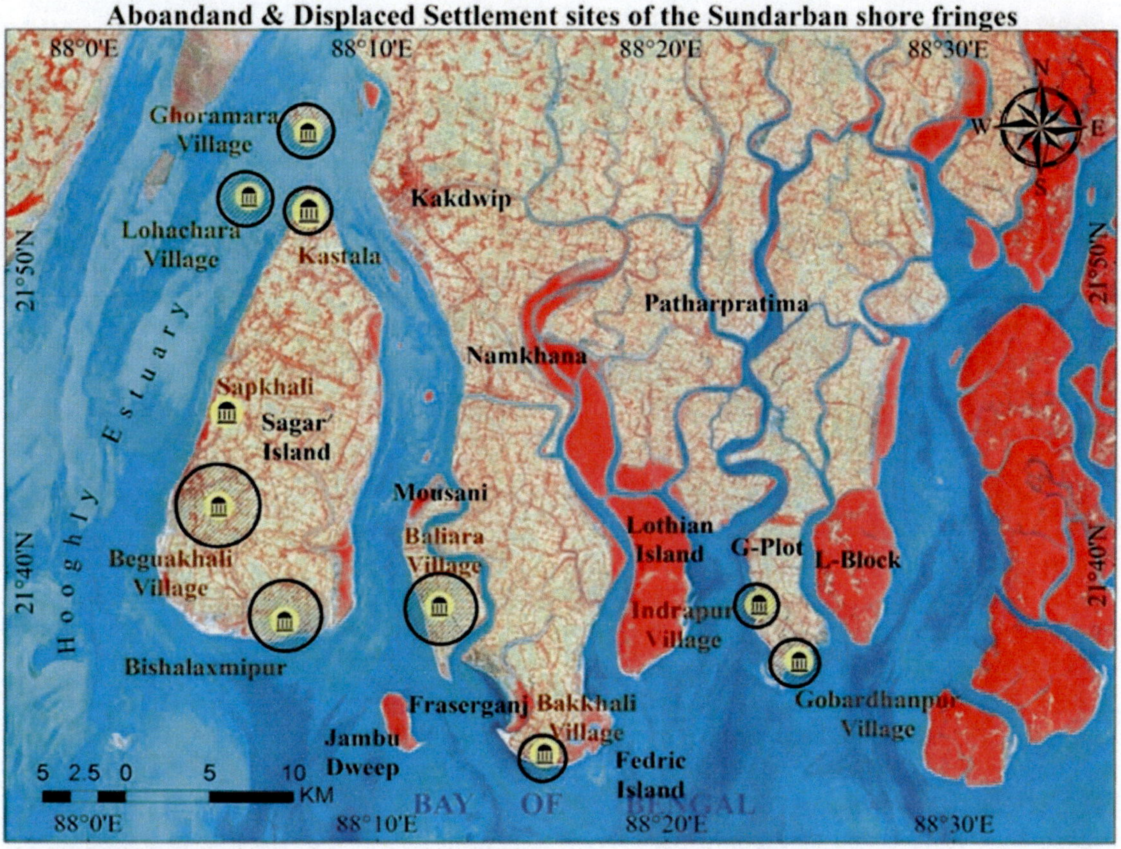

Fig. 16.10 The sites of village settlement abandoned on the shoreface and estuary bank of the western Sundarban (indicative of sea level rise)

Fig. 16.11 Damaged settlements of Boatkhali village on the sea face of Sagar Island

Conclusions

Sundarban coastal communities of the reclaimed tracts along the sea-facing areas are highly vulnerable to cyclone landfalls, storm surges, tidal waves, high rate of shoreline retreats, climate variabilities, and sea level rise process. During the decades of 1970s and 1980s of the previous century, the high embankments and inner island road spaces were utilized by the people affected by tidal waves and cyclones for their temporary shelters. A large number of displaced people was also rehabilitated in the decade of 1990s nearby the open grounds of island fringes after the events of high-magnitude storms. Later on with the cyclone landfalls accompanied with storm surge inundations in 2007 and 2009, the displaced people of the Sundarban were suffering from economic stress due to the losses of agricultural prospects and culture of inland fisheries under village tanks, thus starting migration inland for other options of livelihood activities. The large-scale damages took places, however, in the year 2019 to 2020 cyclones in the coastal fringes and along the river bank margins. Lots of people have also decided to relocate their houses inland further but mostly intended to migrate in the peri-urban areas of the coastal districts of South Bengal after the losses of land and housing properties in the disaster-affected areas. The advancing sea is eating up the land margins gradually along the sea face and also along the estuary banks. More and more people will be displaced in the near future in the coastal Sundarban. Over 10,000 people have been migrated from the Sagar Island administrative block since 1970s.

The fringe areas have been degraded due to loss of the buffers, but attempt is needed to restore the mangroves along the sea faces and estuary banks to reduce the rapid impacts of sea level rise.

Acknowledgments I am indebted to Dr. Sujit Dasgupta of GSI (Former Deputy Director) and Dr. Koushik Gangopadhyay of the University of Calcutta (Archaeology Department) for strongly motivating me in the research on abandoned coastal settlements. Mr. Joydeb Saradar of Remote Sensing and GIS Department of Vidyasagar University is acknowledged for rendering immense assistance during data analysis for the present work. Finally, we are grateful to the students of Coastal Management of Geography Department of Vidyasagar University for their assistance in different stages of the field works (1971, 1918, and 1919) in the western Sundarban.

References

Aila Cyclone (2009). Cyclonic Storm Aila over north-west & adjoining central Bay of Bengal, IMD, Sphere India Secretariat, New Delhi (May 25, 2009) pp. 1–12, https://reliefweb.int/sites/reliefweb.int/files/resources/58AD8134921A2ECD852575C40063743B. Accessed June 2020.

Bulbul Cyclone (2019). Cyclone warning in India standard operation procedure (09-Nov-2019, India Meteorological Department, Ministry of Earth Sciences, Govt. of India, https://mausam.imd.gov.in/imd_latest/contents/pdf/cyclone_sop.pdf, pp. 1–229 Accessed March 2020.

Brooks, S., Pollard, J., & Spencer, T. (2020). Contemporary challenges for shoreline change analysis. In EGU General Assembly Conference Abstracts (p. 8153).

David Bahr (USA) et al. 2018; IPCC Working Committee report on Sea level change. https://www.ipcc.ch/site/assets/uploads/2018/02/WG1AR5_Chapter13_FINAL.pdf

Jana, S. (2020). Micro-level coastal vulnerability assessment in relation to post-Aila landscape alteration at the fragile coastal stretch of the Sagar Island. *India. Regional Studies in Marine Science, 33*, 100908. https://doi.org/10.1016/j.rsma.2019.100908

Karpytchev, M., Ballu, V., Krien, Y., Becker, M., Goodbred, S., Spada, G., & Khan, Z. (2018). Contributions of a strengthened early Holocene monsoon and sediment loading to present-day subsidence of the Ganges-Brahmaputra delta. *Geophysical Research Letters, 45*(3), 1433–1442.

Kumar, K. V., SPaltt, A., Chakrabortt, A. K., Bhan, S. K., Chowdhury, B., & Sanyal, T. (1994). Change detection study of islands in Hooghly estuary using multi-date satellite images. *Journal of the Indian Society of Remote Sensing, 22*(1), 1–7.

Milliman, J. D., Yun-Shan, Q., Mei-e, R., & Saito, Y. (1987). Man's influence on the erosion and transport of sediment by Asian rivers: The Yellow River (Huanghe) example. *The Journal of Geology, 95*(6), 751–762.

Nicholls, R. J., Leatherman, S. P., Dennis, K. C., & Volonte, C. R. (1995). Impacts and responses to sea-level rise: Qualitative and quantitative assessments. *Journal of Coastal Research*, 26–43.

Paul, A. K. (2002). *Coastal geomorphology and environment* (pp. 348–360). ACB Pub, Cal.

Paul, A. K. (2000). Cyclonic storms and their impacts on West Bengal Coasts. In: G.V. Rajamanickam and Michael, J.Tooley (eds.) Quaternary Sea-level Variation, Shoreline Displacement, and coastal environments. New Academic Publishers, Delhi, pp. 8–31.

Paul, Ashis Kumar. (2018) Studies on Coastal morphometry with total station survey in shore face of Sagar Island to assess the impacts of Sea level Rise. Page 460 INCA Volume-38 ISSN: 0927-8392,2019.

Paul, A. K., & Bandyopadhyay, M. K. (1987). Morphology of Sagar Island, a part of Ganga

Delta. *The Journal of Geological Society in India. Bangalore, 29*(4), 412–423.

Paul, A. K., Subrata, J., Amrit, K., Abhinanda, B., Farhin, S., Suman, S., & Dipankar, M. (2016). Problems of Some Erosion Affected Areas of Sagar Island and Proposal for Mangrove Ecotourism Development Sites in a Sustainable Manner. *Eastern Geographer, XXII*(1), 569–582. ISSN 0973–7642.

Paul, A. K., Ray, R., Kamila, A., & Jana, S. (2017). Mangrove degradation in the Sundarbans. In Coastal wetlands: alteration and remediation (pp. 357–392). Springer, Cham. https://doi.org/10.1007/978-3-319-56179-0.

Paul, S., Ghosh, S., Oglesby, R., Pathak, A., Chandrasekharan, A., & Ramsankaran, R. A. A. J. (2016). Weakening of Indian summer monsoon rainfall due to changes in land use land cover. *Scientific Reports, 6*(1), 1–10.

Sahana, M; Rehman, S., Paul, A.K., & Sajjad, H. (2019). Assessing socio-economic vulnerability to climate change-induced disasters: Evidence from Sundarban biosphere reserve, India. Geology, Ecology and Landscapes, 1–13. https://doi.org/10.1080/24749508.2019.1700670.

Umpoon Cyclone (2020). Super Cyclonic Storm "AMPHAN" over the southeast Bay of Bengal (16th – 21st May 2020), Regional specialized Meteorological Centre tropical cyclones, New Delhi, India Meteorological Department, pp. 1–57. https://internal.imd.gov.in/press_release/20200614_pr_840.pdf, Accessed 13-Jun-2020.

Warrick, R. A., & Oerlemans, H. (1990). Sea level rise. In J. T. Houghton, G. J. Jenkins, & J. J. Ephraums (Eds.), *Climate change, the IPCC scientific assessment* (pp. 257–281).

17 Analyzing the Vehicle-Induced Air Pollution and Its Impact in Azadpur Mandi, Delhi

Ekta Raman, Subhash Anand, I. L. Meitei, Piyush Gupta, and Usha Rani

Introduction

Improvement and urbanization over the previous decade have prompted fast increment in the number of inhabitants in Delhi, the metropolitan city of India. Subsequently, there has been a colossal increment in the number of vehicles, which are causing exceptionally significant levels of air contamination. Vehicular outflows are turning out to be the most prevalent wellspring of air contamination in Delhi. The climate can influence air quality, air quality can influence environmental change, and both can legitimately or, by implication, influence well-being. Vehicular contamination is a significant supporter of air contamination in Delhi (Rizwan et al., 2013) wherein the commitment of transport division was assessed to be as high as 72 percent (Goyal et al., 2006). The two primary impacts of environmental change on air quality are weakening of evacuation measures (dispersal, precipitation) and an expansion in the concoction structure of the atmosphere. It is unequivocal that anthropogenic activities such as transportation and industries increments in the all-around blended ozone-harming substances have significantly upgraded the greenhouse effect, and the subsequent driving keeps on expanding (Myhre et al., 2013). Emissions, transport, dispersal, chemical transformation, and deposition of pollutants can be influenced by meteorological variables such as temperature, humidity, wind characteristics, and vertical mixing. It was assessed that around 3000 metric huge amounts of air toxins were radiated each day in Delhi, with a significant commitment from vehicular contamination (67 percent), trailed by coal-based warm force plants (12 percent). There was a rising pattern from 1989 to 1997 as checked by the Central Pollution Control Board (CPCB) (Rizwan et al., 2013). In another study, it has also been reported that approximately 70–80 percent of air pollution in developing countries is caused only by vehicular emission particularly from improper infrastructure of road, low quality of fuel, and large numbers of old vehicles with low vehicle maintenance (Bigazzi & Fighiozzi, 2014). The atmosphere can influence air quality, air quality can influence environmental change, and both can straightforwardly or in a roundabout way influence well-being. The two main effects of climate change on air quality are deterioration of removal processes (dispersal, precipitation)

E. Raman · S. Anand (✉) · U. Rani
Department of Geography, Delhi School of Economics, University of Delhi, Delhi, India

I. L. Meitei
Department of Geography, D.M. College of Arts, Dhanamanjuri University, Imphal, Manipur, India

P. Gupta
Department of Geography, Atarra Post Graduate College, Bundelkhand University, Jhansi, Uttar Pradesh, India

A. R. Siddiqui, A. Sahay (eds.), *Climate Change, Disaster and Adaptations*, Sustainable

and an increase in the chemical composition of the atmosphere. Azadpur Mandi, Delhi, is Asia's most abundant wholesale fruits and vegetable market. It has been announced as the Azadpur market of national importance in India on January 7, 2004. This market is very much crucial for national significance. It provides the distribution center for vegetables like a potato, cauliflower, onion, garlic, and ginger and fruits like banana, orange, apple, mango, etc.

This market was constructed in 1977 by Delhi Development Authorities (DDA). The DDA had allotted the shops to traders. Every day, more than 5000 trucks with millions of tons of fresh fruits and vegetables arrive from all over India (APMC, 2018). The problem of transport in the study area ranges from the inefficiency of roads and transport infrastructure. Lack of mass transportation systems, illegal encroachments on roads, lack of pedestrian facilities, and weak traffic management systems are the major issues of concern in Azadpur Mandi area. Increasing vehicle traffic is creating many parking issues in Mandi. Due to lack of parking space in Azadpur Mandi, people used to park their vehicle on the road; it narrows down the road width. This chapter attempts to analyze the relationship of vehicular population and air pollutants and the associated health problems with air pollution in Azadpur Mandi.

Study Area

The study area is Azadpur Mandi which exists in Delhi, and Delhi has unique position with regard to its location (Anand, 2010). Azadpur Mandi comes under the northwest district of Delhi (Fig.17.1). As per the Census of India 2011, the population of this district is 883,418. Population density is 14, 973 persons per square kilometers.

Data Set and Methodology

The study has used both primary and secondary databases. Qualitative and quantitative methods have been used to fulfill the research. Two hundred respondents were surveyed through a survey questionnaire in Azadpur Mandi. The selected samples included all the working-class section of Azadpur Mandi. That is, from the wealthy class to poor class, youth above 20 years to living older people, men and women, everyone's response is essential. Stratified random sampling technique proofed the ideal for this purpose. For an interview, specific selections of the intellectual person from Agriculture Produce Marketing Committee (APMC) and Town and Country Planning Organization (TCPO) government office are conducted to know their opinion about the traffic issues and to study traffic congestion in Azadpur Mandi, Delhi. In this research area, the secondary data is the registered number of vehicles such as truck, tempo, rickshaws, etc. It has been analyzed in detail and presented using descriptive statistical technique with the help of SPSS software, MS office, and GIS software. All stakeholders including residents, laborers, transport service providers, customers, and shopkeepers have been taken into consideration during the survey to achieve metadata to fulfill the objective requirement. All concerned and service providers/controllers such as Delhi Transport Corporation (DTC), Central Pollution Control Board (CPCB), Municipal Corporation of Delhi (MCD), Delhi Development Authority (DDA), Center for Science and Environment (CSE), Delhi Police, and School of Planning and Architecture (SPA) have also been taken into consideration during preparation of impact-related metadata. Other archive sources like journal, books, newspaper, and reports are also have been consulted and analyzed for the completion of the study.

Results and Discussion

The research mainly discusses the relationship of the total number of registered vehicles and the air quality parameters of Azadpur Mandi. This research also considers the people's perception on the condition of traffic and problems related to air pollution. People have also given perception on health problems due to traveling or working in Azadpur Mandi.

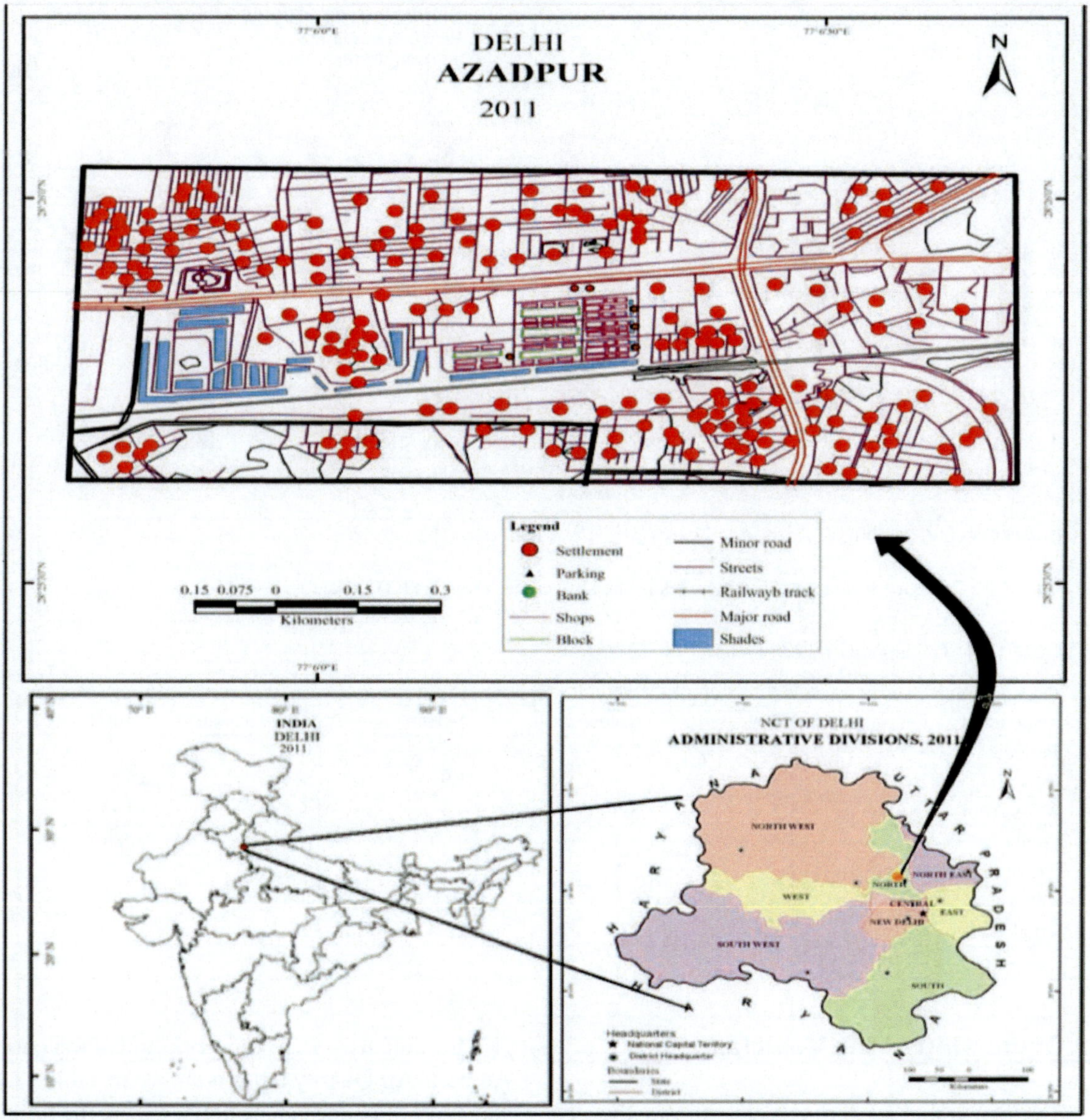

Fig. 17.1 Road map of Azadpur Mandi

Growth of Registered Vehicles

In Azadpur Mandi, more than 5000 vehicles enter every day (APMC, 2018). The numbers of vehicles are increasing day by day. There are many reasons for this increase, and one is that the population growth has risen to 21.6 percent between 2001 and 2011 (Census of India, 2011). People are coming to Delhi for different purposes. Azadpur Mandi provides a wide range of job opportunities to the people of Delhi. There is a yearly development and dissemination of class insightful vehicles in the study area where a light engine vehicle has expanded since 2011, and there has been a fast increment in the quantities of the bikes, the low cost of little vehicles. Direct financing components has elevated individuals to buy four-wheeler and three-wheeler vehicles rather than bikes (Fig. 17.2).

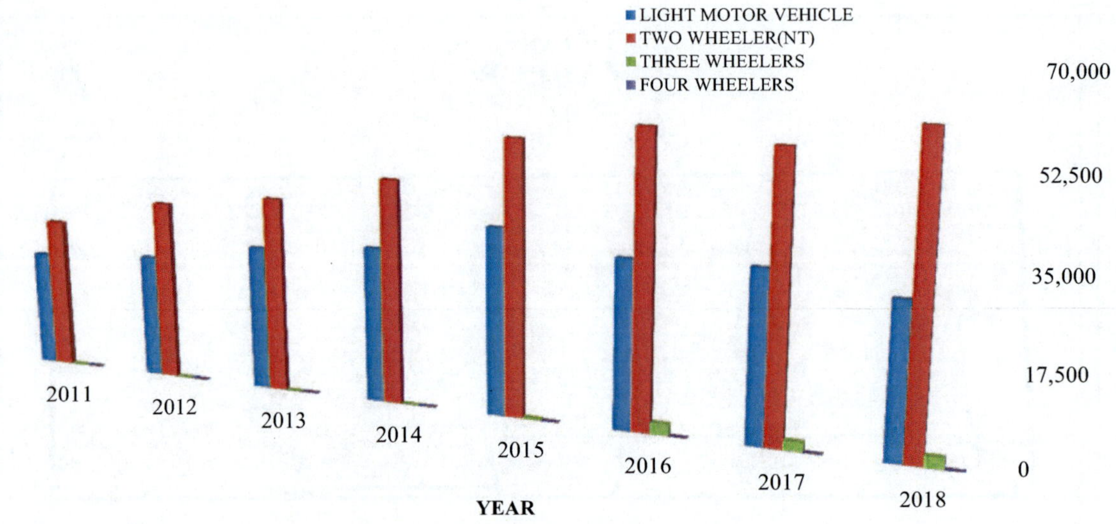

Source: RTO, 2018

Fig. 17.2 Category-wise registered vehicles in the study area. (*Source*: RTO, 2018)

Table 17.1 Ownership of the vehicles

Occupational status of respondent	Ownership of the vehicles (%)				
	Nil	Below 5	5–10	More than 10	Total
Palledar	8.5	5.5	0	0	14
Loader	3	2.5	1.9	1.5	8.9
Massakhor	5.5	3.5	5.5	4.5	19
Aadhti	5.5	5.5	9	8.6	28.6
Truck driver	2	3.5	2	2	9.5
Others	6	6.5	4	3.5	20
Total	30.5	27	22.4	20.1	100

Source: Primary Survey, 2017

Ownership of the Vehicles

The ownership of the vehicles shows the economic conditions of the respondent. With the distribution of the ownership of vehicle based on the occupation of respondents, the research found that Aadhti has the highest number of vehicle in Mandi as they got the highest income. Loaders and Massakhor had many vehicles as they were also high-earning people of Mandi (Table 17.1). There were 20 percent respondents who were having more than 10 vehicles, 22.4 percent of the respondents occupied the vehicle ownership between 5 and 10, 30 percent of the respondents have nil vehicles, and 27 percent of the respondents have vehicle ownership below five in number.

The annual average concentration of the entire Ambient Air Quality index is shown in Table 17.2. The level of ozone can be noticed from the figure, and the highest concentration of ozone in the atmosphere is found in the year 2015, while the lowest concentration of ozone is located in the year 2011 (36.97 μg/m^3).

In Fig. 17.2, with the concentration of PM10, it can be seen that the highest concentration of PM10 in the atmosphere is found in the year 2017 (558 μg/m^3) while the lowest level is found in the year 2015 (337 μg/m^3). This shows an increasing trend of a pollutant from the year 2015, which is recorded at its peak in the year 2017. With the concentration of PM2.5, it can be seen that the highest level of PM2.5 in the atmosphere is found in the year 2017 (378.2 μg/

Table 17.2 Ambient Air Quality Standard (AAQS)

Pollutants	Time weighted average	Concentration in ambient air	
		Industrial area	Residential, rural, and other area
Sulphur dioxide ($SO2$)	Annual average	80 µg/m³	
	24 hours	120 µg/m³	80 µg/m³
Oxide of nitrogen (NOx)	Annual average	80 µg/m³	60 µg/m³
	24 average	120 µg/m³	80 µg/m³
Particulate matter (PM 10)	Annual average	360 µg/m³	140 µg/m³
	24 average	500 µg/m³	200 µg/m³
Particulate matter (PM 2.5)	Annual average	120 µg/m³	60 µg/m³
	24 average	150 µg/m³	100 µg/m³
Carbon monoxide (CO)	8 hours	5.0 µg/m³	2.0 µg/m³
	1 hour	10.0 µg/m³	4.0 µg/m³

Source: CPCB, 2016

m³) while the lowest concentration is located in the year 2015 (202.58 µg/m³). This show an increasing trend of a pollutant from the year 2015, which is recorded at its peak in the year 2017. With the concentration of NOx, it can be seen that the highest level of NOx in the atmosphere is found in the year 2017 (378.2 µg/m³) while the lowest concentration is found in the year 2015 (202.58 µg/m³). These show an increasing trend of pollutant from the year 2015, which is recorded at its peak in the year 2017 (Fig. 17.3).

It can be seen that the highest level of PM2.5 in the atmosphere is found in the year 2017 (251 µg/m³) while the lowest concentration is found in the year 2014 (86.5 µg/m³). These show an increasing trend of a pollutant from the year 2015, which is recorded at its peak in the year 2017. In Fig. 17.3, with the concentration of sulfur dioxide (SO_2), it can be seen that the highest level in the atmosphere is found in the year 2017 (27.6 µg/m³) while the lowest concentration is found in the year 2014 (11.73 µg/m³).

These show an increasing trend of a pollutant from the year 2015, which is recorded at its peak in the year 2017. The main reason of the high level of air pollution at Azadpur Mandi was the increasing number of the vehicles in Azadpur Mandi (Table 17.3), and the ownership of private vehicle creates noxious air emissions in the atmosphere.

Relationship Between Vehicular Pollution and Registered Vehicles

There exists a positive relationship between the number of vehicles entered and the pollution with $r = 0.5727$, but without applying the test of significance, we cannot generalize this positive relationship as the no. of years taken is very few. So the test is carried out as $t = (r\sqrt{n} - 2)/(\sqrt{1} - r2) = 2.030$ (Fig. 17.4).

The tabulated value of t for 10 (12–2) degrees of freedom is 3.17 at 1 percent, 2.33 at 5 percent, and 1.81 at 10 percent. Since the computed value is larger than 10 percent, it can be concluded that the correlation coefficient is significant and may not be considered as zero. So there has been a significant relationship between the vehicular population and pollution.

Thus, the hypothesis that the vehicular population plays a significant role in air pollution is true in all sense. Figure 17.5 shows the regression analysis of the two variables (concentration of pollutants and the no. of vehicles). The coefficient (3025.4) shows the mean increase in pollution with the increase in no. of vehicles. It is evident that with the increase in 1 lakh of vehicles, the pollution of the area is increased by 3025.4 µg/m which is a serious condition for this place as the study area is the busiest place, and there has been an increase in the number of vehicles. The value of R2 (86 percent) also suggests strong goodness of fit which means the model can be used to calculate the prediction of the future pollution of the study area.

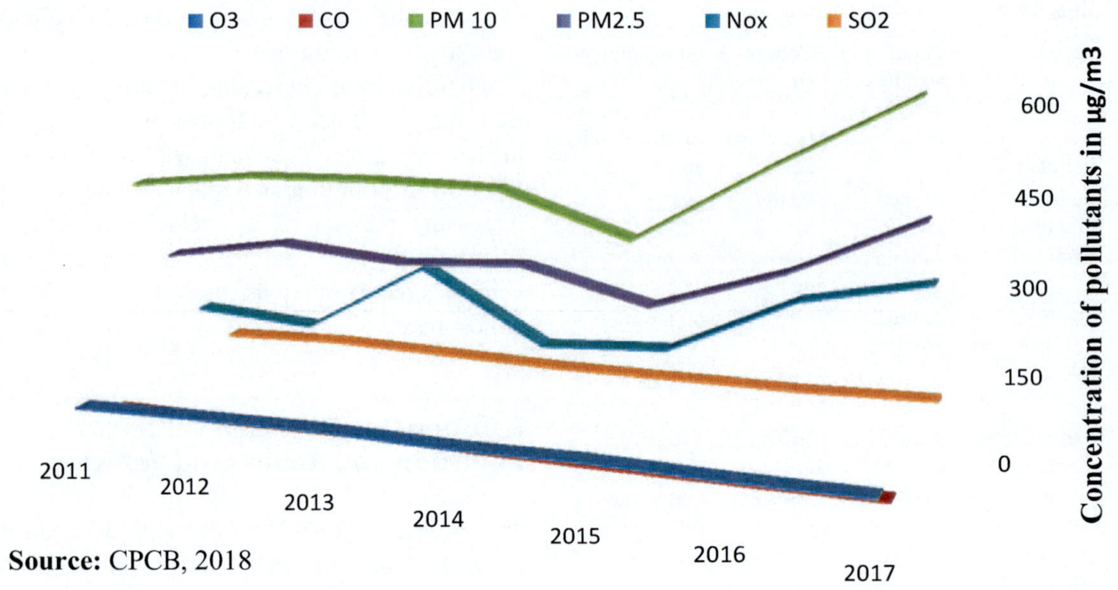

Fig. 17.3 Ambient air qualities at Azadpur Mandi. (*Source*: CPCB, 2018)

People's Perception

People's opinions and perception play a very significant role for the ground truthing. Table 17.4 shows the seasonal distribution of air pollution as per the perception of the people in Azadpur Mandi. As analyzed, the majority of the respondents gives their consent to the higher pollution level during the winter season and a least in the monsoon seasons. They all accepted that one of the reasons for high pollution in winter is the religious ceremony, which is very much prevalent in winter. One such example as noted by the majority of the respondent is the Diwali and other festivals as Gurpurab, which is performed from 11:55 pm to 12:30 am for New Year and Christmas festivities, which results in the use of enormous amount of firecrackers that pollutes the area and results in the creation of haziness in Azadpur Mandi. Besides, the use of firewood and charcoal also plays a major role in contributing to the air pollution. The burning of crop residue and the heavy influx of vehicles during the winter especially the vegetable trucks also play a major factor in increasing the pollution in the study area. It happens due to some atmospheric conditions such as winter inversion where cooler air is trapped under the warm air and forms a kind of atmospheric lid during winter season. Other discretionary particles, for instance, sulfate particles, cool the air and advance disintegrated cloud affiliations (Shindell et al., 2013). However, in monsoon season, rain occurs and all air pollutant comes down with raindrops like PM, etc. PM2.5 can similarly impact the environment. The residue (dim carbon) can ingest heat, accordingly growing close by temperatures (Bond et al., 2013). Close surface ozone is another assistant toxin made by the association of precursor blends with sunshine, including UV radiation (Jacob and Winner, 2009). This joins the ability to direct the effects of natural change by diminishing buildup levels similarly as phony cooling particles (Ming et al., 2014).

In the age group of below 20 years, 5.5 percent of the respondent's opined that air pollution was severe in the winter seasons, while 7 percent opined that they are facing air pollution in the summer months. In the age group of 20–40 years, about 12 percent of the respondents opined that in winter months, they found difficulty in taking a breath because of the severity of air pollution. They also added that they found eye burning in this season because of air pollution. In the age group of 40–60 years, 12 percent of the respondent's opined that they found that summer is also affecting their health severely because of heat waves.

Table 17.3 Vehicles in lakhs and concentration of pollution in μg/m³ (2011–2015)

No. of vehicles entered in lakhs							Concentration of pollution in μg/m³ (2011–2015)						
	2011	2012	2013	2014	2015	Total	SO2	PM10	PM2.5	O3	CO	NO_X	Total
January	0.08	0.09	0.09	0.07	0.13	0.46	16.81	378.46	230.28	33.79	2.55	131.06	792.95
February	0.07	0.08	0.07	0.05	0.12	0.39	17.93	291.15	150.44	44.59	1.81	116.33	622.25
March	0.07	0.08	0.07	0.06	0.12	0.4	20.89	233.95	111.82	51.58	1.49	89.01	508.74
April	0.08	0.07	0.07	0.07	0.09	0.38	28.53	275.24	112.25	66.78	1.64	119.67	604.11
May	0.08	0.08	0.08	0.07	0.12	0.43	21.99	304.51	123.05	64.74	1.46	110.49	626.23
June	0.07	0.07	0.08	0.06	0.03	0.31	13.41	223.73	99.64	48.26	1.33	60.81	447.19
July	0.08	0.07	0.08	0.06	0.05	0.34	13.57	131.18	69.85	35.36	1.46	64.28	315.69
August	0.07	0.06	0.08	0.05	0.03	0.29	15.60	124.59	56.72	37.05	1.23	82.48	317.68
September	0.05	0.09	0.06	0.05	0.07	0.32	19.08	155.75	69.95	41.63	1.23	81.72	369.36
October	0.11	0.09	0.08	0.08	0.09	0.45	23.49	283.71	155.60	66.42	1.84	158.62	689.69
November	0.1	0.08	0.12	0.07	0.14	0.51	21.13	481.45	286.57	51.80	2.80	232.21	1075.97
December	0.05	0.05	0.05	0.04	0.09	0.28	19.47	422.68	248.77	35.41	2.91	207.93	937.17
Total	0.93	0.91	0.93	0.72	0.25								

Source: RTO, 2018; CPCB, 2018

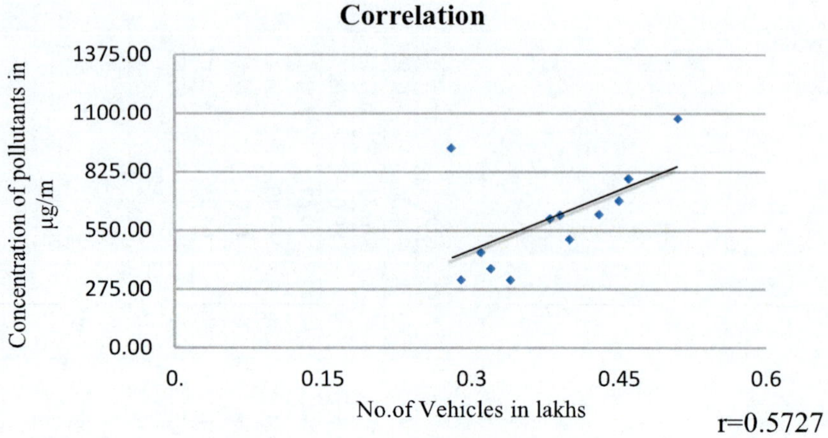

Fig. 17.4 Correlation between vehicle number and air pollutants

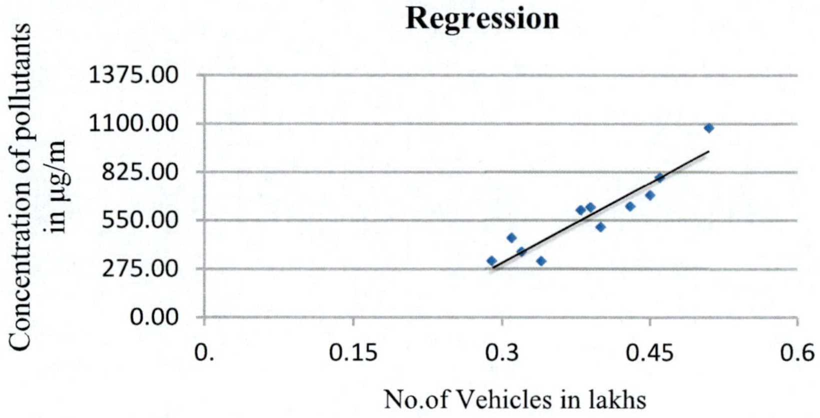

Fig. 17.5 Regression between vehicle number and air pollutants

Effects of Air Pollution on Human Health

Table 17.5 shows air pollution problem where respondents had opinion that which location is more problematic inside the Mandi. Among the age group of between 20 and 40 years, 19 percent of the respondents opined that they are facing problems inside Mandi.

Among the age group of above 60 years, about 9 percent of the respondents opined that they are facing issues inside Mandi. Among the age group of between 20 and 60 years, about 22 percent of the respondents are facing problems outside Mandi. In the age group of above 60 years, 2 percent of the respondents says that they are facing problems at G. T. (Grand Trunk) road.

About 10 percent of the respondents below 20 years of age are suffering from some respiratory illness (Table 17.6). Air contamination is a result of industrialization just as urbanization has risen as a significant hazard factor for human well-being and beyond (Reddy and Roberts, 2019). Introduction to particulate matter (PM) from traffic can cause unfavorable well-being risks (Singh et al., 2019).

In the age group of 20–40 years, 29 percent of the respondents are facing some form of respiratory problems. Among all age groups, there is 35.9 percent of the respondents facing heart-related problems. In the age group above 60 years, about 10 percent of the respondents are suffering from some form of respiratory problems (Table 17.6). So it shows that air pollution really affects the health of respondents, and it creates an extra financial burden of health-related expenditure. Many respondents were working as a laborer so they cannot afford money on health. This again affects their life expectancy and creates an economic burden.

Table 17.4 Seasonal distribution of air pollution (%age)

Age of the respondent (years)	Seasons		
	Winter	Summer	Monsoon
Below 20	5.5	7.1	2.7
20–40	11.7	8.0	4.0
40–60	20.3	12.0	9.0
60 above	9.5	6.2	4.0
Total	47.2	33.3	19.7

Source: Primary Survey, 2017

Table 17.5 People's perception of breathing problem due to air pollution

Age of the respondent (years)	Where do you get unbearable to live in (%)		
	Inside Mandi	Outside Mandi	G. T. road
Below 20	3.0	7.5	10.6
20–40	19.0	12.0	5.0
40–60	10	9.5	4.0
60 above	8.9	8.5	2.0
Total	40.9	37.5	21.6

Source: Primary Survey, 2018

Table 17.6 Distributions of diseases

Age of the respondent (year)	Type of respiratory health issues (%)		
	Lung related	Heart related	Others
Below 20	3.5	0.9	1.5
20–40	9.2	15	4.5
40–60	20.8	20	11.0
60 above	10.3	5	3.3
Total	43.8	35.9	20.3

Source: Primary Survey, 2018

Conclusion

From the above discussion, it is evident that the vehicular population plays a significant role in the increment of the pollution level of the study area which results in the increment of various lung- and heart-related issues, especially in the adults. In order to resolve the problems up to an extent, the encouragement of electric vehicles at the subsidized rate, the introduction of air filter, the large user of snake plants, and the plantation of roadside trees are a must. In Azadpur Mandi, the encroachment of road by roadside vendors should be stopped so vehicles can move easily. There should be provision of more parking spaces for vehicles so illegal parking on roadside will stop. It is a matter of fact, and these suggestions will help reduce traffic and extra emission of air pollutants for urban sustainability.

Acknowledgments The authors acknowledge and thanks the Department of Geography, Delhi School of Economics, University of Delhi and Delhi University Library System for providing institutional facilities. We acknowledge Mr. *Tarun Kapoor*, Dy., Commissioner of Transport Department and Delhi Police, for providing the information. We are greatly thankful to the officials and staff of TCPO. Thanks to Mr. K. K. Joddar, Chief Planner, APMC, DTC, SPA; Dr. M.P. George, Scientist in Delhi Pollution Control Committee (DPCC) Air Lab and IEG Faculty-In-Charge of Computer center; and Prof. *Chandra Sekhar Rao* Nuthalapati for their insightful guidance.

References

Anand, S. (2010). *Solid waste management*. Mittal Publications. Available at https://books.google.co.in/books?id=djHAA6CqOU4C&pg=PA54&source=gbs_selected_pages&cad=3#v=onepage&q&f=false. Accessed on 2 Sept.

APMC, (2018). MIS Report of APMC, Azadpur Mandi, pp. 1–11.

Bigazzi, A. Y., & Fighiozzi, M. A. (2014). Review of urban bicyclists intake and uptake of traffic related air pollution. *Transport Reviews, 34*(2), 221–245.

Bond, T. C., Doherty, S. J., Fahey, D. W., Forster, P. M., Berntsen, T., & DeAngelo, B. J. (2013). Bounding the role of black carbon in the climate system: A scientific assessment. *Journal of Geophysical Research: Atmosphere, 118*(11), 5380–5552. https://doi.org/10.1002/jgrd.50171

Census of India. (2011). *District census handbook Delhi* (pp. 1–1222). Government of India.

CPCB, (2016). *Environment, effluent, emission and air pollution standards & guidelines, Delhi*. p. 1–119.

CPCB (2018). *Ambient air quality data of Delhi Stations, Central Pollution Control Board, New Delhi.*

Goyal, S. K., Ghatge, S. V., Neema, P., & Tamhane, S. M. (2006). *Understanding urban vehicular pollution problem vis-a-vis ambient air quality-case study of a megacity (Delhi, India), Environmental Monitoring and Assessment* (pp. 557–569). Springer.

Jacob, D. J., & Winner, D. A. (2009). Effect of climate change on air quality. *Atmospheric Environment, 43*(1), 51–63. https://doi.org/10.1016/j.atmosenv.2008.09.051

Ming, T., Richeter, R., Liu, W., & Caillol, S. (2014). *Fighting global warming by climate engineering: Is the earth radiation management and the solar radiation management any option for fighting climate change? Renewable and Sustainable Energy Review* (pp. 792–834). Elsevier.

Myhre, G., Shindell, D., Breon, F. M., Collins, W., & Fuglestvedt, J. (2013). *Contribution of working group I to the fifth assessment report of the intergovernmental panel on climate change, The physical Science* (pp. 659–740). Cambridge University Press.

Reddy, K. S., & Roberts, J. H. (2019). Mitigating air pollution: Planetary health awaits a cosmopolitan moment. *The Lancet Planetary Health, 3*(1), e2–e3.

Rizwan, S. A., Nongkynrih, B., & Gupta, S. K. (2013). Air pollution in Delhi: Its magnitude and effects on health. *Indian Journal of Community Medicine, 38*(1), 4–8. https://doi.org/10.4103/0970-0218.106617

RTO, (2018). Department of Transport, Government of NCT of Delhi, 2017–2018.

Shindell, D. T., Lamarque, J. F., Schulz, M., Flanner, M., Jiao, C., & Chin, M. (2013). Radiative forcing the ACCMIP historical land future climate simulation. *Atmoshpheric Chemistry and Physics, 13*(6), 2939–2974. https://doi.org/10.5194/acp-13-2939-2013

Singh, V., Biswal A., Kesarkar, A. P., Mor, S., & Ravindra, K. (2019). High resolution vehicular PM10 emissions over megacity Delhi: Relative contributions of non-exhaust sources. *Science of the Total Environment*. https://europepmc.org/article/med/31683208, available at. https://europepmc.org/article/med/31683208

18

Human Mobility Response to Natural Disasters and Environmental Change

L. Manawadu and V. P. I. S. Wijeratne

Introduction

Every year, a large number of people in the world migrate from a place to another place involuntarily due to natural disasters such as floods, droughts, landslides, cyclones, earthquakes, tsunami, etc. Most of the disasters destroy natural rhythm of human lives, abolishing shelters and property and finally causing human displacement. At present, climate change-related natural disasters such as precipitation extremes, sea level changes, droughts, etc. generate enormous pressure on livelihoods and access to food and water, which may force the victims to take decisions to move to more sustainable living environments. Additionally, anthropogenic climate change is highly influencing human migration and other forms of people moving to manage the disaster risk (International Organization for Migration, IOM, 2020). However, the mechanism of the disaster-induced migration is very complex. In most of the cases, it is very difficult to understand the most significant environment-related pull factors that determined the migration because not only physical or environmental factors but also socioeconomic, political, and cultural factors contributed with environmental factors to determine the people's intention to move in one place or another. These aspects of sociocultural factors are difficult to measure and understand although these factors mainly lead to the creation of human pressure about the disaster migration. Due to those reasons, different types of disaster-induced migration can be seen in the world such as planned migration, travel and relocation, government-forced resettlements, displacements, etc. The common migration patterns in the world are travel and relocation, resettlements, and displacements. Therefore, disaster-induced migration can be identified as a form of environmental adaptation and management of disaster risk in the world.

In the present world, climate change-related natural disasters are one of the biggest challenges for human societies. Mbaye and Zimmermann (2015) pointed out that although environmental disasters are mostly considered as an important driver of migration in today's world, it is not new and has probably existed since the beginning of human history. Today a huge population than ever can be seen in the world and according to the 2015 world statistics, the number of international migrants will exceed 244 million due to many reasons. However, the push factor of the majority of migrants was different kind of disasters. Natural disasters are less predictable, and early assessment of disaster's magnitude is also quite challenging. Also, predicting its impact on migration is difficult because the social, economic, political, and environmental factors underlying

L. Manawadu (✉) · V. P. I. S. Wijeratne
Department of Geography, University of Colombo, Colombo, Sri Lanka
e-mail: lasan@geo.cmb.ac.lk

A. R. Siddiqui, A. Sahay (eds.), *Climate Change, Disaster and Adaptations*, Sustainable

migration are complex and varied (McLeman et al., 2014). The nature of the disaster-induced migration is depending on the type of disasters, direct versus indirect effects, and magnitude and severity of the disasters. For instance, temperature variation and rainfall pattern (precipitation and temperature anomalies) do not have direct effect on migration. Moreover, temperature will increase migration or displacements indirectly but only for agricultural lands. Some disasters (floods and droughts) increase migration, but it is not common in the middle- and low-income countries (Beine & Parsons, 2017). Most of the disasters are expected to increase as a result of global warming. Statistics of WHO (2020) have shown that "every year natural disasters kill around 90,000 people and affect nearly 160 million people worldwide. Indeed, the natural disaster frequency and severity of many disasters are already be increasing." In 1990, the Intergovernmental Panel on Climate Change (IPCC) noted that the greatest single impact of climate change could be on human migration with millions of people displaced by shoreline erosion, coastal flooding, and agricultural disruption (International Organization for Migration, 2008).

Since the 1930s, natural disasters have led to a large number of internal migration in many countries over the world. Drought and dust bowl in 1930s (Great Plains of the United States and Canada) cause human migration of nearly three million people. Also, Hurricane Andrew in 1992 and the hurricane of 2004 led to a large number of human migration. Furthermore, Katrina and Riata of Louisiana and Mississippi in 2005 displaced about two million people. Nearly ten million people have migrated in Africa's Sahel Region and Bangladesh, Philippines, India, etc., and it can be seen in the same situation like in floods and earthquake (Alexeev et al., 2010). A significant number of people living in Asia and Pacific regions migrated every year due to natural disasters such as floods, droughts, soil degradation, cyclone, and typhoons (Asian Development Bank, 2013). A significant proportion (85%) of current human displacements were attached to the extreme weather events rather than geophysical events like earthquake, volcano eruptions, and tsunami (Wilkinson et al., 2013). The number of disaster-induced migration increased significantly year on year because the frequency of natural disasters also increases significantly and cannot be predicted precisely. In some cases, although the magnitude of the disaster is low, the impact of such disasters has increased ever than before. For instance, 9% of disaster displacements were associated with small- or medium-scale events in the period of 2008 to 2015 (UNHCR, 2019). Some people are crossing the country borders to reach to a very safe area, and mostly, this trend can be seen in Africa and Central and South America (UNHCR, 2019). Migration-associated disasters can be seasonal, long term, or permanent, and there are no reliable global estimates of past and current migration flows in response to disasters. However, how significant the global trend of forced displacement in the world can be seen in Fig. 18.1. The internally displaced population was less than 25% until 2013 and increased nearly 40% in 2018.

Mostly, disaster-induced internal migration from different developing countries has influenced many issues in the economy and the society. Not only socioeconomic but also it leads to cultural, political, and education landscape in the countries. The impact of climate change is numerous, and its cause to limit natural resources such as drinking water and disasters can also induce conflicts linked to the food security (UNHCR, 2019). People are trying to adapt to these environmental changes, but some are being forcibly displaced from their living places by the effect of climate change-related natural disasters or are relocating in order to survive. Some of disaster-induced displacements can be very sudden, short distance, and temporary because they no longer have options or sustainable solutions than to move from their living places shortly to seek safety and protection. Disaster-induced migration can push people into lower rugs in the social ladder without considering where they were in the social ladder before the disaster event. As an example, the major floods in Pakistan in 2010 led to displacement of nine million people from the risk areas, but after 1 year, most of them

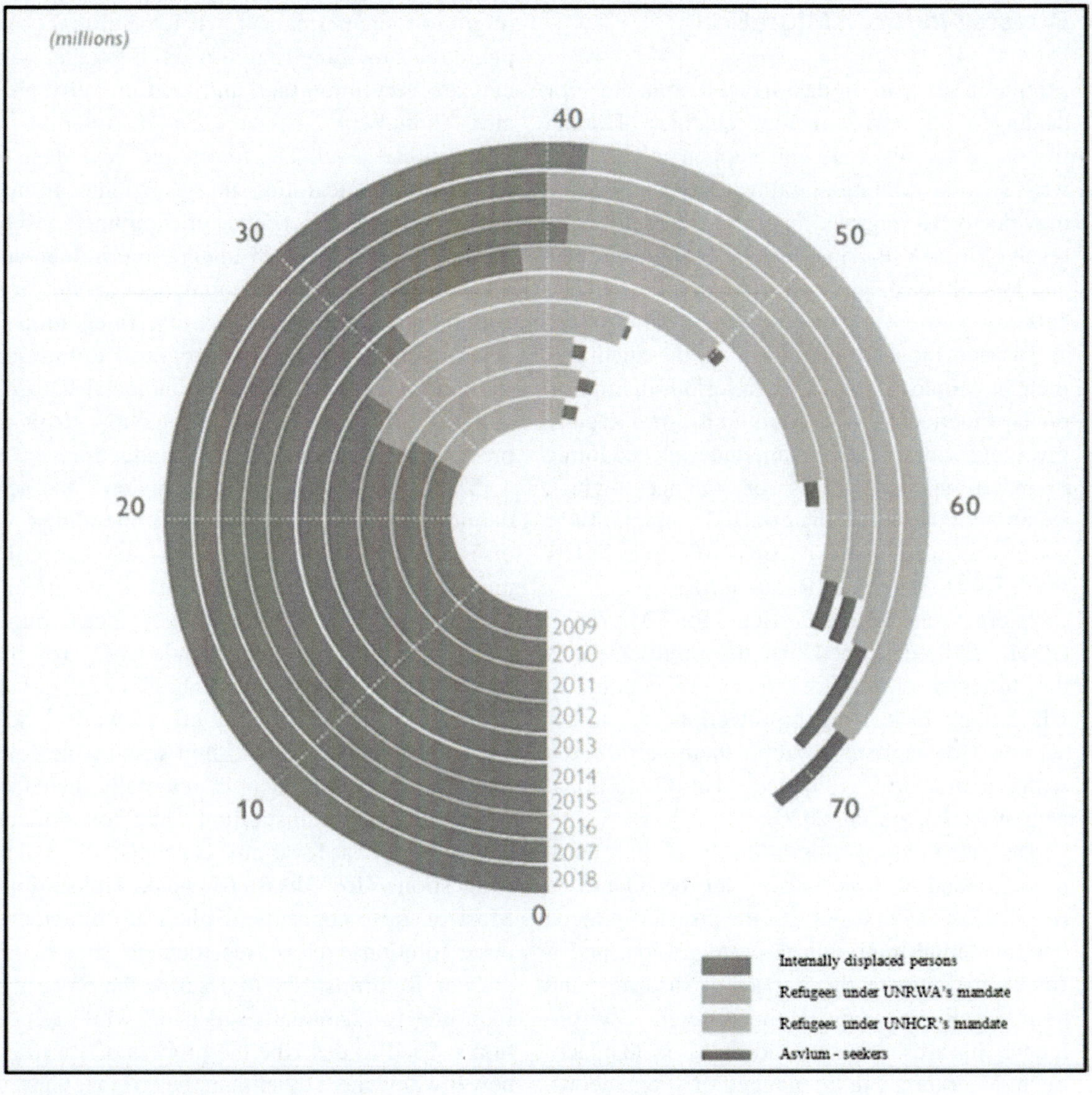

Fig. 18.1 Trend of global displacement (2009–2018). (Source: UNHCR, 2019)

returned back due to such socioeconomic issues faced in new settlement areas. Therefore, the migrations due to natural disasters have received high attention in recent years.

Recently, Sri Lanka is also vulnerable to many of natural disasters such as floods, droughts, and landslides which are very common with current climate change-related rainfall variability. According to the statistics of Internal Displacement Monitoring Center (IDMC, 2019), the total number of disaster-induced displacements of Sri Lanka were 87,000 in 2019. The total displacements of the 2004 tsunami were around 553,000 in 14 of 25 districts in Sri Lanka. "Sri Lanka was among the three most affected counties in the 2017 estimate in the terms of weather-related loss events, ranking second highest on the Climate Risk Index which measures fatalities and economic losses occurring as a result of extreme weather" (Eckstein et al., 2019; UNDRR, 2019). This chapter mainly focuses on the disaster-induced migration in Sri Lanka, issues, human adaptation strategies, and related policies. Understanding the characteristics of the disaster-induced human mobility will help to identify the effective policies, adaptation plans, and investments.

Disaster-Induced Migration

Migration is one of the ancient strategies for dealing with environmental changes. Human responses for disasters can happen in different ways such as migration and resettlement. People may decide to migrate, displace, or resettle as a result of loss of housing, loss of living resources, and loss of social and cultural resources (IPCC, 1990).

Disaster-induced migration is the displacement of people as a result of "a serious disruption of the functioning of a community or a society involving widespread human, material, economic or environmental losses or impacts, which exceeds the ability of the effected community or society to cope using its own resources" (UN office for Disaster Risk Reduction).

International Organization for Migration (IOM, 2019) described that the resettlement is the "transfer of refugees from the country in which they have sought protection to another state that has agreed to admit them as refugees with permanent residence status" (Global Migration data portal, 2020).

Disaster-induced migration is a long-term process, and a few stages can be identified (Fig. 18.2). The first stage of the disaster-induced environmental migration is premigration, and it causes to mitigate the impact of disasters and help humans to adapt to the different environmental hazards. The second stage is displacement. Migration can be planned or spontaneous, individuals, families, or community. It can also be internal or international. Most of the people move shorter or longer distances to find new places and safe environment as an internal migration. Sometimes, it can be international, and people are willing to resettle in any other countries. According to the nature of the disaster event, people are forced to move from one location to another, or it can occur under emergency circumstances. The third stage of the life cycle involves return or resettlement in another location different from the original location. Mostly, human displacement can be temporary, with most migrants expecting to return to their home after some time. However, sometimes disaster-induced migrations are permanent, when the migrants are unable or unwilling to return. In this stage, policies are very important, and land property policies should be the concern in internal or international level. The fourth and final stage of the cycle is integration into the home or new location. The policy frameworks outlined will be the key determinants of integration, influencing the access of displaced populations to housing, livelihoods, safety, and security. Integration is also affected by plans and programs to mitigate future dislocations from environmental hazards, coming full circle on the life cycle to a focus on prevention, adaptation, and risk reduction.

There are a vast amount of reviews and academic- or professional-oriented literatures on disaster-induced displacements and resettlements. Most of the literatures have mainly focused on "forced migration," "legal framework," "protection and attendance," "return," "resettlement," and "reintegration" in relation to the disaster-induced migration. However, there are no acceptable ways to find sustainable solutions for the resettlement of internally displaced persons. Scholars, institutions, and governments have put forward various concepts, ideas, and suggestions for disaster-induced relocations. Most of these concepts or ideas are directed to three solutions: return, resettlement, and reintegration. Return is used to describe the process of returning to "habitual residence," while relocation is used to describe the process of starting a new life anywhere other than the original place of residence but still in the same country. The United Nations guiding principle 28:1 states that the competent authorities have the primary responsibility and obligation to create conditions and provide means to allow internally displaced persons to return voluntarily to their homes or places of habitual residence in conditions of safety and dignity or voluntarily settle in other parts of the country (Sangasumana, 2018). Reintegration is used to describe the social, economic, cultural, and political structures in which former internally displaced persons reenter their original or new communities (Consortium of Humanitarian Agencies, 2003).

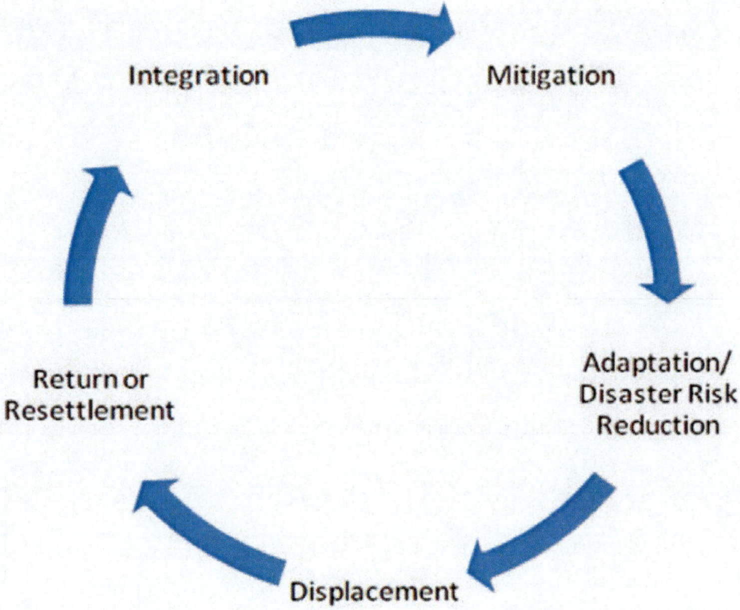

Fig. 18.2 Life cycle of disaster-induced migration. (Source: Martin, 2010)

Most of migrations due to climate change-related disasters can be internal and affected people seeking to find more habitable locations, with greater economic opportunities, within their own countries (Martin, 2010). According to the archaeological findings, human settlement patterns have been developed or existed by responding to the environmental changes, and people think that migration is an important mechanism to response to disasters since the past. However, the disaster-induced migration is complex, but the distinction between intensive and extensive risks is a useful starting point in characterizing the relationship between climate risk and human mobility. Movements of people in response to intensive risks are very different from those in the context of extensive risks. These different types of movement can be categorized as (1) migration, (2) displacement, and (3) planned relocation (Warner et al., 2014; Wilkinson et al., 2013). Figure 18.3 further illustrates those three concepts.

In addition, other main categories of the disaster-induced migration is forced displacements and voluntary migration. In most of the cases, voluntary migrants move to urban centers expecting better urban facilities and job opportunities. Usually, the severity of the disaster and the environmental condition are the primary drivers of disaster-induced displacements, and other socioeconomic factors can be discussed as secondary factors (Black et al., 2011).

According to Fig. 18.4, disaster-induced environmental changes influence different environmental, social, economic, demographic, and political drivers. These factors that drive migration cause the decision of the people, and sometimes, people are not decided to move at all. Most of the people in the world are not willing to move due to lack of resources and socioeconomic problems (Black et al., 2011).

Disaster-Induced Migrations in Sri Lanka

"Sri Lanka's geographic and climatic diversity exposes it to several types of natural hazards, with the most frequently occurring events being floods, droughts and cyclones" (Lavell and Ginnetti, 2014). Increasing temperature, changing rainfall and monsoon patterns, and tropical storms have caused large number of disaster-induced human migrations (Mombauer, 2018). In recent decades, disaster losses in Sri Lanka have increased significantly (Vijekumara,

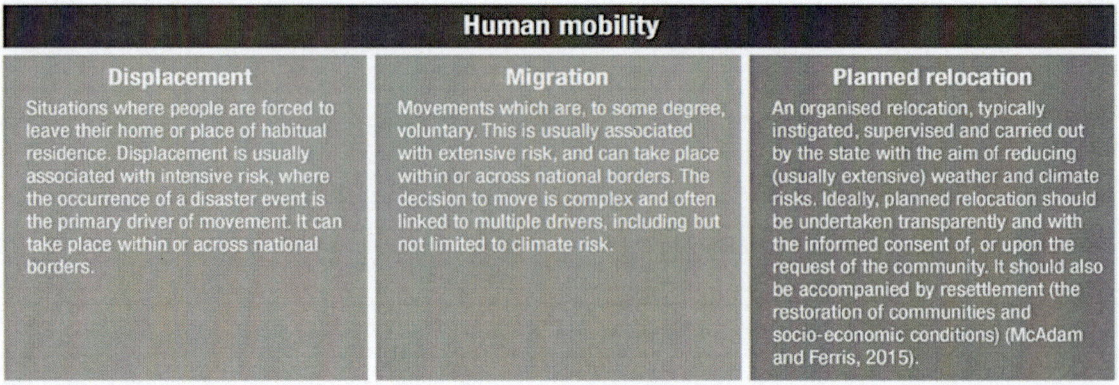

Fig. 18.3 Different types of human mobility. (Source: Advisory Group on Climate Change and Human Mobility, 2014)

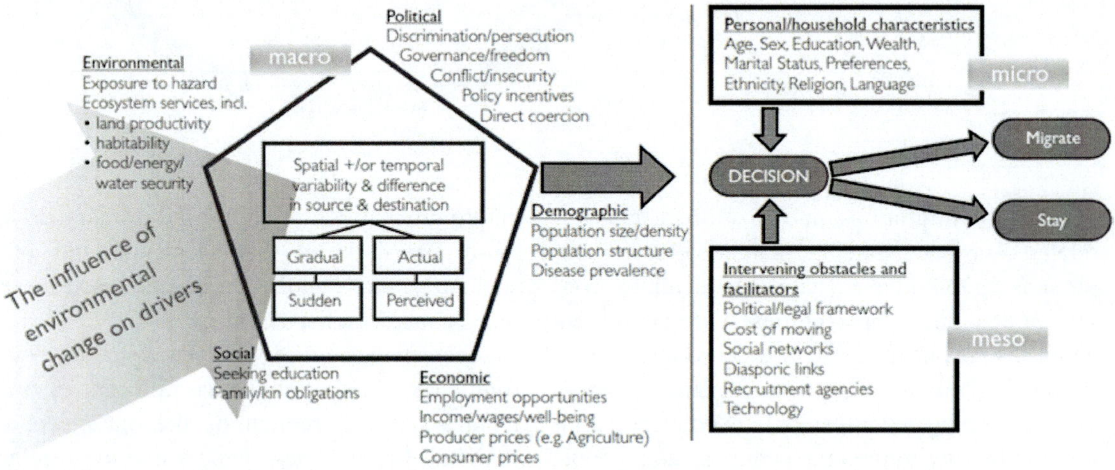

Fig. 18.4 Drivers of migration and the influence of environmental changes. (Source: Black et al., 2011)

2015). During the past decade, most natural disaster-related displacement has occurred in the western, middle, and southern parts of the country. Approximately 95 percent of the damages and destructions to housing since the 1970s has been caused by cyclones and monsoon floods, and the tsunami in 2004 further caused to displace more than a million people (Lavell and Ginnetti, 2014). The types of disaster occurrences between 1990 and 2018 period in Sri Lanka can be seen in Fig. 18.5.

In addition, it can be considered that the frequency of occurrences of natural disasters has increased in the recent past. According to the National Physical Planning Policy (NPPP) of Sri Lanka 2030, "more than two-thirds of the population of Sri Lanka is exposed to natural disaster conditions" (National Physical Planning Department [NPPD], 2010). Therefore, migration or force displacements are the vital strategies for minimizing the impact of natural disasters, and people can move from the risk areas to the safer areas. Mainly, migration or relocation will be a disaster preventive strategy to reduce exposure to natural disasters and thereby to prevent risk on lives and properties (Rathnasiri, 2018).

As IDMC (2020) pointed out, the average expected number of displacements in Sri Lanka per year is 20,548, and the large part of the displacement can happen due to floods. However, flood-induced displacements are mostly not a permanent displacement. The average expected annual flood displacements in Sri Lanka is nearly 18,800, and displacements in 2019 were 87,000

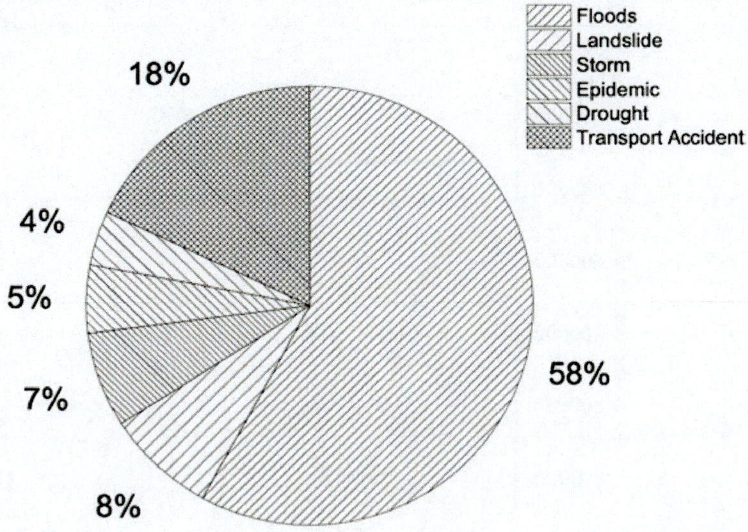

Fig. 18.5 Disaster occurrences (1990 to 2018). (Source: CRED, 2019)

(IDMC, 2020). After floods, the landslide disaster is the most common natural disaster that causes high number of people displacements in Sri Lanka, and it has been the cause for the largest number of permanent displacements in Sri Lanka. Out of the total land extent of the country, nearly about 30 percent of land is prone to landslides in the wet zone (Sangasumana, 2018). Therefore, it can be mentioned that the risk of natural disasters in Sri Lanka has increased in the recent past (Lavell and Ginnetti, 2014). Figure 18.6 shows the total number of annual disaster displacements in Sri Lanka from 2008 to 2019. The number of displacements shows a decreasing trend with some fluctuations. The highest number of displacements have been reported in 2011 (685,000) and the lowest in 2015 (66,000). Although, man-made disasters (terrorist attacks) have influenced on the higher number of disaster displacements before 2009 in Sri Lanka, after 2009, displacements were limited to floods and landslide disasters.

Most Sri Lankans are trying to acclimate with the environmental changes and reluctant to migrate from their residence areas. However, in some instance, many are being forcibly displaced from their homes by the authorities and conducted resettlement projects to relocate disaster victims. Most of the disaster-induced displacements in Sri Lanka are internal and due to wiliest floods and landslides especially the 1978 cyclone and 2004 tsunami that also caused to resettle many people. Landslide is a very common natural disaster in Sri Lanka which leads to permanent internal human displacements in the country.

During the past few years, many people have been affected by landslides in Sri Lanka. For example, in 2016, 34,033 people from 9620 families were affected by landslide in Kegalle District, and most of the affected people have evacuated to safer places. By the year 2017, the National Building Research Organization (NBRO) has identified 14,680 families living in landslide-prone areas from nine administrative districts (Table 18.1). Also, more recently, a few major landslides are recorded in Kotapola-2003, Hanguranketha and Walapane-2007, Galahawatte-2011, Meeriyabedda-2015, Aranayaka-2016, Bulathsinhala-2017 and Nivithigala-2017.

According to NBRO, Galle, Monaragala, Colombo, Kurunegala, and Gampaha districts have also been identified as high-risk areas during the last couple of decades. Landslide disaster resettlements were implemented in different areas of Sri Lanka. For instance, the resettlement of families affected by the Ratnapura flood and landslides in 2003, the Hanguranketha and Walapane landslide resettlement plans in 2007, the Galahawata landslide resettlement project in 2011, and the Meeriyabedda resettlement plan in

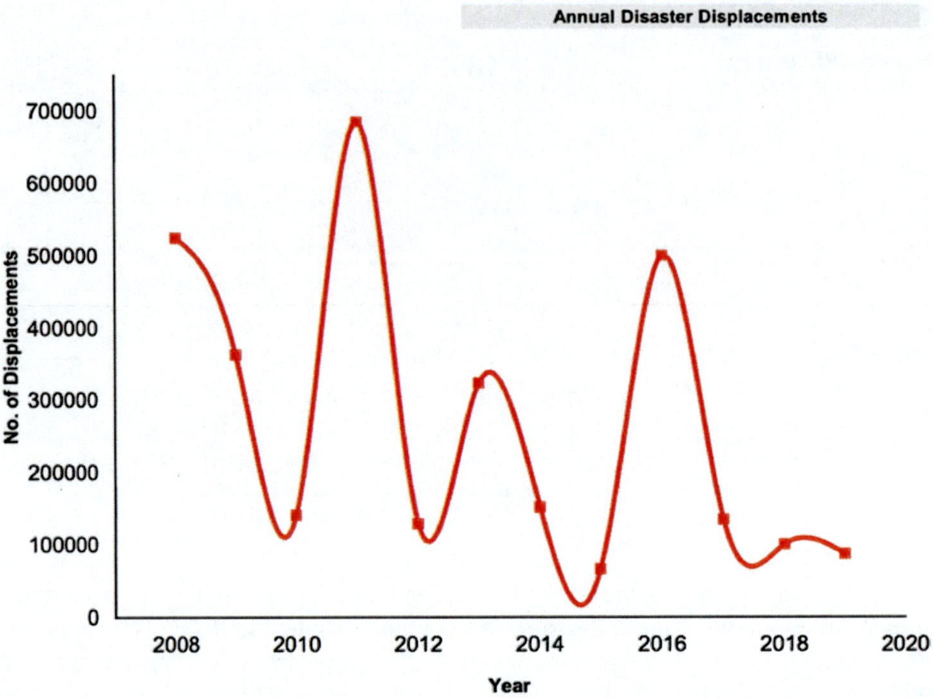

Fig. 18.6 Annual disaster displacements in Sri Lanka from 2008 to 2019. (Data source: Internal Displacement Monitoring Center (IDMC), 2020. Prepared by author)

Table 18.1 Families living in high-risk landslide areas in Sri Lanka

District	No. of families living in high-risk areas
Badulla	6418
Nuwara Eliya	3496
Kandy	1292
Matale	210
Kegalle	824
Kalutara	929
Matara	591
Ratnapura	757
Hambantota	343
Total	*14,860*

Source: The National Building Research Organization (NBRO), 2017

2015 are some of the major landslide resettlement schemes implemented locally (Vijekumara and Karunasena, 2016).

The 2004 Indian Ocean tsunami led to an unprecedented process of relocation in Sri Lanka, moving thousands of homelands beyond the coastal buffer zone set by the government (De Silva, 2018). As a result of tsunami, approximately 1100 km out of the 1583 km coastline of Sri Lanka, including west, south, east, and north coastlands, were affected, depending on the coastal configuration after Tsunami. It has been estimated that more than 98,000 permanent houses had to be rebuilt in other places. These displacements of landslides and tsunami are permanent and long term, while human mobility due to floods, droughts, or other natural disasters are temporary, and after risk period, people will return to their original locations. Also, such human mobility is seasonal, and displacement time may depend on the magnitude and severity of the disaster that occurred.

Disaster-Induced Displacement Issues and Human Adaptation

Relocation caused by disasters has received great attention from the governments, the international organizations, and the emergency management agencies during the recent past. They are particularly concerned about effective impact assessment and the development of management models to cope with disasters, which may cause the need for large-scale forced relocation. In forced migration, "people are forced to leave their home or homes for any reason and are designated to specific areas established in the country, providing at least a minimum of facilities and services to restore their lives" (De Silva, 2018).

Relocation is one of the long-term strategies of any disaster management plan. This does not mean providing land and housing but helping to rebuild the lives of the displaced people. In any case, it will be a great challenge to bring any relocation plan to a satisfactory level. It is obvious that there are some problems in the long-term relocation policy implemented by the government of Sri Lanka. However, several issues can be identified in relation to relocation process, and most of the people could not adapt to the new environment due to different economic, health, and social distraction and infrastructure and management issues. One of the most important sociological aspects of relocation is decelerating human networks which are very significant in rural Sri Lankan communities. Every person shares a common territory, resources, and equal culture with others who are living his/her origin place, and such a group can be considered as a social group. This is a very common social network in rural areas, and all people have adapted to their origin places. When disasters happened, social networks are also interrupted, and this might lead to serious results such as social disorders and social conflicts. The normal pattern of social system is permanently suspended due to disaster displacements. Rural Sri Lankans are bounded with a very strong social networks and trust their neighbors, sharing their day-to-day social events and sometimes food and clothes. This kind of social networks, behaviors, and attitudes can be destroyed by a relocation. After relocation, many people suffer vicarious trauma because of their relationship breakdown, bereavement, and losing the previous social connections and contacts.

Although government and nongovernment institutions are playing a very significant role in the post-disaster resettlement process in Sri Lanka, still there can be seen some kind of gaps and dissatisfactions due to some socioeconomic issues faced by displaced people. First, people who faced the unexpected disaster displacement faced more difficulties when living in temporary shelters due to different disasters happening in Sri Lanka. However, after that, they also faced different socioeconomic issues due to the slow and inefficient resettlement process. Different issues faced by post-disaster resettlement is clearly shown in Fig. 18.7.

It has been observed that there is very poor involvement of all the stockholders for pre-disaster management processes due to lack of studies in disaster risk analysis while a relatively better involvement for post-disaster management in Sri Lanka (Sangasumana, 2018). Therefore, people are suffering with their inability to receive the basic needs of daily life, and they are often reminded of their comfortable life spent before disaster incident. Disaster-induced migration in Sri Lanka is different from other south Asian countries, and a very few cases were reported as self-migrations or displacements. For example, Badulla is high landslide hazard-prone area, and the total number of high-risk families were 3776. However, only a few families were willing to self-migrate to different areas at their discretion (16 families in Soranatota), 225 families were totally refused any kind of displacement, and all other families were willing to displace if the government provides some kind of support and incentives. Galle district also is a high disaster-prone district, and the resettlement plans are being prepared for both flood and landslide disasters. A total of 407 families has to be resettled in Galle district from the following DS divisions: Nagoda (145), Thawalama (60), Baddegama (41), Neluwa (38), and Welivitiya Divithura (35). Out of 407, 75 families totally disagreed for any kind of displacement and are not willing for any self-displacements.

In general, a large proportion of the disaster-affected people have different problems with relocation, and they are not willing to move from their homelands. Therefore, the government forced them to relocate and implemented a large-scale resettlement program in several districts in

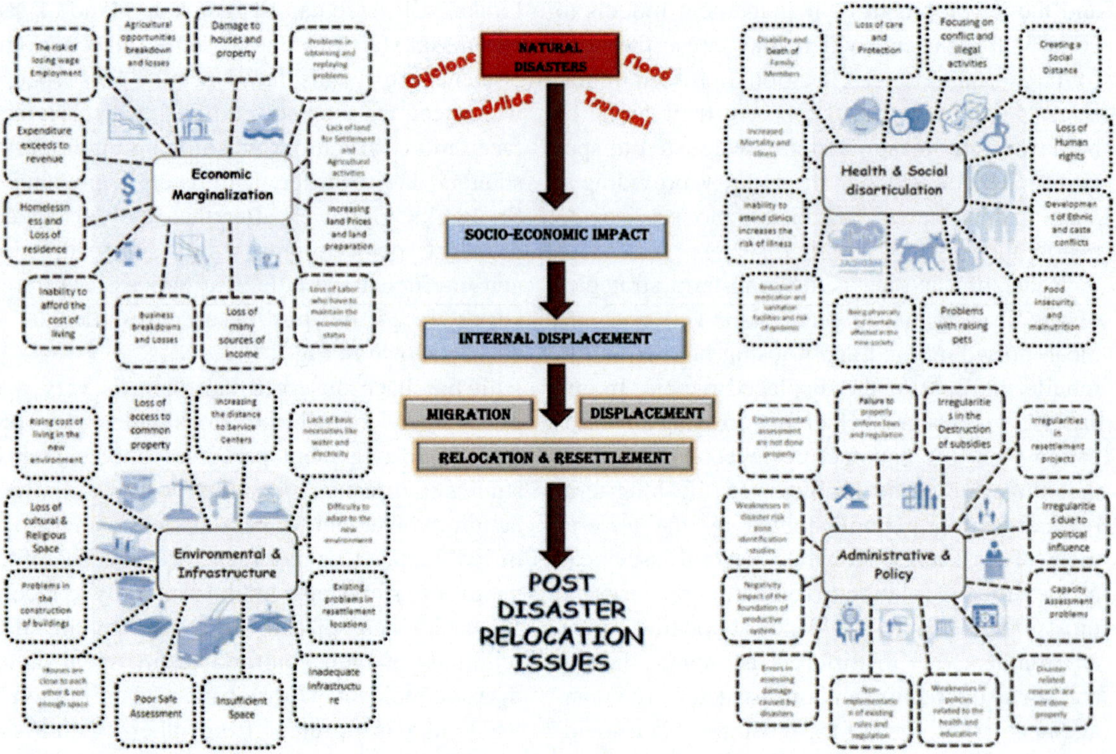

Fig. 18.7 Post-disaster relocation issues in Sri Lanka

Sri Lanka. However, displaced people were experiencing different issues and challenges in adopting to the new environment. Specifically, relocation is highly influencing social and cultural landscape of the affected people. Recently, the Sri Lankan government has introduced a new relocation procedure with the following compensation packages:

- If the affected family has their own land, the government will provide compensation of Rs. 400,000 for the land.
- The affected family can buy a land using government financial aid of Rs. 400,000 and build a house with the government aid of Rs. 1,200,000.
- The affected family can buy a house using government financial aid of Rs. 1,600,000.
- They can resettle in a government land, and they will receive Rs. 1,200,000 to build a house.

However, during the initial stages of disaster relocation, people were facing many issues in relation to the overdue government involvement to release the allocated money. According to the semi-structured interviews held in Galle and Badulla, some victims criticized that even though they have received their first installment for house construction, still (more than 1 year), they have not received the second installment, and they have to stay at temporary shelters and relative's houses with limited facilities. Also, they were given a notice that before the commencement of constructions, they should have received a valuation report, but it took a very long time due to lack of expertise in the government officers. Therefore, in most of the cases, victims were in uncertain environment for a long period, and dissatisfaction and poverty were increasing day by day.

Even though victims were migrated, adoption for the new environment is one of the very serious problems they are facing. One of the main challenges commonly they are facing is inability to adapt to the new economic environment. As new livelihood opportunities are sought, this leads to increased anxiety and despair. For instance, "In 2016 Aranayaka landslide was a high impact landslide and that regions had been normally generated US$ 160,000 annually from their home gardens and plantations (tea, rubber, etc.). Still the responsible authorities were not able to rebuild the earlier economic environment in the new resettled areas" (Sangasumana, 2018). Most of the tsunami-affected population in Sri Lanka were engaged in fishing industry. However, after the relocation, people settled far away from the original locations, and most of them had to seek different income sources to meet their basic needs. Therefore, the relocated community had been enrolled with informal economy activities as well as illegal income sources. Also due to this problem, the number of unemployed person has increased in these areas.

The other common problem is inadequate facilities in the resettled areas, specifically inadequate provision for the education of children, poor road systems, lack of water resources, hospitals, supermarkets, etc. Moreover, the facilities provided are not proportionate to the number of displaced people. For example, the resettlement program provides the same amount of compensation for all the victims without drawn attention to the number of members of the family, and the facilities of new houses are insufficient for some families.

Another important issue faced by the relocated victims is people's perception of the host community. Normally, the host population attitudes are not very favorable on the resettled population, and in most cases, the host population were not in a position to welcome victims. Some resettled areas have created a segregation for different races, ethnicity, cultural and economic status, etc. Also, poor social relationships and broken social networks can be seen due to separation from previous relatives and neighborhoods. Less interactions and poor social relationships with host community and some social conflicts are happening for natural resources. The label of "displaced community" and poor communication with host community normally creates very unsatisfactory and uncomfortable life for displaced community.

Abandonments of newly built houses is another crucial issue behind the resettlement programs due to various reasons. In most of the resettlement programs, after several months, a significant number of resettled families have returned to their original locations. Adaptation failure, economic losses, shelter problems, and bad attitudes of host community are some of the reasons for such return migration.

Resettlement Policies, Plans, and Recommendations

Disaster mobility is part of the disaster management cycle, which is included in the recovery phase. However, most studies have shown that disaster resettlement goals are often not achieved, and opportunities for community development are lost due to the inefficient resettlement process (Vijekumara, 2015). In 2001, the cabinet approved the National Involuntary Resettlement Policy to address the shortcomings of the land acquisition act, which provides protection for people displaced by development projects. However, Sri Lanka has no specific policy to address the needs of people displaced by disasters, which is entirely handled by Divisional Secretary as a government agent and nongovernmental agencies. There are no systematic and well-defined procedures or methods used in any disaster-induced resettlement project (Dissanayake et al., 2018; Vijekumara, 2015).

Especially with donor-driven approaches of resettlements, they are often criticized for failing to ensure the built environment and people's resilience due to two reasons. First of all, it involves project-related failures, mainly poor planning, implementation, coordination, and participation. Second, it involves outcomes that fail to meet the needs and expectations of victims. Disasters sometimes make land unsuitable for human habitation. After such disasters, for determining the risk of such disasters by changing their topography, economy, or population structure, Sri Lanka government agencies such as the

Ministry of Irrigation, the NBRO, and the Sri Lanka Land Reclamation and Development Corporation (SLLRDC) declare the restoration or limitation of the land concerned. Shortly after the declaration, the government or nongovernment agencies of Sri Lanka often began to resettle to protect people living in these areas (Sridarran et al., 2018). However, Sri Lanka has formulated resettlement policies framework and guidelines to address the disaster relocation in several times but could not implement properly. For example, after tsunami, the government introduced housing policy for resettlement, and the tsunami resettlement conditions deteriorated after the announcement of the "buffer zone policy," which, depending on coastal conditions, restricts access to land between 300 and 500 m from the beach. This policy has made many people homeless. Finally, according to the high population in the area, it was reduced between 100 and 200 meters. Although the government implemented a policy, it does not succeed in practice. In addition, they introduced a disaster resettlement housing guideline for flood and landslide disasters in 2017. However, as mentioned in the previous section, it also brings different issues when implemented in different disaster-affected areas.

Further, "The National Council for Disaster Management was established in 2005 under the Sri Lanka Disaster Management Act, No 13 of 2005, following the Tsunami. In 2007, the Resettlement Authority was founded under the Resettlement Authority Act, No 09 of 2007. The National Housing Development authority also has granted powers and functions in terms of resettlement, displaced and potentially displaced persons. The National Housing Department (MHD), The National Housing Development Department (NHDD), The Greater Colombo Economic Commission (GCEC), The Urban Development Authority (UDA) and some agencies also retained some authorities to formulate housing policies and organize housing projects on behalf of the government for targeting disaster-induced population. Despite the regulations, a number of local and international guidelines on resettlement have been developed by these agencies" (Sridarran et al., 2018). However, these policies and guidelines do not succeed. It is important to have a specific policy to guide the mechanism for the implementation of preventive resettlement programs and to standardize the process of minimizing anomalies in field practice.

Additionally, different suggestions can be made on the success of post-disaster resettlement in Sri Lanka. First, before planning a resettlement, it is very important to conduct a comprehensive study and analysis of natural disasters in the resettlement area. Secondly, it is also necessary to identify the government and nongovernmental organizations that are most relevant to disaster management and can actively participate. Otherwise, priority should be given to all human needs and their interests in the relocation plans. In addition, in order to successfully resettle, socioeconomic and cultural information should be collected as a baseline study from the affected communities and analyzed for future planning. In particular, individual resettlement options will be the most appropriate choice for communities with different characteristics. All these decisions should be based primarily on the results of socioeconomic and cultural studies. Also, it is better if the new land is close to former settlement since most of family livelihoods are based on former location. Finally, community consultation and participation can be used to make resettlement policies in disaster-induced displacements.

Conclusion

Normally, most people are migrated or displaced around the world due to the pressure of natural disasters. Although environmental disasters are mostly considered as an important driver of migration in the present world, it is not new and has probably existed since the beginning of human history. Human responses for disasters can happen in different ways such as migration and resettlement. This study mainly focuses on the identification of the nature of human mobility responses and human migration responses to natural disasters in Sri Lanka, and finally, the consequences and human adaptations for disaster-induced human mobility in Sri Lanka are discussed. Relocation is one of the long-term strategies of any disaster management plan. This does not mean providing land and housing but helping to rebuild the lives of the displaced population. In any case, it will be a great challenge to bring any relocation plan to a satisfactory level. There are many issues that can be identified in post-disaster resettlements in Sri Lanka, and it has highly influenced the social and cultural behavior of affected people. People cannot adapt easily with the new environment, and they are facing different challenges such as inability to adapt the new economic environment, lack of social network, losing previous job, lack of natural resources, etc. It is obvious that there are some problems in the long-term relocation policy implemented by the government of Sri Lanka. Sri Lanka has no specific policy to address the needs of people displaced by disasters, which is entirely dependent on different government and nongovernmental agencies. There are no systematic procedures or methods used in any resettlement project. Therefore, different suggestions can be made to minimize the post-disaster resettlement issues, and community consultation should be used for making permanent disaster resettlement policies in Sri Lanka.

References

Alexeev, A., Good, D. H., & Reuveny, R. (2010). *Weather-related disasters and international migration*. Indiana University School of Public and Environ-mental affairs [Mimeo]. Retrieved 11 Oct 2011, from http://umdcipe.org/conferences/Maastricht/conf_papers/Papers/Effects%20of%20Natural%20Disasters.pdf.

Asian Development Bank. (2013). The rise of natural disasters in Asia and the Pacific. *Asian Development Bank*.

Beine, M., & Parsons, C. R. (2017). Climatic factors as determinants of international migration: Redux. *CESifo Economic Studies, 63*(4), 386–402. https://doi.org/10.1093/cesifo/ifx017

Black, R., Adger, W. N., Arnell, N. W., Dercon, S., Geddes, A., & Thomas, D. (2011). The effect of environmental change on human migration. *Global Environmental Change, 21*(SUPPL. 1), S3–S11. https://doi.org/10.1016/j.gloenvcha.2011.10.001

CRED. (2019). EM-DAT: The Emergency Events Database - Université catholique de Louvain (UCL) - CRED,. Brussels: Université catholique de Louvain.

Consortium of Humanitarian Agencies (2003). Human Security Report 2003, CHA. 1–27.

De Silva, M. (2018). ScienceDirect ScienceDirect ScienceDirect Involuntary disaster relocation and its impact on children: a case Involuntary disaster relocation and its impact on children: a case study in Galle, Sri Lanka study in Galle. Sri Lanka, 00(2017), 1–8.

Dissanayake, P., Hettiarachchi, S., & Siriwardana, C. (2018). Increase in Disaster Risk due to inefficient Environmental Management, Land use policies and Relocation Policies. Case studies from Sri Lanka ScienceDirect Science Direct Increase in Disaster Risk due to inefficient Environmental Management, Land use policies and Relocation Policies. Case studies from Sri Lanka. *Procedia Engineering, 212*(January), 1326–1333. https://doi.org/10.1016/j.proeng.2018.01.171

Eckstein, D., Hutfils, M.-L., & Winges, M. (2019). *GLOBAL CLIMATE RISK INDEX 2019 Who Suffers Most From Extreme Weather Events? Weather-related Loss Events in 2017 and 1998 to 2017*. Germanwatch.

Global Migration Data Protal. (2020). https://gmdac.iom.int/global-migration-data-portal Accessed: 4th September 2020.

IDMC. (2019). Global Report on Internal Displacement Summary GRID 2019. *Internal Displacement Monitoring Centre, 5*, 16. http://www.internal-displacement.org/global-report/grid2019/.

IDMC. (2020). Country Information. https://www.internal-displacement.org/countries/sri-lanka (Accessed: 5th September 2020).

International Organisation for Migration – IOM. (2019). World Migration Report 2020. https://publications.iom.int/books/world-migration-report-2020

International Organization for Migration – IOM. (2008). World Migration 2008. https://publications.iom.int/books/world-migration-report-2008-managing-labour-mobility-evolving-global-economy

IPCC. (1990). ipcc_far_wg_I_full_report.pdf. In Climate Change.

Lavell, C., & Ginnetti, J. (2014). The risk of disaster-induced displacement in South East Asia and China Technical paper. 56. http://www.internaldisplacement.org/publications/2014/the-risk-of-disaster-induced-displacement-in-south-east-asia-and-china/

Martin, S. (2010). *Climate Change and International Migration, Climate Change and Migration*. The German Marshall Funds of the United States.

Mbaye, L., & Zimmermann, K. (2015). Environmental disasters and migration. IZA Discussion Papers, 9349.

McLeman, R., Obokata, R., & Veronis, L. (2014). Empirical research on international environmental migration: A systematic review. *Population and Environment, 36*(1), 111–135. https://doi.org/10.1007/s11111-014-0210-7

Mombauer, D. (2018). Moving Grounds: Climate Migration and Refugees, https://slycantrust.org/post/

moving-grounds-climate-migration-and-refugees. Accessed 4th September.

National Building Research Organization (NBRO). (2017). Annual Report, https://www.nbro.gov.lk/images/ongoing/OPERATIONAL-HIGHLIGHTS-2018.pdf Accessed: 5th September, 2020).

National Physical Planning Department [NPPD] (2010). National Physical Planning Policy and Plan - Sri Lanka 2030. April.

Rathnasiri, C. (2018). Resettlement of Population at High Risk of Natural Disasters in Sri Lanka Resettlement of Population at High Risk of Natural Disasters in Sri Lanka,8th Annual NBRO Research Symposium, ISBN: 9789550283118.

Sangasumana, P. (2018). Post Disaster Relocation Issues: A Case Study of Samasarakanda Landslide in Sri Lanka. *European Scientific Journal, ESJ, 14*(32), 1. https://doi.org/10.19044/esj.2018.v14n32p1

Sridarran, P., Keraminiyage, K., & Fernando, N. (2018). Acceptance host of a resettlement programme: Programme: Acceptance to be the host literature review literature.

UNDRR. (2019). Disaster Risk Reduction in Sri Lanka: Status Report 2019. Bangkok, Thailand, United Nations Office for Disaster Risk Reduction (UNDRR), Regional Office for Asia and the Pacific.

Vijekumara, P. A. (2015). A study on the resettlement planning process applied in post-landslide disaster resettlement projects in Sri Lanka. *Innovations for Resilient Environment*, 13.

Vijekumara, P. A., & Karunasena, G. (2016). Analysis on Resettlement Process : Landslide Disasters in Sri Lanka. 008.

Wilkinson, T. J., Ali, A. N., Bell, C. J., Carter, F. A., Frampton, C. M., & McKenzie, J. M. (2013). The impact of learning environment disruption on medical student performance. *Medical Education 47*(2), pp. 210–213.

Warner, K., Martin, S., Nassef, Y., Lee, S., Melde, S., Chapuisat, H. E., Frank, M., & Afifi, T. (2014). Changing climate, moving people: framing migration, displacement and planned relocation. *UNU-EHS Publication Series, 9*, 1–52.

19

Temperature and Rainfall Extremes over Southern India (1969–2014): Frequency Distribution and Trends

Mahendra S. Korade and Amit G. Dhorde

Introduction

The fifth assessment report of the Intergovernmental Panel on Climate Change (IPCC, 2014) stated that changes in climate have caused impacts on all natural as well as human systems all over the Earth during the recent decades (see also Jaswal et al., 2015). In the context of warming climate, changes in extremes of temperature are also consistent (Trenberth et al., 2007), and they are more sensitive to climate change than mean values; therefore, extremes have received more attention (Yan et al., 2002; Fan et al., 2012; Wang et al., 2013; Jaswal et al., 2015). Many studies have documented changes in climate extremes around the world (Gajic-Capka and Zaninovk 1997; Easterling et al., 2000; DeGaetano & Allen, 2002; Kürbis et al., 2009; Rodríguez-Puebla et al., 2010; Kothawale et al., 2010; Bednorz, 2011; Guirguis et al., 2011). Although most of the land regions showed warming over the past century and it continued in the present century (Hu et al., 2012; Rummukainen, 2013), the pattern of change has not been spatially uniform (IPCC, 2014). This heterogeneity in the pattern of change results from regional differences in the response of the climate system to increasing radiative forcing and from the climate variability (Horton et al., 2015). Climate variability and response of the climate systems to increasing radiative forcing are dynamic and complex. This complexity substantially increases the challenge of climate study and its attribution at regional and local scales. The studies conducted over China confirm that extremes have remarkably consistent increasing trends for each of the nighttime and daytime temperature extremes, but trends for rainfall indices are spatially less coherent (Fan et al., 2012). Alexander et al. (2006) analyzed different indices of global temperature and rainfall extremes and concluded widespread significant changes. However, rainfall changes were less coherent compared to temperature.

Weather variability and occurrences of extreme events have been amplified around the world, and more or less similar scenario has been observed in India (Revadekar et al., 2012). In India, a few studies have been conducted on the changes of extreme climatic events country-wise (Pai et al., 2004; De et al., 2005; Kothawale et al., 2010; Pal and Al-Tabbaa 2010; Dash & Mamgain, 2011; Jaswal et al., 2015). These studies have shown that the frequency of occurrence of hot (cold) days and hot (cold) nights have widespread increasing (decreasing) trends. Moreover, significant increase was noticed in the frequency, spell

M. S. Korade
Department of Geography, Shri Shiv Chhatrapati College, Junnar, Pune, India

A. G. Dhorde (✉)
Department of Geography, Savitribai Phule Pune University, Pune, India
e-mail: amitdhorde@unipune.ac.in

A. R. Siddiqui, A. Sahay (eds.), *Climate Change, Disaster and Adaptations*, Sustainable

duration, and their spatial coverage. Arora et al. (2005) analyzed time-series of mean annual and seasonal temperatures over a set of 125 Indian stations. They reported an increasing trend over southern and western Indian stations and a decreasing trend at the stations of the north Indian plains. It is also reported that temperature trends were faster over south India compared to north India during both dry and wet seasons (Kothawale et al., 2012). The results of different studies reveal signals of decrease in rainfall and its extremes over central and northeast Indian region (Subash et al., 2011; Subash & Sikka, 2014). Further, Panda and Kumar (2014) added south India to the above regions where wet days and moderate and total rainfall were found to have decreased. Significant changes were noted in monsoon rainfall extremes especially in the period from 1979 to 2003 over Kerala (Pal & Al-Tabbaa, 2011). Goswami et al. (2006) have shown significant increase in the frequency and magnitude of extreme rainfall events and a decrease in moderate events over central India. Dash et al. (2009) have reported a significant increase in short and dry rain spells and decrease in long rain spells.

The damages due to climatic extremes have increased over the years, particularly in developing nations like India. It is necessary to adopt disaster-specific risk reduction measures at both individual as well as policy level to reduce potential impacts. Ecosystems are vulnerable to extreme events, and therefore, there is a need to undertake regional studies of climate extremes (Rodríguez-Puebla et al., 2010). Moreover, for the study of extremes and related changes in their variability, daily data becomes necessary. This research considers extreme temperatures on a daily basis as well as spells of extreme days, periods which will likely have greater socioeconomic and health impacts. In view of all above, an attempt is made in this study to document trend and variability in temperature and rainfall extremes for the period of 1969–2014 based on daily data of selected surface meteorological stations well distributed over the southern parts of India.

Data and Methodology

Study Area

The present study focuses on the changes over southern parts of India; therefore, the study area covers southern India, largely defined by 8° 4′–23° 30’ N and 68° 7′–89° 5′ E, which has been chosen to study the changes in temperature and rainfall extremes. This vast area includes the states of Andhra Pradesh, Goa, Karnataka, Kerala, Maharashtra, Orissa, Tamil Nadu, and the parts of Chhattisgarh, Gujarat, Jharkhand, Madhya Pradesh, and West Bengal (Fig. 19.1). The terrain of study area varies from west to east; it covers western coastal plains, hilly region of Western Ghats, plateau, small hills of Eastern Ghats, and eastern coastal plains. The area has variety of climatic regions from high rainfall regions of west coast to rain shadow region of Western Ghats. High variability in temperature is observed over interior parts, whereas coastal region has less fluctuation in daily temperature.

Data Used

Data were acquired for daily minimum and maximum surface air temperature observed at 62 and rainfall observed at 61 meteorological stations in the study area (Table 19.1). The data set was provided by the National Data Center (NDC) of the India Meteorological Department, Pune.

Quality Checks

RClimDex software package (http://cccma.seos.uvic.ca/ETCCDMI) has been used to clean up the data sets. It helps to remove errors in data like rainfall value below 0 mm and daily minimum temperature more than maximum temperature. To define outliers, four standard deviations as the threshold were chosen for a finer quality control of the data set. Generalized data plots generated in RClimDex were used to visually inspect the data.

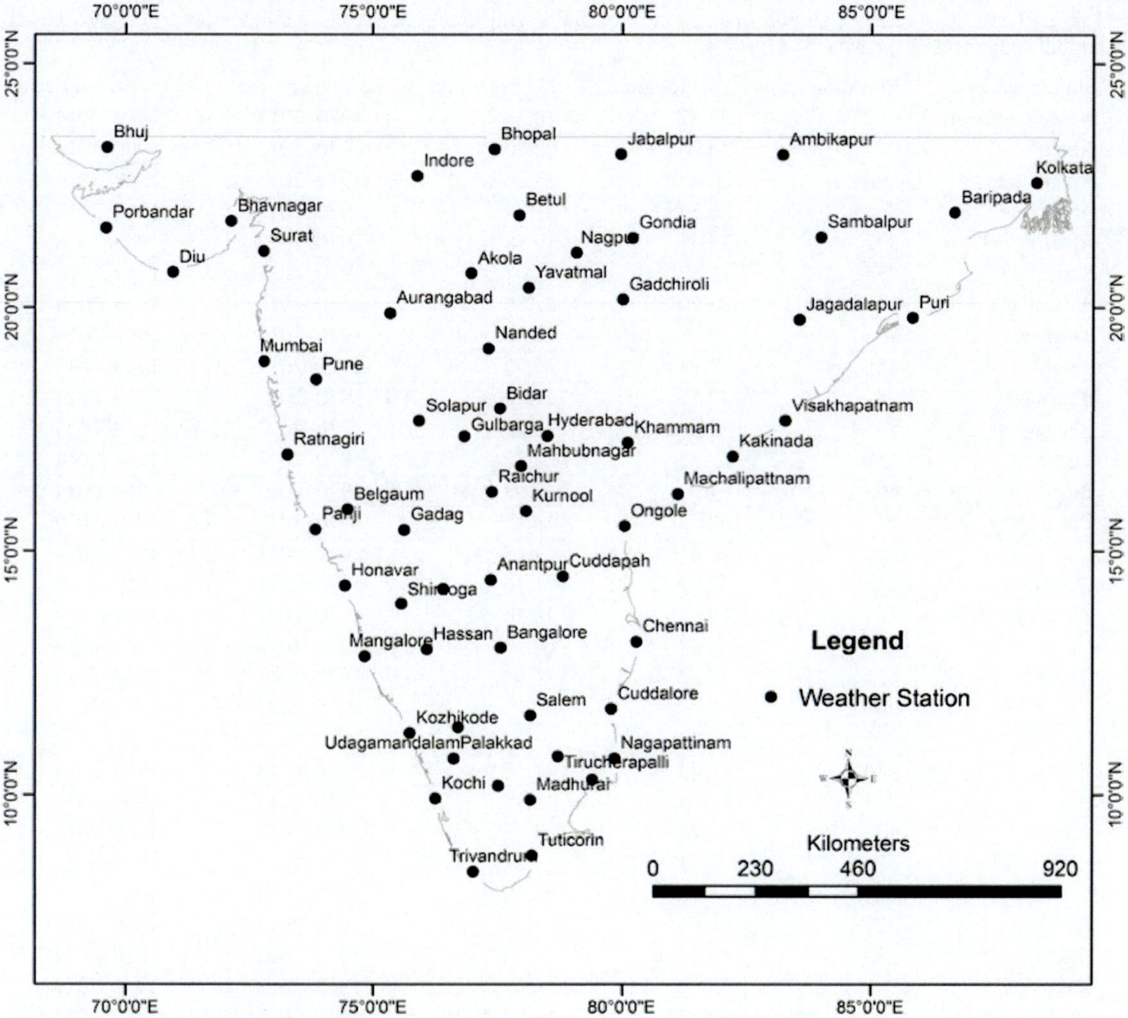

Fig. 19.1 Study area and location of weather stations

Computation of Indices

The joint World Meteorological Organization (WMO), Commission for Climatology (CCl)/World Climate Research Programme (WCRP) project on Climate Variability and Predictability (CLIVAR)/Joint WMO-Intergovernmental Oceanographic Commission (IOC) Technical Commission on Oceanography and Marine Meteorology (JCOMM) Expert Team on Climate Change Detection and Indices (ETCCDI) coordinated the development of a suite of climate change indices which primarily focus on extremes (Peterson et al., 2001). The development of the indices is described at the link http://cccma.seos.uvic.ca/ETCCDMI/. Recently, the ETCCDI revisited its definitions of indices, and presently, a total of 27 indices are considered to be core indices. They are based on daily temperature values and daily rainfall amounts. The ETCCDI is also concerned with improving indices and analysis tools and developed a user-friendly software package called RClimDex that is freely available to the international research community. In the present study, RClimDex software is used to calculate 13 temperature indices and 6 rainfall indices (Table 19.2).

Table 19.1 List of selected stations located in south India with station name, elevation, longitude, latitude, baseline period, and ranges of the data

Station name	Elevation (m)	East longitude	North latitude	Baseline period	Start year–end year
Adirampattinam	6	79.42	10.31	1970–2013	1969–2014
Akola	279	76.98	20.70	1970–2013	1969–2014
Ambikapur	600	83.24	23.14	1970–2013	1969–2014
Anantapur	8	86.75	21.24	1970–2013	1969–2014
Aurangabad	592	75.35	19.88	1970–2013	1969–2014
Bangalore	934	77.57	13.02	1970–2013	1969–2014
Baripada	34	86.72	21.96	1970–2013	1969–2014
Belgaum	766	74.50	15.86	1970–2013	1969–2014
Betul	656	77.94	21.90	1970–2013	1969–2014
Bhavnagar	25	72.15	21.78	1970–2013	1969–2014
Bhopal	504	77.44	23.25	1970–2013	1969–2014
Bhuj	118	69.63	23.28	1970–2013	1969–2014
Bidar	602	77.55	17.93	1970–2013	1969–2014
Chennai	8	80.30	13.15	1970–2013	1969–2014
Chitradurga	718	76.41	14.23	1970–2013	1969–2014
Cuddalore	2	79.79	11.76	1970–2013	1969–2014
Cuddapah	133	78.82	14.48	1970–2013	1969–2014
Diu	3	70.97	20.73	1970–2013	1969–2014
Gadag	651	75.63	15.44	1970–2013	1969–2014
Gadchiroli	209	80.02	20.17	1970–2013	1969–2014
Gondia	305	80.22	21.44	1970–2013	1969–2014
Gulbarga	486	76.84	17.36	1970–2013	1969–2014
Hassan	933	76.09	12.99	1970–2013	1969–2014
Honavar	13	74.45	14.29	1970–2013	1969–2014
Hyderabad	506	78.51	17.37	1970–2013	1969–2014
Indore	564	75.90	22.70	1970–2013	1969–2014
Jabalpur	427	79.98	23.14	1970–2013	1969–2014
Jagdalpur	383	83.56	19.76	1970–2013	1969–2014
Kakinada	7	82.23	16.95	1970–2013	1969–2014
Khammam	102	80.12	17.23	1970–2013	1969–2014
Kochi	7	76.26	9.93	1970–2013	1969–2014
Kodaikanal	602	77.52	10.19	1970–2013	1969–2014
Kolkata	8	88.38	22.56	1970–2013	1969–2014
Kozhikode	3	75.75	11.27	1970–2013	1969–2014
Kurnool	283	78.07	15.83	1970–2013	1969–2014
Machilipatnam	10	81.14	16.18	1970–2013	1969–2014
Madurai	124	78.16	9.90	1970–2013	1969–2014
Mahbubnagar	529	77.97	16.75	1970–2013	1969–2014
Mangalore	19	74.85	12.85	1970–2013	1969–2014
Mumbai	10	72.82	18.90	1970–2013	1969–2014
Nagapattinam	2	79.86	10.75	1970–2013	1969–2014
Nagpur	309	79.10	21.13	1970–2013	1969–2014
Nanded	363	77.32	19.17	1970–2013	1969–2014
Ongole	9	80.06	15.52	1970–2013	1969–2014
Palakkad	77	76.62	10.75	1970–2013	1969–2014
Panaji	20	73.86	15.45	1970–2013	1969–2014
Porbandar	9	69.61	21.64	1970–2013	1969–2014
Pune	557	73.87	18.53	1970–2013	1969–2014

(continued)

Table 19.1 (continued)

Station name	Elevation (m)	East longitude	North latitude	Baseline period	Start year–end year
Puri	9	85.85	19.80	1970–2013	1969–2014
Raichur	395	77.39	16.22	1970–2013	1969–2014
Ratnagiri	23	73.30	16.99	1970–2013	1969–2014
Salem	287	78.16	11.63	1970–2013	1969–2014
Sambalpur	148	84.00	21.45	1970–2013	1969–2014
Shimoga	575	75.57	13.92	1970–2013	1969–2014
Solapur	477	75.93	17.68	1970–2013	1969–2014
Surat	10	72.82	21.17	1970–2013	1969–2014
Tiruchirappalli	85	78.72	10.79	1970–2013	1969–2014
Trivandrum	48	77.01	8.42	1970–2013	1969–2014
Tuticorin	3	78.19	8.75	1970–2013	1969–2014
Udagamandalam	2231	76.71	11.39	1970–2013	1969–2014
Visakhapatnam	104	83.29	17.68	1970–2013	1969–2014
Yavatmal	439	78.13	20.41	1970–2013	1969–2014

Table 19.2 Definitions of 13 temperature and 6 rainfall indices used in the study

Index	Descriptive name	Definition	Units
Temperature indices			
TMAXmean	Annual mean Tmax	Arithmetic mean of monthly mean value of Tmax	°C
TMINmean	Annual mean Tmin	Arithmetic mean of monthly mean value of Tmin	°C
TNn	Coldest night	Lowest value of daily minimum temperature in the year	°C
TXn	Coldest day	Lowest value of daily maximum temperature in the year	°C
TNx	Hottest night	Highest value of daily minimum temperature in the year	°C
TXx	Hottest day	Highest value of daily maximum temperature in the year	°C
TN10p	Cool nights	Percentage of days when TN < 10th percentile	Days
TX10p	Cool days	Percentage of days when TX < 10th percentile	Days
TN90p	Warm nights	Percentage of days when TN > 90th percentile	Days
TX90p	Warm Days	Percentage of days when TX > 90th percentile	Days
CSDI	Cold spell duration indicator	Annual count of days with at least six consecutive days when TN < 10th percentile	Days
WSDI	Warm spell duration indicator	Annual count of days with at least six consecutive days when TX > 90th percentile	Days
DTR			
DTR	Diurnal temperature range	Difference between TX and TN	°C
Rainfall indices			
RX1day	Max 1-day precipitation amount	Monthly maximum 1-day precipitation	mm
CDD	Consecutive dry days	Maximum number of consecutive days with RR < 1 mm	Days
R10mm	Number of heavy precipitation days	Annual count of days when PRCP >= 10 mm	Days
SDII	Simple daily intensity index	Annual total precipitation divided by the number of wet days (defined as PRCP >= 1.0 mm) in the year	mm/day
R95p	Very wet days	Annual total PRCP when RR > 95th percentile	mm
PRCPTOT	Annual total wet-day precipitation	Annual total PRCP in wet days (RR >= 1 mm)	mm

Trend Analysis

The extremes indices generally do not follow a Gaussian distribution, and hence a simple linear least squares estimation of the trend would not be appropriate. Therefore, nonparametric Mann-Kendall (MK) rank statistic-based slope estimators are used to test the significance of long-term trends in the time-series (Subash et al. 2011; Zhang et al., 2011; Del Río et al., 2013). MK test is widely used in environmental science because it is simple and does not assume a distribution of residuals, and it is robust to the effect of outliers in the series and can cope with missing values and values below detection limit (Revadekar et al., 2012). Linear least square method was used to detect the magnitude of trends for all indices (Pal & Al-Tabbaa, 2011) because Mann-Kendall test does not give the rate of change (Subash et al. 2011; Wang et al., 2013). Finally, calculated trends were integrated with GIS environment to represent their spatial distribution.

Results and Discussion

Indices of temperature extremes averaged over southern parts of India tend toward warming for the period from 1969 to 2014. Trends in indices of temperature extremes at individual stations exhibit widespread warming significant at 95 percent confidence level. The magnitudes of the trends in indices based on maximum temperature are generally higher than indices based on minimum temperature. Magnitudes of trends in indices of rainfall extremes are lower compared to indices of temperature extremes and also do not show spatially coherent pattern. Very few stations show statistically significant trends for rainfall.

Trends in Temperature Extremes

Each region is unique, and it gives response to temperature changes with respect to their physical characteristics. Present study region is a combination of different micro regions; hence, changes in temperature were not spatially coherent. Table 19.3 illustrates standard deviations of different warm and cold extremes which help to understand variability of trends. TXn had highest standard deviation within lowest and highest values in both minimum and maximum temperature. CSDI was deviating more than the WSDI index. All cold extreme indices showed high variability than warm extremes except TX90p. This result emphasizes that cold extremes were more vulnerable in the past warming climate.

Spatial Variability

Cold Extremes

Figure 19.2 illustrates the spatial distribution of trends in cold extremes. Figure 19.3 shows the frequencies of trend magnitudes per decade for cold extremes indices at 62 weather stations. The temperature of coldest night (TNn) increased at the rate of 0.12 °C/decade, and about 58% of the stations showed increasing trends. Coldest day (TXn) also indicated a generally increasing trend over the analysis period at 53% stations, but the regional rate was −0.21 °C/decade. This discrepancy is due to higher decreasing magnitude at the stations of Shimoga and Bidar. The stations with larger change magnitudes in TNn and TXn are evenly distributed over the study area (Fig. 19.2). Frequency distributions indicated that TXn of 53% stations ranges from 0 to 1 °C/decade, while TNn of 60% stations ranges from 0 to 2 °C/decade (Fig. 19.3).

The number of cool nights (TN10p) and cool days (TX10p) had decreasing trends at most of the stations. For TN10p, 66% (37%) of the stations, and for TX10p, 79% (48%) of the stations had negative (statistically significant) trends, and about 60% and 76% of stations for magnitudes range from −4 to 0 days/decade for TN10p and TX10p, respectively (Fig. 19.2). Regionally, TN10p had a magnitude of −0.54 night/decade, while TX10p had −1.07 days/decade.

For cold spell duration indicator (CSDI), 66% of the stations had decreasing trends with 60% reporting from −5 to 0 days/decade (Fig. 19.3). All stations over the coast had decreasing trends, but greater magnitudes for CSDI indices were observed along the eastern coast (Fig. 19.2). At

Table 19.3 Standard deviation of trends of temperature and rainfall extremes over study area

Warm extremes	SD	Cold extremes	SD	Rainfall extremes	SD
TMAXmean	0.20	TMINmean	0.35	RX1day	9.93
TXx	0.70	TXn	1.33	CDD	4.99
TNx	0.31	TNn	0.78	R10mm	0.79
TX90p	2.58	TX10p	2.23	SDII	0.48
TN90p	2.54	TN10p	3.29	R95p	32.12
WSDI	3.21	CSDI	5.91	PRCPTOT	47.68

the regional level, CSDI depicted increasing trend, and the magnitude was 0.26 days/decade.

Warm Extremes

The spatial distribution of stations with different trends for warm extremes is shown in Fig. 19.4. About 55% of the stations for hottest night (TNx) and 73% of the stations for hottest day (TXx) had increasing trends, with about 16 and 26% of the stations, respectively, being statistically significant. The trend for TNx had a rate of −0.01 °C/decade, and TXx was 0.22 °C/decade. Almost the entire western part of the study area recorded positive trends in TXx, whereas most of the stations with significant increase were concentrated in southern part of the study area. Mix pattern of trends was observed over study area for TNx. A total of 53 and 70% of stations for TNx and TXx had rate ranging between 0 and 0.5 °C/decade.

For the percentage of days exceeding the 90th percentile (TN90p and TX90p), about 76% (36%) and 84% (60%) of the stations had increasing (significant) trends. Most of the significant increasing stations of TX90p were concentrated in the southern part of the study area and along the west coast (Fig. 19.4). The central part of the study area was dominated by the decreasing trends in TN90p. TN90p of 61% stations ranged from 0 to 3 days/decade, while the frequency of trend magnitudes of 77% stations for TX90p ranged from 0 to 5 days/decade, which is evenly distributed (Fig. 19.5). For the warm spell duration indicator (WSDI), about 74% of the stations had increasing trends, amongst which 44% were statistically significant. Stations with significant increase were concentrated in three pockets, i.e., northern part, central part, and southern coastal region. The regional change rate for WSDI was 2 days/decade. Frequency plots of warm extremes illustrated in Fig. 19.5 demonstrate that the climate is shifting toward a warmer regime.

Trends in Diurnal Temperature Range (DTR)

For DTR, around 61% of the stations exhibited increasing trends, and at nearly 39% of the stations, the trends were statistically significant. The stations with the largest DTR increase are located in the central part of the study area, and except few stations, the entire western coastline showed significant increasing trends (Fig. 19.6). The trends of about 53% stations ranged from 0 to 0.4 °C/decade. Regional change rate was 0.07 °C/decade. Increasing trend in DTR was very gentle, which might be mainly caused by a slight faster increase in maximum temperature than minimum temperature though both increased. At country level, Kumar et al. (2006) showed that minimum temperatures were increasing faster than the maximum temperatures. Reverse condition has been observed for southern India in the present study. However, both studies indicate a warming trend in the climate. For TMAXmean and TMINmean, 86% (58%) and 73% (48%) of stations respectively had increasing (significant) trends. For TMAXmean, 82% of the stations had change magnitudes ranging from 0 to 0.4 °C/decade (Fig. 19.7). For TMINmean, 60% (66%) of the stations had change magnitudes ranging from 0 to 0.3 (0.4) °C/decade. The stations located on the western coastline and southern parts of eastern coastline had greater magnitudes for TMAXmean, whereas high-magnitude stations for TMINmean were concentrated over southern tip of study area and along the eastern coastline. At the regional level, TMAXmean and TMINmean had been increasing, and the magni-

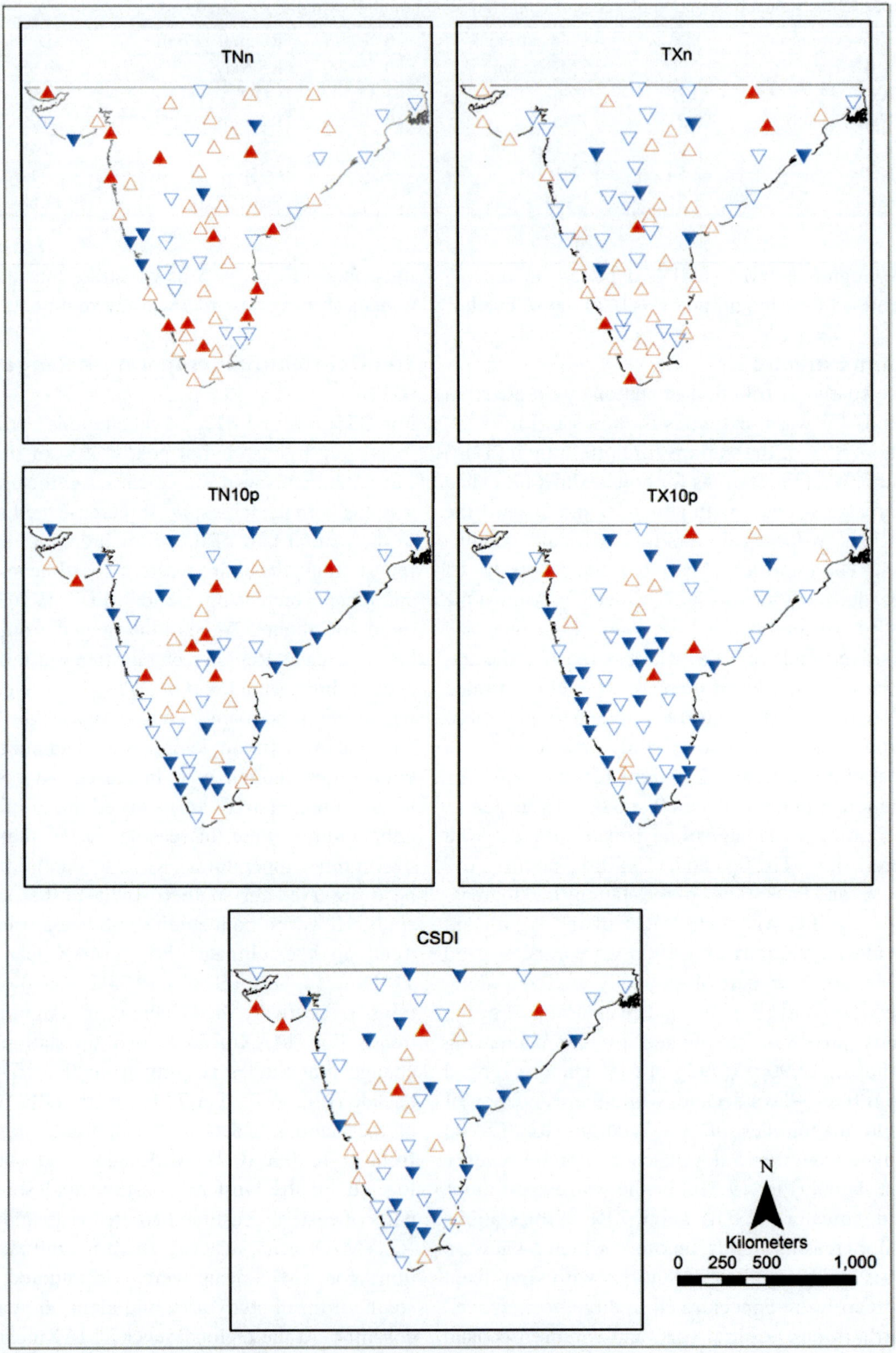

Fig. 19.2 Spatial pattern of trends in cold extremes (upward/downward pointing triangles indicate positive (negative) trends; filled triangles illustrate trends significant at the 0.05 level)

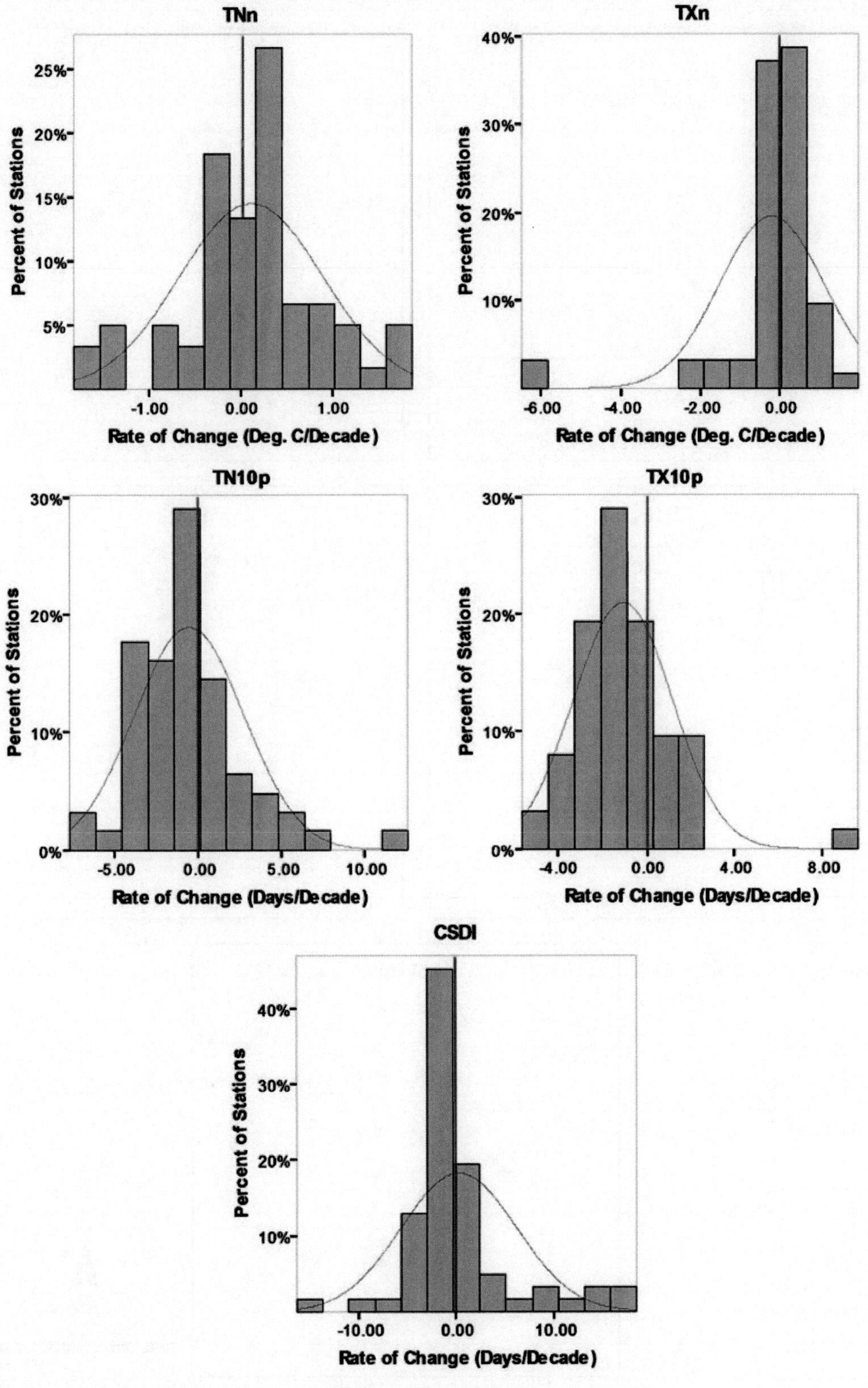

Fig. 19.3 Trend magnitude per decade for cold extreme indices and their frequency

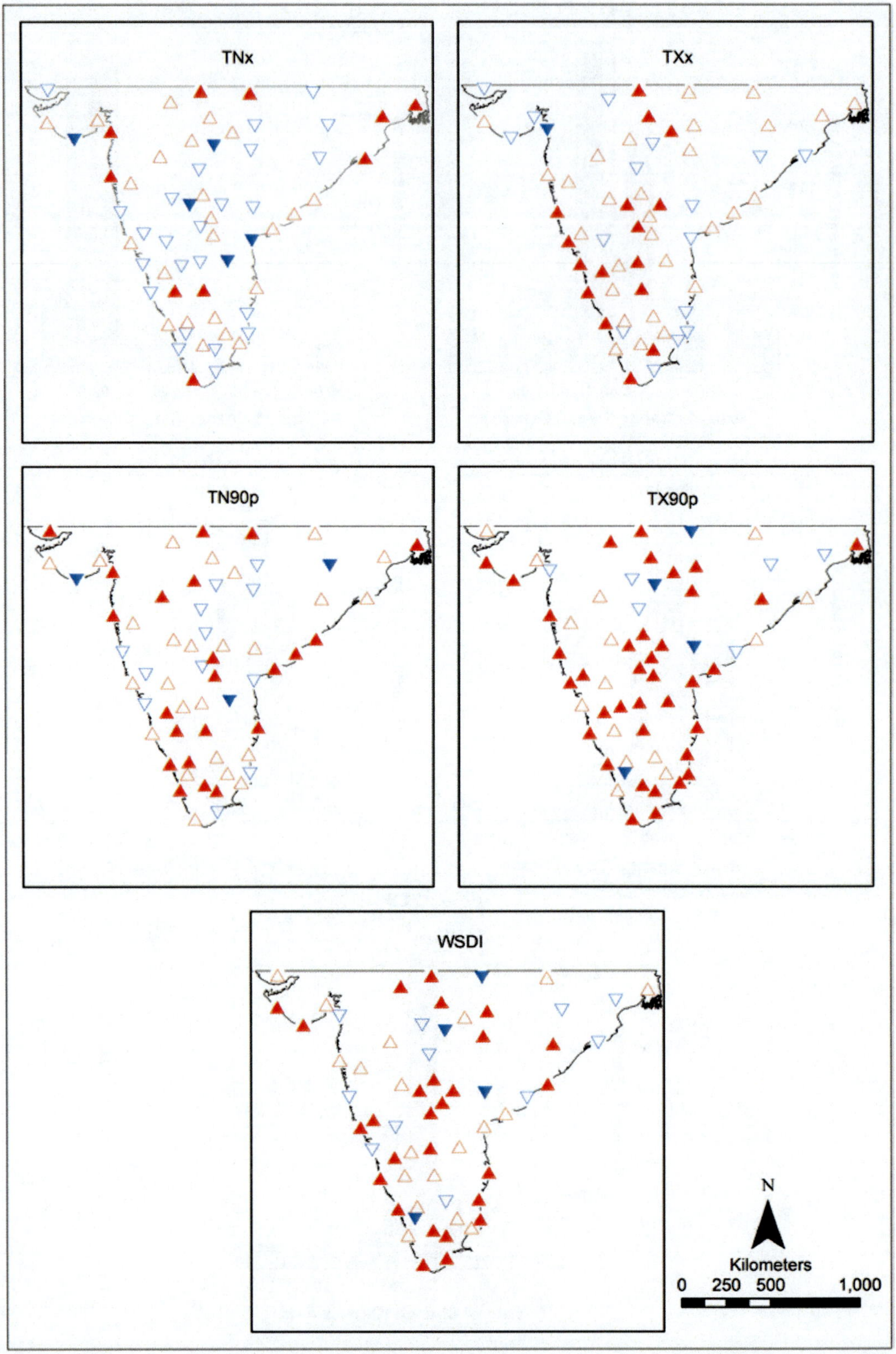

Fig. 19.4 Spatial pattern of trends in warm extremes (upward/downward pointing triangles indicate positive (negative) trends; filled triangles illustrate trends significant at the 0.05 level)

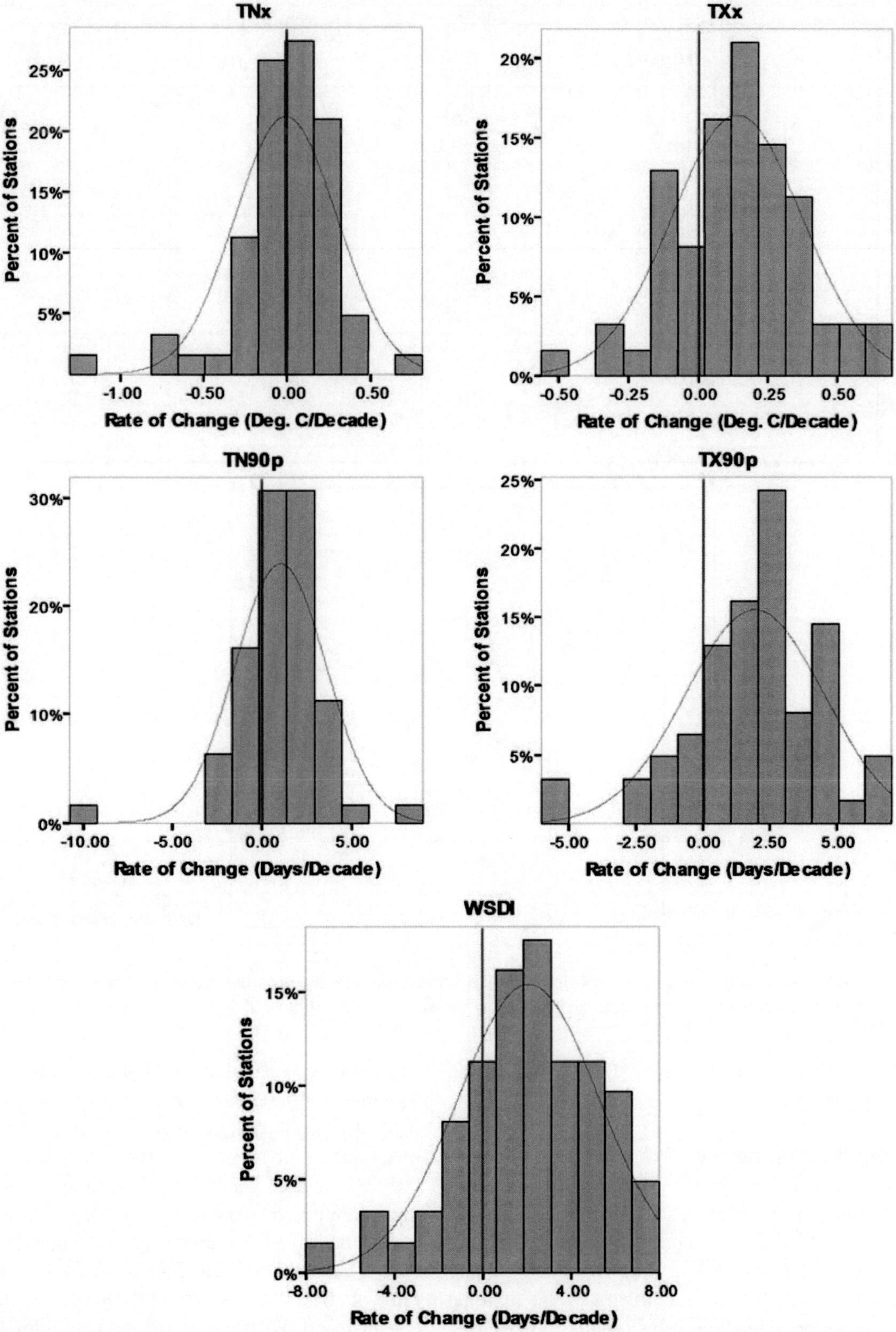

Fig. 19.5 Trend magnitude per decade for warm extreme indices and their frequency

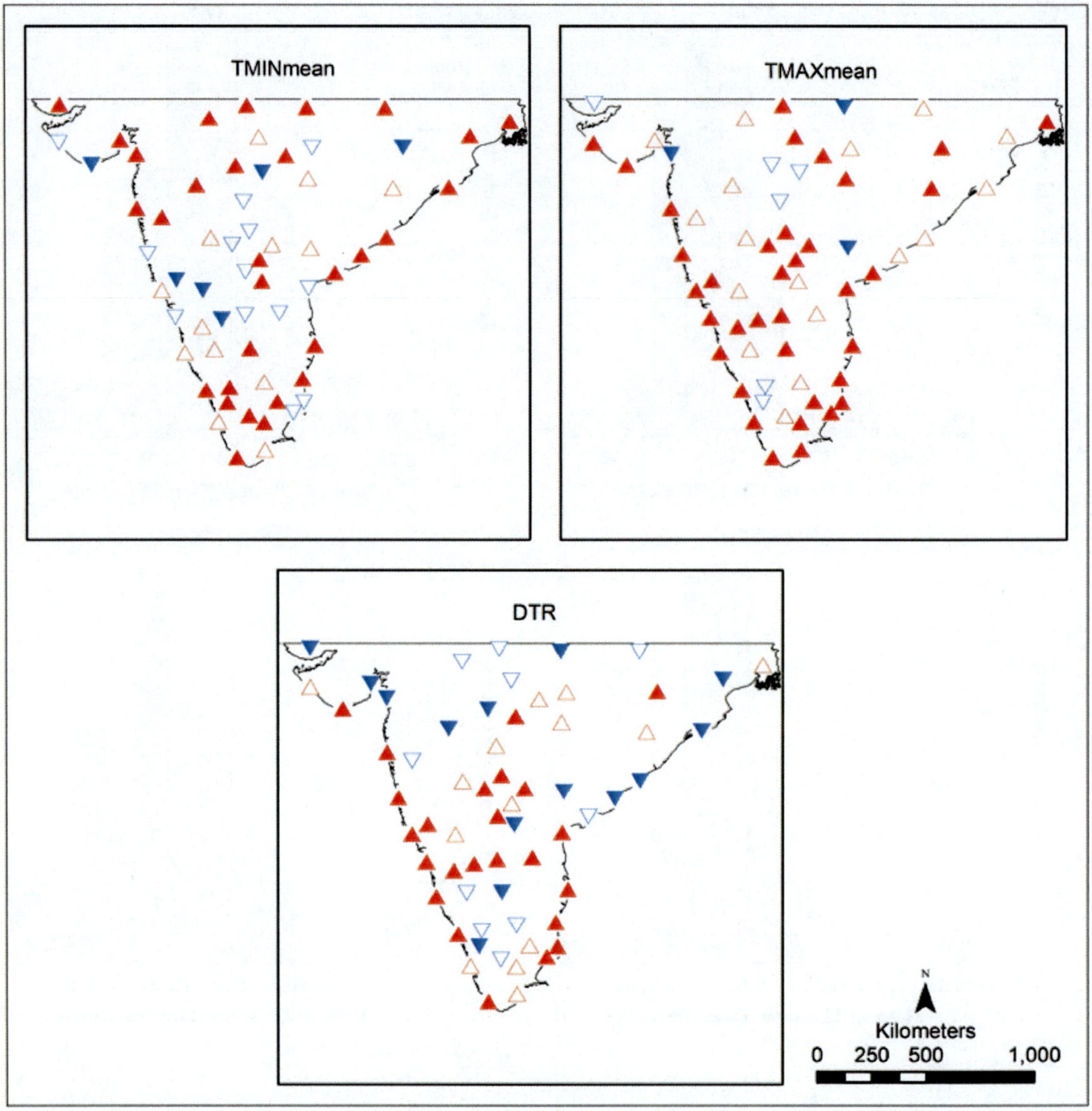

Fig. 19.6 Spatial pattern of trends in TMINmean, TMAXmean, and DTR (upward/downward pointing triangles indicate positive (negative) trends; filled triangles illustrate trends significant at the 0.05 level)

tudes were 0.14 and 0.06 °C/decade, respectively.

Regional Tendency

To calculate the regional average series, arithmetic mean of values at all stations in the study area was used "since the aggregate values represent regional changes" (Klein Tank et al., 2006; Wang et al., 2013). Table 19.4 illustrates the regional changes per decade in warm and cold extremes. All warm indices had increasing trend with high magnitude compared to cold indices except TNx. The mean of maximum temperature increased by 0.14 °C/decade, whereas the rate of change of mean values of minimum temperature was only 0.06 °C/decade. The range of extremes in maximum temperature increased during the study period because TXx increased with the rate of 0.22 °C/decade, whereas TXn decreased with the rate of 0.21 °C/decade. TX90p (1.94 days/decade) and TX10p (−1.07 days/decade) also depicted similar situation in maximum tempera-

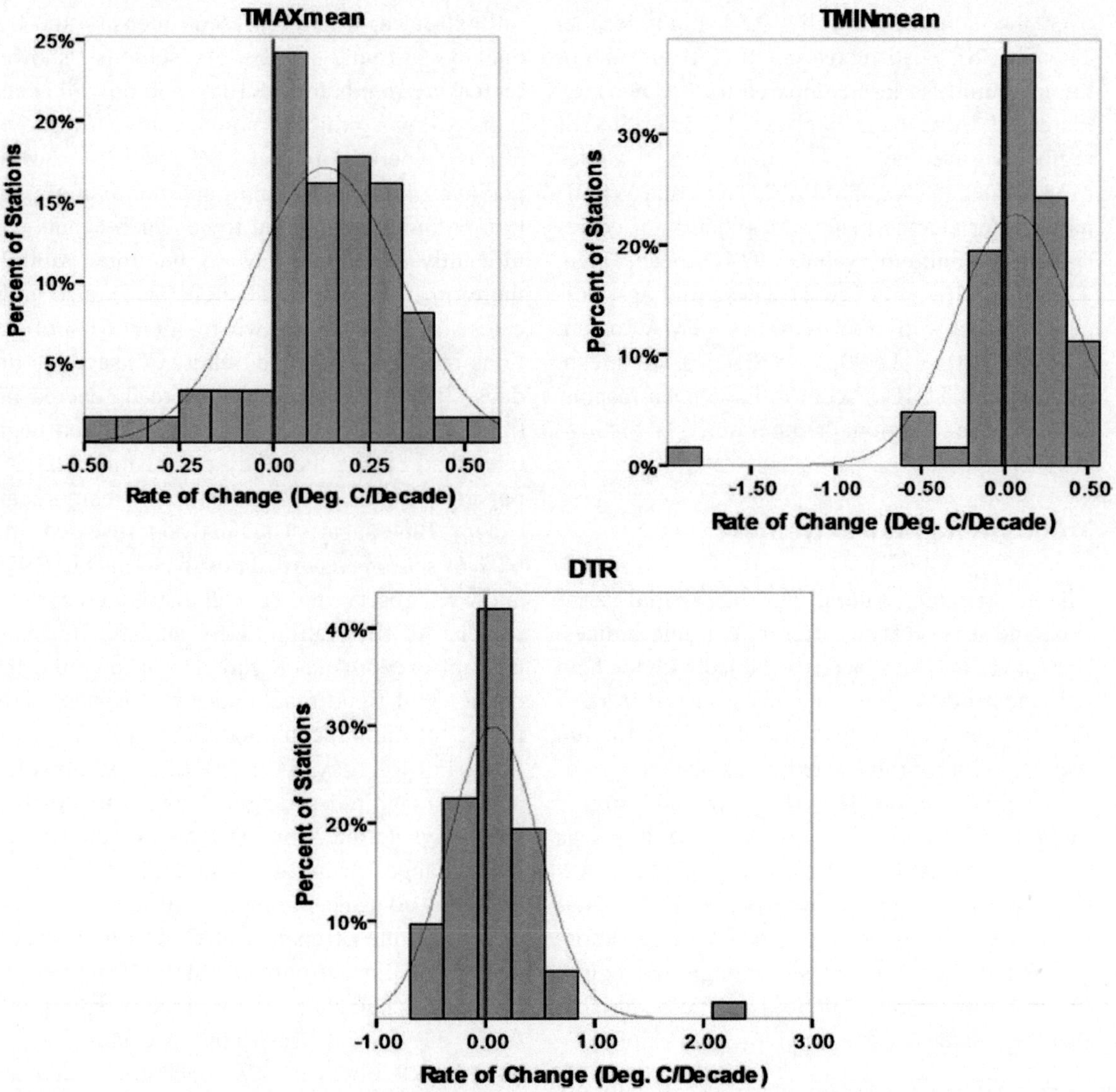

Fig. 19.7 Trend magnitude per decade in TMAXmean, TMINmean, and DTR and their frequency

Table 19.4 Mean regional changes per decade for temperature extremes (average of trends observed at all weather stations)

Regional changes					
Warm extremes	Change/decade	Cold extremes	Change/decade	Rainfall extremes	Change/decade
TMAXmean	0.14 °C	TMINmean	0.06 °C	RX1day	−0.85
TXx	0.22 °C	TXn	−0.21 °C	CDD	−0.83
TNx	−0.01 °C	TNn	0.12 °C	R10mm	0.07
TX90p	1.94 days	TX10p	−1.07 days	SDII	0.17
TN90p	1.06 days	TN10p	−0.54 days	R95p	9.36
WSDI	2.10 days	CSDI	0.26 days	PRCPTOT	8.97

ture. In case of night time temperature, slight variation has been observed. Although TNx had decreasing trend, warm nights had increased by almost 1 night/decade. Cool nights had decreased over the study period by 0.54 night/decade. Though CSDI had increased (0.26 day/decade), the magnitude is less compared to WSDI which increased by 2 days per decade. In terms of regional average, all maximum values (TMAXmean, TXx, TX90p, TN90p, and WSDI) had greater change magnitudes than their corresponding minimum values (TMINmean, TXn, TX10p, TN10p, and CSDI). These findings indicate that most of the extremes (TMAXmean, TXx, TX90p, TN90p, WSDI, TMINmean, TX10p, and TN10p) seem to have been responsible for rise in regional temperature.

Trends in Rainfall Extremes

Like temperature, rainfall also has spatial variation, and it reflects in different extreme indices. Spatial variation has been explained with the help of standard deviation of trends obtained in rainfall extremes over study area (Table 19.3). Region was coherent for rainfall extremes like number of heavy precipitation days (R10mm) and simple daily intensity index (SDII) because it shows very less variability over the region. Very wet days (R95p) and annual total wet-day precipitation (PRCPTOT) had higher standard deviation, and that means it fluctuates from region to region. Overall low-intensity indices are less variable than high-intensity extreme indices of rainfall.

Spatial Variability

Spatial trends in rainfall extremes are shown in Fig. 19.8. Station level indices examined are 1-day maximum rainfall (RX1day), consecutive dry days (CDD), heavy rainfall days (R10mm), simple daily intensity index (SDII), very wet days (R95p), and total rainfall (PRCPTOT). These trends are computed on annual scale, and the baseline period for percentile threshold index is from 1970 to 2013 (for more, see Table 19.1). Compared to the indices of temperature extremes, the trends in indices of rainfall extremes display low magnitude. Very few stations depict statistically significant trends for rainfall extremes with less spatial coherence. Most of the stations across the region show increasing trends in heavy rainfall extremes, which represent intensity and frequency. Around 56% of stations showed increasing trends for RX1day, and only 4(1) stations were significant with positive (negative) trends, whereas around 54% stations showed positive trends in R10mm, and not a single station recorded significant trend. Such events significantly contribute toward the total rainfall; therefore, total rainfall (PRCPTOT) also indicates similar tendency, where, about 61% of stations registered positive trends. Consecutive dry days (CDD) showed widespread decreasing trends. Almost 61% of stations confirmed negative trend during the study period for CDD. All negative trends were statistically insignificant except Bhavnagar. The analysis revealed that 62% of stations recorded positive trends in R95p, and were observed to be well distributed over the region. All the stations were statistically insignificant except four. Figure 19.9 shows the frequencies of trend magnitudes per decade of the indices of rainfall extremes at the 61 weather stations. For RX1day, 54 (39)% of the stations had change magnitudes ranging from 0 to 20 (0 to −20) days/decade. For CDD, 57% of the stations had change magnitudes ranging from 0 to −10 days/decade. Compared to other rainfall extremes, rate of change of CDD was normally distributed. For R10mm and SDII, 44 and 66% of the stations had change magnitudes ranging from 0 to 1 days/decade, respectively. Change magnitude for R95p was mostly concentrated between −50 and 50 mm/decade. Change magnitude for PRCPTOT ranges from −100 to 100 mm/decade except outliers. For R95p and PRCPTOT, 51 and 48% of the stations had change magnitudes ranging from 0 to 50 mm/decade, respectively.

Regional Tendency

Table 19.4 illustrates the regional changes per decade in rainfall extreme indices. In terms of regional average, RX1day and CDD showed decreasing trend by −0.85 mm/decade and −0.83 days/decade, whereas R10mm, SDII, R95p, and PRCPTOT had increased by 0.07 mm/decade, 0.17 days/decade, 9.36 mm/decade, and 8.97 mm/decade, respectively. The proportion of stations with positive (statistically significant)

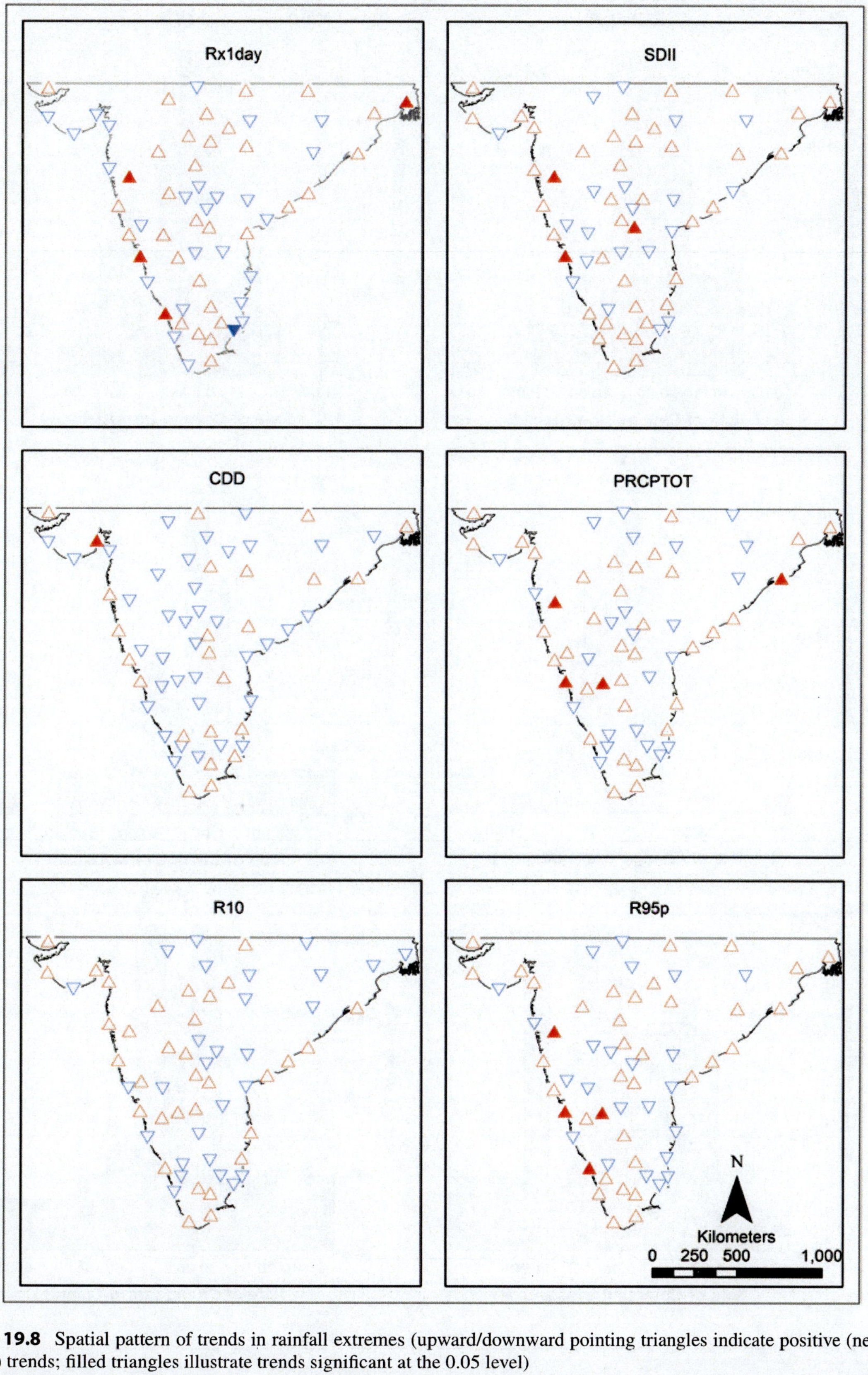

Fig. 19.8 Spatial pattern of trends in rainfall extremes (upward/downward pointing triangles indicate positive (negative) trends; filled triangles illustrate trends significant at the 0.05 level)

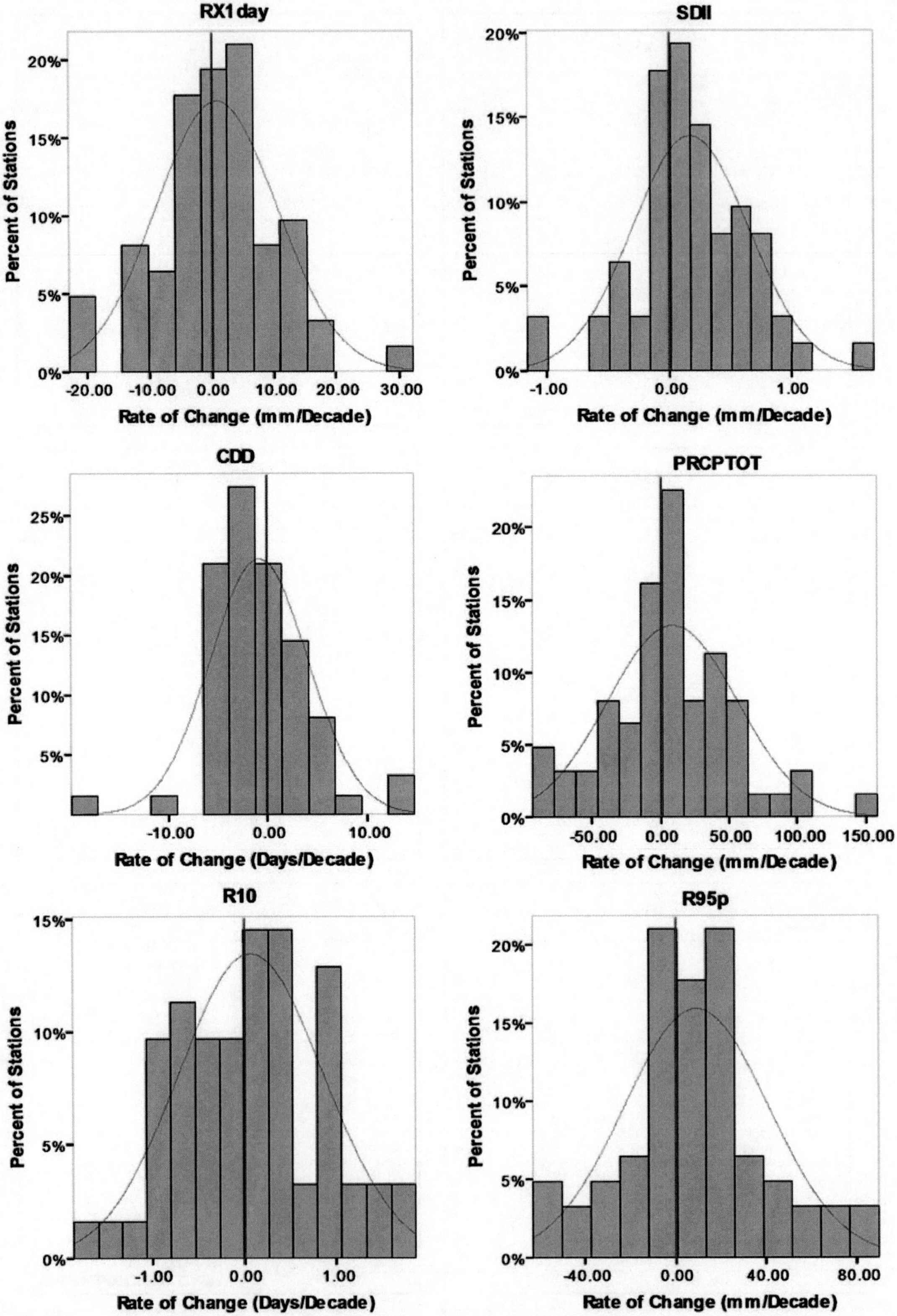

Fig. 19.9 Trend magnitude per decade for rainfall indices and their frequency

trends for these six indices was 55% (6.5%), 39% (2%), 53% (00%), 68% (5%), 61% (6%), and 60% (6%), and approximately 66, 65, 82, 92, 82, and 87% of stations had trend magnitudes from −50 to 50 mm/decade, −10 to 10 days/decade, −1 to 1 days/decade, −1 to 1 mm/decade, −50 to 50 mm/decade, and −80 to 80 mm/decade, respectively. These results indicate that although the rainfall extremes have decreased over south India in recent years, these changes seem to be insignificant.

Elevation and Climate Dependency

The relationship between trends in climate extreme (magnitudes) at individual surface stations and elevation of those respective stations was analyzed (Table 19.5). Weak correlations between trend and elevation were observed, but not a single index had statistically significant correlation. The relationship between elevation and cold extremes shows negative pattern (Wang et al., 2013) except in TX10p and TMINmean, and in warm extremes, this relationship is positive (except TMAXmean). The relationship between trends in rainfall extremes and elevation was negative though the magnitude was less. The trends of R10mm, R95p, and PRCPTOT displayed a weak decrease with increasing altitude. The results of relationships between climate extremes and elevation are not statistically significant but are indicative of the relationship. Cold extremes (TN10p, TNn, TXn, and CSDI) have decreased with increase in elevation, whereas warm extremes (TN90p, TNx, TX90p, TXx, and WSDI) have increased with elevation. Thus, the results indicate that stations with high elevation have been warming faster through both warm and cold extremes compared to nearby stations with lower elevation. The relations found for TX10p, TX90p, WSDI, TXx, TNn, and TNx with elevation are similar to the study carried out by Revadekar et al. (2013) for South Asia stations. Liu and Chen (2000) found a tendency for the warming trend to increase with the elevation in the Tibetan Plateau and its surrounding areas. This tendency of the warming at high-elevation

Table 19.5 Correlation (r) between climate extremes (magnitude unit/decade), mean climate (unit/decade), and elevation (meters)

Indices		Mean climate (mean temperature)	Elevation
Cold extremes	TN10p	−0.00558	−0.00743
	TNn	0.190902	−0.2459
	TX10p	−0.19583	0.269602
	TXn	0.097054	−0.22076
	CSDI	0.023834	−0.03162
	TMINmean	−0.00647	0.080183
DTR	DTR	0.026409	−0.10188
Warm extremes	TN90p	−0.08928	0.146918
	TNx	−0.23593	0.254859
	TX90p	−0.06279	0.003761
	TXx	−0.18079	0.218118
	WSDI	−0.08993	0.1107
	TMAXmean	−0.02815	−0.02054
Rainfall extremes	RX1day	0.056467	−0.03738
	CDD	0.100789	−0.06692
	R10mm	−0.19475	−0.00436
	SDII	−0.13624	−0.07975
	R95p	−0.07014	−0.00376
	PRCPTOT	−0.18653	−0.00918

sites was more pronounced than at low elevations, and the same has been reported by many other regional studies (Beniston et al., 1997; Rangwala & Miller, 2012). The multimodel-based ensemble indicates warming rates in mountains will be enhanced relative to non-mountain regions (Rangwala et al., 2013). The relationship between trends of rainfall extreme and elevation shows apparent pattern, with negative relations of all indices.

The relationships between trend (magnitudes) and mean temperature are also analyzed, and results are shown in Table 19.5. Negative correlations between warm extremes and mean temperature were observed. After careful inspection, it was observed that faster warming occurs at lower temperature though this relationship is not significant. Similar conclusions have been derived by You et al. (2010) for Tibetan Plateau and Wang et al. (2013) for arid area of northwestern China. For cold extremes, the mix pattern of relationships has been found, but this relationship is also not significant. Cool days and nights and TMINmean had negative correlation with mean temperature, whereas TNn, TXn, and CSDI increased with mean temperature. RX1day and CDD has increased with mean temperature, whereas reverse relationship has been observed for R95p, PRCPTOT, R10mm, and SDII.

Conclusion

The analysis done in this study indicates that most of the stations show an increase in percentage of warm nights and a corresponding decrease in cold nights. The trends in maximum temperature extremes had greater magnitudes than the trends in minimum temperature extremes. For temperature extremes, mixed pattern has been observed. The trends in TNn and TXx were positive, indicating a shift in mean temperature towards high values. In case of TNx, around 56 percent of stations showed increasing trend, whereas about 59 percent of stations showed positive trend in TXn, but magnitudes of negative trends were greater than the positive. Trends in WSDI were higher than CSDI. The frequency distributions of cold and warm extremes suggested climate moving over to the warmer side. For example, coldest day and coldest night today are warmer than their predecessors during 1960s and 1970s.

In case of rainfall indices, few percent of stations showed statistically significant changes. For CDD, single station showed significant change. Extreme rainfall indices like RX1day and CDD had a regional decreasing trend; however, very few trends at individual stations were statistically significant.

Elevation dependency was observed in temperature and rainfall extremes. Cold extremes decreased with the increase in elevation, whereas warm extremes showed increasing pattern with increase in elevation. Rainfall extremes had negative relation with elevation. Mean climate dependency of temperature extremes was exactly opposite to the elevation dependency. Rainfall extremes depicted negative correlation with mean climate. Though both elevation and mean climate were negatively correlated with rainfall extremes, the magnitude of relation with mean climate was higher than elevation.

The results of this study suggest that south India can expect changes in temperature extremes that are generally consistent with broad-scale warming. However, changes will be different because each location is influenced by local as well as regional factors, and hence, future changes in temperature extremes may be less coherent. For rainfall, very less changes are expected over study region.

Acknowledgments The authors would like to thank India Meteorological Department for supplying the required data. This work has been the output of the major research project supported by University Grants Commission (UGC). Therefore, the authors are grateful to UGC.

References

Alexander, L. V., et al. (2006). Global observed changes in daily climate extremes of temperature and precipitation. *Journal of Geophysical Research – Atmospheres, 111*, 1–22. https://doi.org/10.1029/2005JD006290

Arora, M., Goel, N. K., & Singh, P. (2005). Evaluation of temperature trends over India / Evaluation de tendances de température en Inde. *Hydrological Sciences Journal, 50*, 37–41. https://doi.org/10.1623/hysj.50.1.81.56330

Bednorz, E. (2011). Occurrence of winter air temperature extremes in Central Spitsbergen. *Theoretical and Applied Climatology, 106*, 547–556. https://doi.org/10.1007/s00704-011-0423-y

Beniston, M., Diaz, H. F., & Bradley, R. S. (1997). Climatic change at high elevation sites: an overview. *Climatic Change, 36*, 233–251.

Dash, S. K., & Mamgain, A. (2011). Changes in the frequency of different categories of temperature extremes in India. *Journal of Applied Meteorology and Climatology, 50*, 1842–1858. https://doi.org/10.1175/2011JAMC2687.1

Dash, S. K., Kulkarni, M. A., Mohanty, U. C., & Prasad, K. (2009). Changes in the characteristics of rain events in India. *Journal of Geophysical Research, 114*. https://doi.org/10.1029/2008JD010572

De, U. S., Dube, R. K., & Rao, G. S. P. (2005). Extreme Weather Events over India in the last 100 years. *Journal of the Indian Geophysical Union, 9*, 173–187.

DeGaetano, A. T., & Allen, R. J. (2002). Trends in Twentieth-Century Temperature Extremes across the United States. *Journal of Climate, 15*, 3188–3205.

Del Río, S., Anjum Iqbal, M., Cano-Ortiz, A., Herrero, L., Hassan, A., & Penas, A. (2013). Recent mean temperature trends in Pakistan and links with teleconnection patterns. *International Journal of Climatology, 33*, 277–290. https://doi.org/10.1002/joc.3423

Easterling, D. R., Meehl, G. A., Parmesan, C., Changnon, S. A., Karl, T. R., & Mearns, L. O. (2000). Climate extremes: Observations, modeling, and impacts. *Science, 289*, 2068–2074. https://doi.org/10.1126/science.289.5487.2068

Fan, X., Wang, Q., & Wang, M. (2012). Changes in temperature and precipitation extremes during 1959-2008 in Shanxi, China. *Theoretical and Applied Climatology, 109*, 283–303. https://doi.org/10.1007/s00704-011-0577-7

Gajic-Capka, M., & Zaninovk, K. (1997). Changes in temperature extremes and their possible causes at the SE boundary of the Alps. *Theoretical and Applied Climatology, 57*, 89–94.

Goswami, B. N., Venugopal, V., Sengupta, D., Madhusoodanan, M. S., & Xavier, P. K. (2006). Increasing trend of extreme rain events over india in a warming environment. *Science, 314*, 1442–1445. https://doi.org/10.1126/science.1132027

Guirguis, K., Gershunov, A., Schwartz, R., & Bennett, S. (2011). Recent warm and cold daily winter temperature extremes in the Northern Hemisphere. *Geophysical Research Letters, 38*, 2–7. https://doi.org/10.1029/2011GL048762

Horton, D. E., Johnson, N. C., Singh, D., Swain, D. L., Rajaratnam, B., & Diffenbaugh, N. S. (2015). Contribution of changes in atmospheric circulation patterns to extreme temperature trends. *Nature, 522*, 465–469. https://doi.org/10.1038/nature14550

Hu, Y., Maskey, S., & Uhlenbrook, S. (2012). Trends in temperature and rainfall extremes in the Yellow River source region, China. *Climatic Change, 110*, 403–429. https://doi.org/10.1007/s10584-011-0056-2

IPCC. (2014). Summary for policy makers. Climate Change 2014 Impacts, Adapt. Vulnerability - Contrib. Work. Gr. II to Fifth Assess. Rep., 1–32. https://doi.org/10.1016/j.renene.2009.11.012.

Jaswal, A. K., Rao, P. C. S., & Singh, V. (2015). Climatology and trends of summer high temperature days in India during 1969–2013. *Journal of Earth System Science, 124*, 1–15.

Klein Tank, A. M. G., et al. (2006). Changes in daily temperature and precipitation extremes in central and south Asia. *Journal of Geophysical Research, 111*, 1–8. https://doi.org/10.1029/2005JD006316

Kothawale, D. R., Revadekar, J. V., & Kumar, K. R. (2010). Recent trends in pre-monsoon daily temperature extremes over India. *Journal of Earth System Science, 119*, 51–65. https://doi.org/10.1007/s12040-010-0008-7

Kothawale, D. R., Kumar, K. K., & Srinivasan, G. (2012). Spatial asymmetry of temperature trends over India and possible role of aerosols. *Theoretical and Applied Climatology, 110*, 263–280. https://doi.org/10.1007/s00704-012-0628-8

Kumar, K. R., Sahai, A. K., Kumar, K. K., Patwardhan, S. K., Mishra, P. K., Revadekar, J. V., Kamala, K., & Pant, G. B. (2006). High-resolution climate change scenarios for India for the 21st century. *Current Science, 90*, 334–345.

Kürbis, K., Mudelsee, M., Tetzlaff, G., & Brázdil, R. (2009). Trends in extremes of temperature, dew point, and precipitation from long instrumental series from central Europe. *Theoretical and Applied Climatology, 98*, 187–195. https://doi.org/10.1007/s00704-008-0094-5

Liu, X., & Chen, B. (2000). Climatic warming in the Tibetan Plateau during recent decades. *International Journal of Climatology, 20*, 1729–1742. https://doi.org/10.1002/1097-0088(20001130)20:14<1729::AID-JOC556>3.0.CO;2-Y

Pai, D. S., Thapliyal, V., & Kokate, P. D. (2004). Decadal variation in the heat and cold waves over India during 1971–2000. *Mausam, 55*, 281–292.

Pal, I., & Al-Tabbaa, A. (2010). Long-term changes and variability of monthly extreme temperatures in India. *Theoretical and applied climatology, 100*(1-2), 45–56.

Pal, I., & Al-Tabbaa, A. (2011). Monsoon rainfall extreme indices and tendencies from 1954-2003 in Kerala. *Indian Climate Change, 106*, 407–419. https://doi.org/10.1007/s10584-011-0044-6

Panda, D. K., & Kumar, A. (2014). The changing characteristics of monsoon rainfall in India during 1971–2005 and links with large scale circulation. *International Journal of Climatology, 34*, 3881–3899. https://doi.org/10.1002/joc.3948

Peterson, T. C., Folland C., Gruza G., Hogg W., Mokssit A., & Plummer N. (2001). Report on the activities of the Working Group on Climate Change Detection and Related Rapporteurs 1998–2001. World Meteorological Organization Rep.WCDMP-47, WMO-TD 1071, Geneva, Switzerland, p. 143.

Rangwala, I., & Miller, J. R. (2012). Climate change in mountains: a review of elevation-dependent warming and its possible causes. *Climatic Change, 114*, 527–547. https://doi.org/10.1007/s10584-012-0419-3

Rangwala, I., Sinsky, E., & Miller, J. R. (2013). Amplified warming projections for high altitude regions of the northern hemisphere mid-latitudes from CMIP5 models. *Environmental Research Letters, 8*. https://doi.org/10.1088/1748-9326/8/2/024040

Revadekar, J. V., et al. (2013). Impact of altitude and latitude on changes in temperature extremes over South Asia during 1971–2000. *International Journal of Climatology, 33*, 199–209. https://doi.org/10.1002/joc.3418

Revadekar, J. V., Kothawale, D. R., Patwardhan, S. K., Pant, G. B., & Rupa Kumar, K. (2012). About the observed and future changes in temperature extremes over India. *Natural Hazards, 60*, 1133–1155. https://doi.org/10.1007/s11069-011-9895-4

Rodríguez-Puebla, C., Encinas, A. H., García-Casado, L. A., & Nieto, S. (2010). Trends in warm days and cold nights over the Iberian Peninsula: Relationships to large-scale variables. *Climatic Change, 100*, 667–684. https://doi.org/10.1007/s10584-009-9721-0

Rummukainen, M. (2013). Climate change: Changing means and changing extremes. *Climatic Change, 121*, 3–13. https://doi.org/10.1007/s10584-013-0888-z

Subash, N., & Sikka, A. K. (2014). Trend analysis of rainfall and temperature and its relationship over India. *Theoretical and Applied Climatology, 117*, 449–462. https://doi.org/10.1007/s00704-013-1015-9

Subash, N., Sikka, A. K., & Mohan, H. S. R. (2011). An investigation into observational characteristics of rainfall and temperature in Central Northeast India-a historical perspective 1889–2008. *Theoretical and Applied Climatology, 103*, 305–319. https://doi.org/10.1007/s00704-010-0299-2

Trenberth, K. E., et al. (2007). Observations: Surface and atmospheric climate change. In S. Solomon et al. (Eds.), *Climate change 2007: The physical science basis* (pp. 235–336). Cambridge University Press.

Wang, H., Chen, Y., Xun, S., Lai, D., Fan, Y., & Li, Z. (2013). Changes in daily climate extremes in the arid area of northwestern China. *Theoretical and Applied Climatology, 112*, 15–28. https://doi.org/10.1007/s00704-012-0698-7

Yan, Z., et al. (2002). Trends of extreme temperatures in Europe and China based on daily observations. *Climatic Change, 53*, 355–392. https://doi.org/10.1023/A:1014939413284

You, Q., Kang, S., Pepin, N., Flügel, W., Yan, Y., Behrawan, H., & Huang, J. (2010). Relationship between temperature trend magnitude, elevation and mean temperature in the Tibetan Plateau from homogenized surface stations and reanalysis data. *Global and Planetary Change, 71*, 124–133. https://doi.org/10.1016/j.gloplacha.2010.01.020

Zhang, Q., Li, J., Chen, Y. D., & Chen, X. (2011). Observed changes of temperature extremes during 1960–2005 in China: Natural or human-induced variations? *Theoretical and Applied Climatology, 106*, 417–431. https://doi.org/10.1007/s00704-011-0447-3

Index

A. R. Siddiqui, A. Sahay (eds.), *Climate Change, Disaster and Adaptations*, Sustainable

D

K

L

M

N